Sudarshan:
Seven Science Quests 2006

Journal of Physics: Conference Series Volume 196

Austin, Texas, USA
6-7 November 2006

ISBN: 978-1-61738-281-9

Printed from e-media with permission by:

Curran Associates, Inc.
57 Morehouse Lane
Red Hook, NY 12571

Some format issues inherent in the e-media version may also appear in this print version.

Printed by Curran Associates, Inc. (2010)

For permission requests, please contact the Institute of Physics
at the address below.

Institute of Physics
Dirac House, Temple Back
Bristol BS1 6BE UK

Phone: 44 1 17 929 7481
Fax: 44 1 17 920 0979

techtracking@iop.org

Additional copies of this publication are available from:

Curran Associates, Inc.
57 Morehouse Lane
Red Hook, NY 12571 USA
Phone: 845-758-0400
Fax: 845-758-2634
Email: curran@proceedings.com
Web: www.proceedings.com

TABLE OF CONTENTS

QUEST 3: SPIN AND STATISTICS

QUEST 4: QUANTUM COHERENCE

QUEST 5: QUANTUM ZENO EFFECT

QUEST 6: TACHYONS

QUEST 7: OPEN SYSTEMS

REFLECTIONS

CONGRATULATORY MESSAGES

This volume contains a unique collection of papers contributed by experts with a long association with George Sudarshan in seven different areas of physics. Each paper recognizes the seminal contributions to physics made by George Sudarshan, and acknowledges their impact on their own research. They were presented at the conference *Sudarshan: Seven Science Quests* held at the University of Texas at Austin in November 2006. Participants spanned George's 5 decades of research and represented three of the 4 inhabited continents. Our only regret is that due to the vast breadth of George's quests, some topics (e.g. Quantum Field Theory), had to be left out. The conference was truly a celebration of Physics itself, due to the participation and contributions of a galaxy of stellar physicists who are leaders in their chosen fields.

The focal point of the conference was to showcase George Sudarshan's breakthrough initiation of these seven areas of physics, and to celebrate his 75th birthday. Many of his former and current students, colleagues, friends and family gathered in Austin, Texas for 2 days and presented their research that was based on his prior and current works, and beautiful personal memories.

Papers presented traced the history of the origin of the seven physics fields in which George's role as the originator/initiator has often been overlooked or misrepresented, and the status of these fields in current times. Even many of his colleagues who have worked with him on one or two of the quests were unaware of the impressive breadth and depth of his contributions to the other quests. We feel that the conference achieved its objective of creating a unified showcase of some of his major contributions in a single volume. Of course, knowing George, his quests are ongoing. We wish him many more years of productive pursuit of science.

This meeting would not have been possible without the financial support of several institutions. We are deeply indebted to the Office of the Vice President of Research, the Physics Department, and the Center for Electromagnetic Materials and Devices of The University of Texas at Austin; and to MetaMaterials LLC, Austin,Texas.

Finally, we would like to thank the participants, and particularly George's family, for their contribution to the wonderful atmosphere achieved during the Conference. We would like also to acknowledge the authors of the papers collected in the present volume, the members of the Scientific Committee for their guidance and support, and the referees for their generous work.

E C G Sudarshan: quantum catalyst

Rodger M Walser

The University of Texas at Austin

E-mail:walser@metamaterials.com

Abstract. A survey of Sudarshan's works is presented to show that his brilliant role in advancing science on many fronts has been like a catalytic agent which, by definition, plays a singular role in increasing the rate of a process, but remains hidden, and is unchanged at the process end. As might be necessary to remain catalytically active for a lifetime, his broad and seminal contributions to science have often gone unappreciated, and rarely given due honor – unfortunately at a great cost to scientific progress.

Help rendered another cannot be measured by the extent of assistance imparted. Its real measure is the recipient's worthiness. (Kural 105, *Thirukkural,* Thiruvalluvar, First Century BCE)

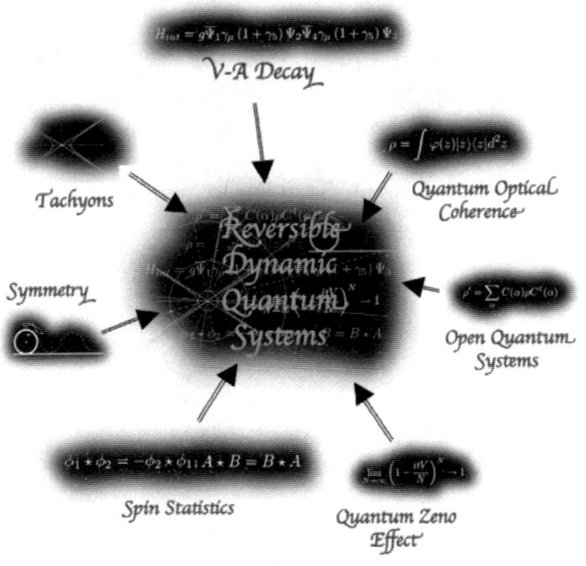

The idea for the "Sudarshan: Seven Science Quests Symposium" was born in 2005 shortly following the announcement, on October 4 by the Royal Swedish Academy of Sciences, that a part of the 2005 Nobel Prize in Physics was awarded to Professor Roy J. Glauber of Harvard University "for his contribution to the quantum theory of optical coherence". Many of George Sudarshan's colleagues around the world, well aware that he was the first to formulate this theory, were stunned by the failure of the Academy to neither recognize his seminal and singular role in the development of this theory, nor to include him in the award. They were also quite familiar with an earlier injustice in which George and Bob Marshak's formulation [1,2] of the V-A theory for the weak nuclear interactions was erroneously assigned to Richard Feynman and Murray Gell-Mann [3]. I will return to these incorrect attributions later.

The Symposium organizers puzzled over the question of why none of the singular contributions of one of the greatest scientists the world has ever known, has been recognized, or honored with a major international award. Could it be, they asked, because his many seminal papers are on a variety of subjects, and he may appear to be the great *originator*, rather than the best follower or expositor, of new ideas? George initiates new ideas with one or at most a few papers (although he has published more than 400) in which he manages to give by far its most elegant formulation and also the complete, basic, and generally the most succinct solution of the key problem. In other words, he finds the core kernel of the problem and brings it out for everyone to see. After that, to others, he may seem to be generally disinterested in exploring all the manifold effects of his own key discovery. He moves to another new problem. His colleagues understand, however, that all of his work has a common theme; how can irreversible processes and transformations (including decay, measurement, etc) be understood in terms of strictly reversible dynamical quantum systems.

George has dedicated his scientific life to the pursuit of this "Holy Grail" problem but, along the way, found his progress blocked by a number of important, unresolved enigmas of physics; the most significant of these have been identified in the diagram on the first page as his Seven Quests.

He addressed these problems, not for their closure, but because of their relationship to his main concern. If George had continued to make contributions to further advances in these areas they would, combined with his seminal resolution of the principal enigmas, have earned him prestigious honors for each.

The composite picture of George is therefore one of a scientist; **(1) whose work initiates a field of inquiry and accelerates the rate at which it develops, (2) whose vital contributions seem to resist acknowledgement, and (3) whose primary goal is unchanged by the profound consequences of the results of his individual quests**. By analogy, this is a fair description of a catalytic agent that, by definition, plays a singular role in increasing the rate of a process, but remains hidden, and is unchanged at the process end. In every field of human endeavor, the catalytic "agent" is a genius, best defined as *someone who sees what everyone else sees - but thinks what nobody else thinks*. Metaphorically speaking, it could be said that a genius is the agent catalyzing an enormous acceleration in the transition of humankind from a state of ignorance to one of enlightenment.

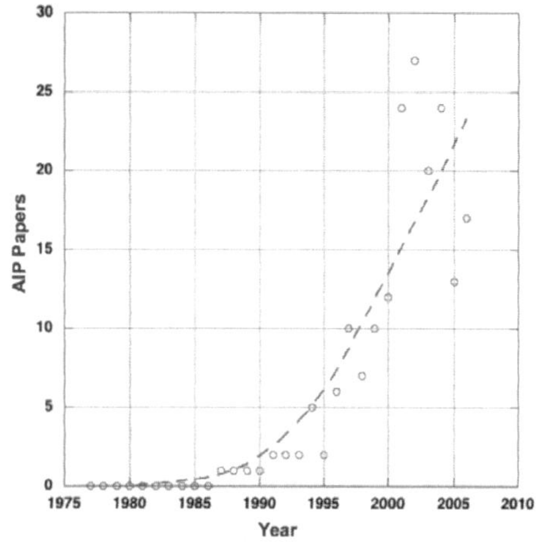

(1) Initiation and Acceleration of Field of Inquiry: To exemplify, note the exponential rise in the number of papers published annually by the AIP on the quantum Zeno effect (QZE), that cite Misra and Sudarshan's seminal 1977 paper [4].

This transition followed George Sudarshan's catalytic role in initiating an understanding of the non-decay of quantum states (quantum non-collapse). Also note the other tell tale of a seminal contribution; a continued increase in citations over a long period (30 years in this case) indicating that this work continues to set the agenda for progress. In his Symposium paper [5], Wayne Itano observes that *"interest in the QZE continues to be high, partially due to the possibility of practical applications in quantum information processing"*.

Rajiah Simon has also addressed this theme in his masterful review [6] of George's role in formulating the theory of Quantum Optical Coherence:

"Commenting on the information theory which came into being, in one stroke, through the remarkable 1948 work of Shannon, A.I. Kinchin writes ….."rarely does it happen in mathematics

that a new discipline achieves the character of a mature and developed scientific theory in the first investigation devoted to it." Shannon's work set the agenda for the classical information theory community. The agenda has not changed in spite of the enormous progress. ...... ***A similar description seems to fit perfectly Sudarshan's seminal work of 1963 which founded the field of Quantum Theory of Optical Coherence."***

The discontinuity in papers on Quantum Optical Coherence that followed the publication of Sudarshan's seminal 1963 paper [7] is seen in this table.

PERIOD	# Papers Citing QUANTUM OPTICAL COHERENCE
1898 - 1962	0
1963**	93
1964 - 2006 (After GS paper**)	30,800 (average = 715 /year)

*(Inspec Database)

The acceleration in the rate of research contributions that followed these breakthrough publications can be understood in analogy with the transition state theory of catalysis. As shown in the following diagrams, prior to the breakthrough, only a very small fraction of the population would be able to overcome the conceptual barrier.

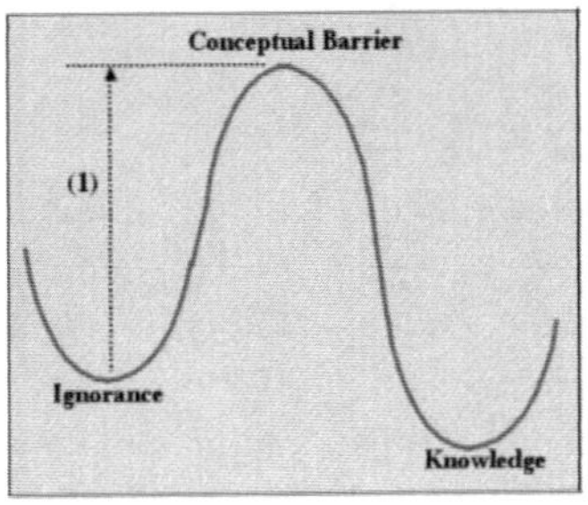

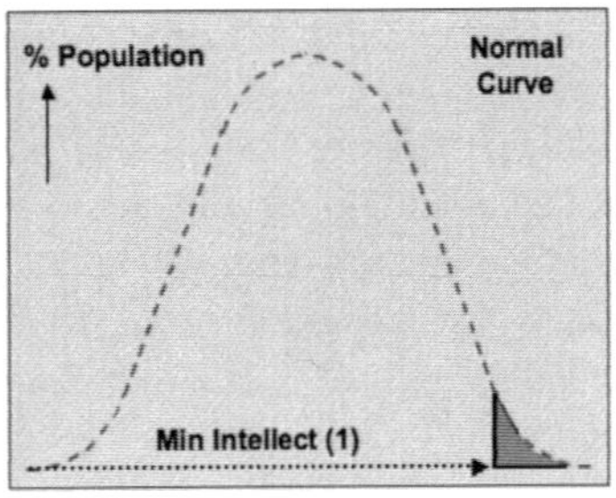

The role of the genius (catalytic agent) is to lower this barrier and make it possible for many more to contribute to further progress as depicted in the modified diagrams.

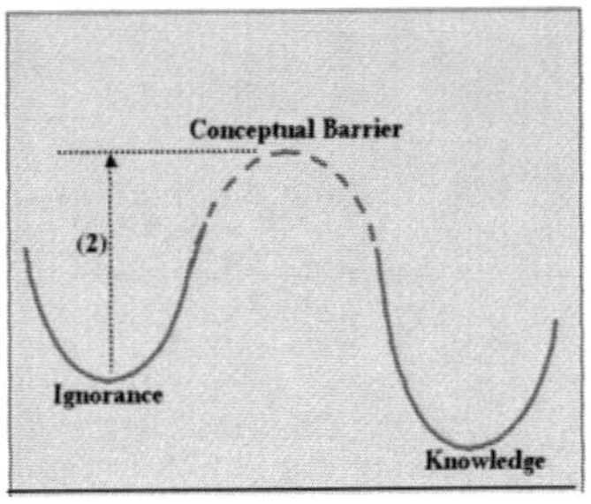

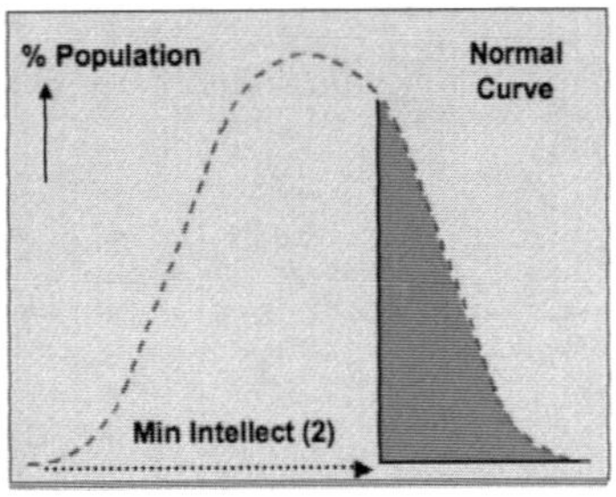

(2) Vital contributions are hidden and unacknowledged. It is readily established that George has catalyzed profound changes in the understanding of a number of scientific issues. But has his vital role in producing these changes remained hidden? Let me examine two spectacular, hard-to-believe examples (formulation of the theories of Quantum Optical Coherence and Weak Nuclear Interactions). In both cases, George's seminal role has been mis-attributed.

George's 1963 paper [7] was the first formulation of Quantum Optical Coherence, and is believed to still set the agenda for this field, now largely concentrated on practical quantum information systems. Some might argue that this transition was generated by R. J. Glauber's earlier publication [8] that is very frequently cited as the first theory of Quantum Optical Coherence. This is clearly erroneous, since this paper did not address *quantum* optical coherence, and his later 1963 paper [9] actually cites Sudarshan's 1963 paper [7].

On November 3, 2005, several of the Symposium organizers sent a letter to the Nobel Committee on Physics of the Royal Swedish Academy, appealing ***" the incorrect attribution of the discovery of the "quantum-mechanical theory of optical coherence" in the citation for the 2005 Nobel Prize in Physics"***. The following is an excerpt from it:

In a communication dated 4 October 2005 posted on its website entitled "Advanced information on Nobel Prize in Physics 2005", the Royal Swedish Academy of Sciences announced the award of half the Nobel prize to Prof. Roy J. Glauber for the discovery of the "quantum-mechanical theory of optical coherence". We, the undersigned, point out that:

1. The actual first formulation of the "quantum-mechanical theory of optical coherence" was given solely by Prof. E.C.G. Sudarshan in his 1963 paper [7].

2. Sudarshan's paper formulates the quantum theory in a manner that allowed its comparison with classical coherence theory. This facilitates the differentiation of specifically quantum effects from classical effects.

3. Quantum effects are manifested only through the non-positivity of the weight function f(z) in the Sudarshan diagonal coherent state representation.

4. All of the subsequently discovered inherently quantum effects, such as anti-bunching and anti-correlation of intensities, can only be understood with Sudarshan's representation.

5. In contrast, in Glauber's original paper [8] the "probability" $p(a_k)$ in eq.4 did not contain any non-positivity. Therefore, it is not the same as Sudarshan's weight function f(z), and cannot represent any strictly quantum phenomena that require non-positivity.

6. Hence, Glauber did not formulate the "quantum-mechanical theory of optical coherence". In fact, in a subsequent paper, Glauber cites Sudarshan's earlier work, criticizes it as being in error (in Sec X), but also adopts it as his own – calling it the "P-representation" (in Sec VII).

7. Accordingly, Sudarshan is the sole discoverer of the "quantum-mechanical theory of optical coherence." The only original, true quantum representation is the ***Sudarshan representation*** *- not the "P-representation" or "Glauber-Sudarshan representation" or the "Glauber representation".*

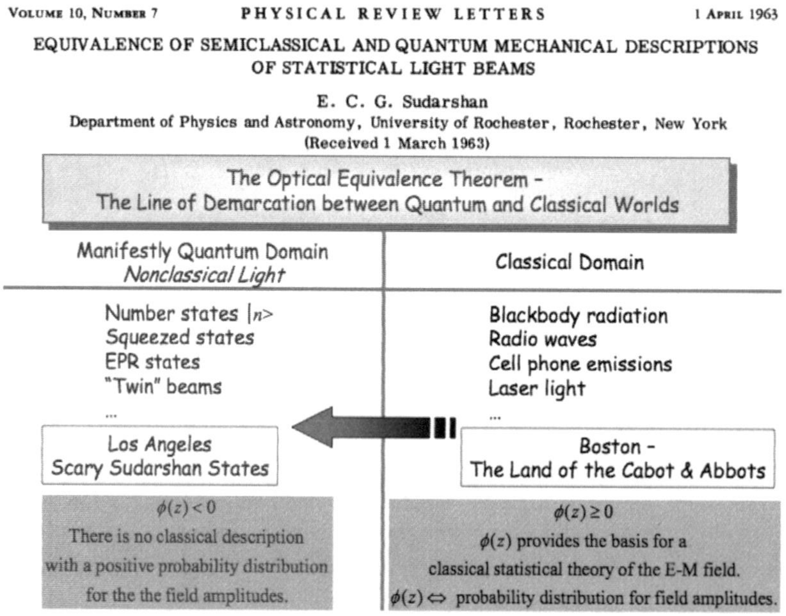

Professor Jeff Kimble, who was the first to experimentally verify Sudarshan's theory of quantum optical coherence [10], provides further support in his historical account [11] in the session on Quantum Optical Coherence.

*"**Two very important papers (were) published in 1963 by George and by Glauber,** .... one had a language and formulas to understand things which had only been discussed in more qualitative terms before....**one had the full quantum description ... and in the lab one had the tools to describe what earlier could be described with classical theories of optical coherence - and that's the Optical Equivalence Theorem which George introduced in his paper.**"*

In his Symposium paper [11], Professor Kimble uses the above figure to further make it clear that Glauber's 1963 paper describes classical optical states, whereas Sudarshan's 1963 paper describes manifestly quantum optical states.

In their deliberations, the Symposium organizers were not concerned with Prof. Glauber's selection for this award, for his contributions to the theory of quantum optical coherence are substantial, meritorious, and worthy. Instead, they focused on trying to understand why a singular contribution of one of the greatest scientists the world has ever known, was being attributed to another scientist.

To Sudarshan, this had to be a cruel repeat of a previous injustice surrounding the attribution of the theory of the Weak Nuclear Interactions. The V-A theory of the weak interactions was first formulated by George and his Ph.D. professor, Bob Marshak, but erroneously attributed to Richard Feynman and Murray Gell-Mann. This egregious error has been acknowledged by both Richard Feynmann [12]:

*The Sudarshan-Marshak paper on the theory and prediction of the (V-A) universal Fermi interaction (UFI), although submitted at the same time as Feynman and Gell-Mann submitted theirs to Physical Review, was published with considerable delay in the Proceedings of the Padua-Venice International Conference (1958) ......, their principal work was overshadowed by the publication of the Feynman – Gell-Mann paper..... **and no mention of the work of Sudarshan and Marshak had been made at the April 1957 Rochester Conference, by which time – several months before the Feynman – Gell-Mann paper was ready – Sudarshan and Marshak had arrived at their principal conclusions. Richard Feyman, when he became aware of this complex situation tried to make amends to Sudarshan and Marshak by making the following remarks at the conclusion of a conference on weak interactions: 'So I would like to say that we have a conventional theory of weak interactions invented by Marshak and Sudarshan, published by Feynmann and Gell-Mann........"***

and by Murray Gell-Mann [13]:

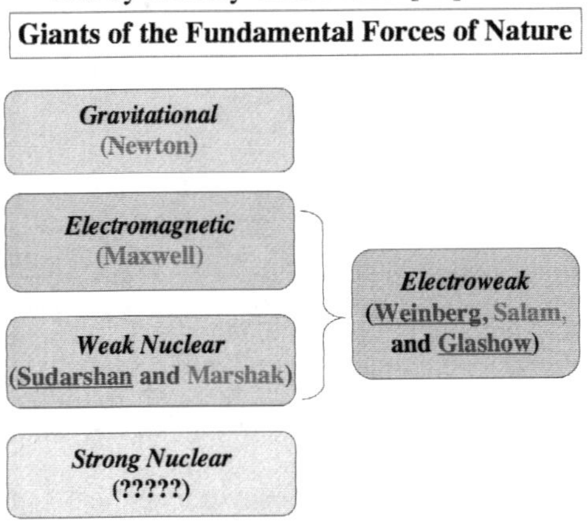

*"But for years everyone cited Gell-Mann and Feynman ... and ignored Sudarshan and Marshak. As late as 1955, Marshak had to publicly remind Feynman at an American Physical Society conference that V-A was not the Feynman - Gell-Mann theory. And he was hurt that it was not until 1970 that **Murray approached him to apologize for the widespread misunderstanding**."*

Their exposition of the Weak Nuclear Force, should have propelled the names of Sudarshan and Marshak alongside those of Newton and Maxwell as suggested in this. This chart not only dramatizes the spectacular recognition they were denied, but shows that this Symposium was important because *all of the living members (names underlined)* who first contributed to our understanding of the fundamental forces participated in it, were afforded an opportunity to clarify this story and set the record straight, **and that they did**.

In his Symposium talk [14], Steven Weinberg observed:

*At that time I did not understand how important that breakthrough (the Sudarshan-Marshak V-A theory) would prove to be. **It opened up the door** to a line of theoretical work, which in less* *than a decade led to a thorough understanding of things that had earlier, when I was a graduate student, seemed like a mélange of miscellaneous facts that had no coherence and certainly no beautiful theoretical interpretations.*

Professor Weinberg makes the case that Sudarshan and Marshak were the first to formulate the V-A theory when in his paper he writes:

"perhaps because Feynman and Gell-Mann were already justly famous for other work, and George, after all, was just beginning his career, people tended to refer to this as the Feynman and Gell-Mann Theory. ***They don't now.*** *I think that with the passage of time, the history has gotten sorted out and it's now generally realized* ***that the earliest statement of the V – A theory was due to Marshak and Sudarshan****".* He further states; *"weak interactions are produced by both vector and axial vector currents, as with great courage and physical insight and élan,* ***Marshak and Sudarshan were the first to propose in 1957****."*

The only other living member of this exclusive club is Sheldon Glashow who shared the 1979 Nobel Prize in Physics with Steven Weinberg and Abdus Salam for their theory unifying the weak and electromagnetic forces. Professor Glashow was scheduled to participate in the symposium, but a medical issue preventing him from attending. In the message he sent to the Symposium [15] he categorically stated for the record that:

"The Sudarshan-Marshak paper (SM) [2] was submitted practically at the same time as the Feynman--Gell-Mann (FG) [3], 'Theory of the Fermi Interaction.' The two papers cover much of the same ground. They offer somewhat different arguments leading to the same V-A theory, and both identify the same conflicting experiments. ***However, several reasons underlie my belief that Sudarshan and Marshak deserve priority in this matter****.*

(1) The FG paper includes the footnote: 'A universal VA interaction has also proposed by Sudarshan and Marshak (to be published).' ***This surely suggests that FG had acess to the SM work.*** *In contrast, the SM paper, submitted on the very same day, includes merely the polite acknowledgment, 'We are grateful to Prof. Gell-Mann....for valuable discussions.' These facts are consonant with Sudarshan's later claim 'that Gell-Mann was informed of our work on the V-A theory [by] the first week of July, 1957, at which time our paper (SM) was completed.'*

(2) Richard Feynman, himself a protagonist in the priority issue, made the following public and unpublished remark much later, in 1974: 'We have a conventional theory of weak interactions (the V-A theory) invented by Marshak and Sudarshan (sic), published by Feynman and Gell-Mann and completed by Cabibbo.' ***So much, I would say, for the issue of priority****."*

Professor Glashow's final statement sends chills up my spine each and every time I re-read it:

"Their (Marshak's and Sudarshan's) daring hypothesis was accompanied by a list of four experimental results...."all of these experiments should be redone...if any stand....it will be necessary to abandon our hypothesis." ***This is theoretical physics at its zenith! The experiments were redone and the results confirmed their hypothesis. It was a stunning accomplishment, yet one never recognized with a prize****."*

Dr. Prashant Valanju, a Symposium organizer, and the first to demonstrate the QZE from an analysis of the experimental data on hadron-nucleus collisions [16], believes that George's work is not given the attention it deserves because *"the breadth of his work, and the way he mostly starts new tracks, is probably why only a very few cognoscenti - those whose souls are really troubled at that moment by the specific problem he solves with such clarity, grace, and perfection - can possibly recognize the greatness of his work. Each of us sees only a small part of this giant elephant. The rest do not see him sweating long in the field, so often do not even know that he has worked in the area. His seminal papers are quoted but often unread. In modern times, when much "work" in science is done by huge groups, and valued more by the quantity and glitz of the wrapper than the quality of its core content, someone like George is sure to go unrecognized."*

(3) Goals are unchanged by the profound consequences of his accomplishments.

The unifying theme of George Sudarshan's work is that of resolving the fundamental dilemma of how irreversible processes and transformations (including decay, measurement, etc) could be understood as strictly reversible dynamical quantum systems. In each of his 7 Science Quests, George Sudarshan has brilliantly used his deep understanding of quantum mechanics to catalyze the resolution of physics enigmas related to this goal. Like any catalyst, his role has been masked, which has allowed him to emerge unchanged by each of his quests, to focus his catalytic potency on other important enigmas of physics.

George writes [17]:

"In most of my work I have to recognize that discoveries do not confer credit on their discoverers; and that great and not-so-great scientists can be small men. But most of all I have learned to value those gifts of insight, irrespective of what is the official position. Not with pride but with awe since the insight came to me, they were not my doing; that I was a channel for the vision. ***This is part of my tradition, to be a seer, to be a rishi.***"

The "rishi" is a "seer" or one to whom wisdom is revealed through states of higher consciousness. Awareness begins in an unbounded state with pure consciousness and then cascades until it reaches the physical world. It has been long suspected that two of the greatest mysteries of modern science, quantum mechanics, and consciousness are somehow related. The very statement of this mystery suggests that one must necessarily be a rishi to be an observer of it.

I believe that the sum of all of George's Quests are leading him to address this enigma and that he is destined to play a seminal role in resolving the physics of consciousness. But mentioning "consciousness" is controversial and only scientists who have received the highest peer recognition could safely venture into this arena. If he were younger, to do so would surely jeopardize his effectiveness; but he is now very experienced and battle tested, so the risk is likely small. I hope that science will bestow on George the necessary recognition that encourages him to apply his brilliance to this problem because, if he does not, it may be a very long time before someone with his capacity to attack such issues comes along.

As mentioned, this Symposium - *Sudarshan: Seven Science Quests* was designed to focus attention on his numerous singular contributions, so George would be recognized as a *genius creator*, distinct from other prolific and important *builders*. Some of his colleagues have suggested that George had more than 7 significant Quests. Stanley Deser, for example, asked if George's work in *Quantum Field Theory* (particularly the work George did with him and Walter Gilbert (Nobel Prize in Medicine) [18]) might be included as an 8th Quest. Others mentioned George's Quest into *Axiomatic Field Theory* (e.g. see Johnson & Sudarshan's important paper on " *Inconsistencies of Quantizing the Spin 3/2 Field*" [19]) and another of his Quests into *Indefinite Metric Quantization* [20,21]. Faced with the daunting chore to select a limited number of his Quests for the Symposium, logistical constraints forced the Organizers to choose 7 of George's most prodigious scientific contributions.

All of the participants in the Symposium are distinguished scientists who have collaborated with George and contributed to his 7 Science Quests. I will leave it to them to tell the rest of the story, and for the reader to decide on George Sudarshan's rightful place in the history of science.

It has been more than 50 years since Sudarshan and Marshak formulated the first theory of the weak interaction. It is evident from the remarks of both Professors Weinberg and Glashow, that it is important to note and respect such seminal efforts, since they lead the way for all further progress. The failure to sort out and honor Sudarshan's and Marshak's achievement has undoubtedly impacted progress in understanding the forces of nature and will continue to do so until it is corrected. The same might be said about George's paradigm-shifting theory of quantum optical coherence, and any of his other 7 Science Quests.

George has paid dearly for this neglect, but what has been the cost to science? Alas, as might be necessary to remain catalytically active for a lifetime, his brilliant works have frequently gone undetected or unnoticed, and rarely given the honor they deserve. In this symposium *Sudarshan: Seven Science Quests*, the participants reviewed and examined 7 of George's principal seminal contributions to science, with the goals of illuminating them, and understanding their common thread, without jeopardizing the continuation of his quest. They have succeeded admirably and, in honoring George Sudarshan, brought great distinction to themselves.

References

[1] Sudarshan E C G and Marshak R E 1957 "*The Nature of the Four-Fermion Interaction*", Proc. of the Conference on Mesons and Newly Discovered Particles in Padua-Venice (Sept 1957), N. Zanichelli (ed.), Bologna (1958). This first paper on V-A was presented in the Padua-Venice conference in Sept 1957, and was published in 1958. The results were presented to Gell-Mann in June 1957

[2] Sudarshan E C G and Marshak R E 1958 "*Chirality Invariance and the Universal Fermi Interaction,*" Phys. Rev **109**, 1860-1862

[3] Feynmann R P and Gell-Mann M 1958 "*Theory of the Fermi Interaction,*" Phys. Rev. **109** (1), 193-198

[4] Misra B and Sudarshan E C G 1977 "*The Zeno's paradox in quantum theory,*" J. Math. Phys. **18**, 756

[5] Wayne Itano 2006 "*Perspectives on the Quantum Zeno Paradox,*" in this symposium session on the Quantum Zeno Effect

[6] Simon R 2006 "*From Nonclassical Optics to Quantum Information Science: The Legacy of Sudarshan's Diagonal Representation,*" in this symposium session on Quantum Optical Coherence

[7] Sudarshan E C G 1963 "*Equivalence of Semiclassical and Quantum Mechanical Descriptions of Statistical Light Beams,*" Phys. Rev. Lett. **10**, 277-279

[8] Glauber R J 1963, "*Photon Correlations,*" Phys. Rev. Lett. **10**, 84-87

[9] Glauber R J 1963, "*Coherent and Incoherent States of the Radiation Field,*" Phys. Rev. **131** (6), 2766

[10] Kimble H J and Mandel L 1976, "*Theory of Resonance Fluorescence,*" Phys. Rev. A **13**, 2123

[11] Kimble H J 2006 "*Neoclassical Light: An Assessment of the Voyage into Hilbert Space,*" This symposium session on Quantum Optical Coherence

[12] Mehra J 1994 The Beat of a Different Drum: The Life and Science of Richard Feynman, Clarendon Press Oxford, p. 477 and references 29 and 40

[13] George Johnson 1999 Strange Beauty: Murray Gell-Mann and the Revolution in 20th Century Physics, Vintage Books, Random House, p. 156

[14] Weinberg S 2006 "*V-A was the Key,*" Plenary Talk in this Symposium session on V-A: Universal Theory of Weak Interaction

[15] Glashow S 2006 "*Symposium Message,*" in this Symposium session on V-A: Universal Theory of Weak Interaction

[16] Valanju P M, Sudarshan E C G, and Chiu C 1980 "Spatio-temporal Development of Hadron-Nucleus Collisions," Phys. Rev. **D 21**, 1304

[17] www.keral.com/celebrities/georgesudarshan/georgespeaks3.htm

[18] Sudarshan E C G, Gilbert W, and Deser S 1960, "*Integral Representations of Two Point Functions,*" Phys. Rev **117**, 273-279

[19] Sudarshan E C G and Johnson K 1961 "*Inconsistency of the Local Field Theory of Charged Spin 1/2 Particles,*" Annals of Physics **13**, 126-145

[20] Sudarshan E C G 1961 "Quantum Mechanical System with Indefinite Metric I," Phys. Rev. **123**, 2183-2193

[21] Sudarshan E C G and Schnitzer H J 1961 "Quantum Mechanical System with Indefinite Metric II," Phys. Rev. **123**, 2193-2201

Journal of Physics: Conference Series **196** (2009) 011001 doi:10.1088/1742-6596/196/1/011001

Sudarshan

Prashant M Valanju

The University of Texas at Austin

E-mail:pvalanju@mail.utexas.edu

When one reads George Sudarshan's many seminal papers on a variety of subjects, one becomes aware of how unique he is. On the one hand, he has been enormously prolific, authoring over 400 important papers. His work is credited with elucidating several important topics in physics and impacting many leading physicists. On the other hand, his style appears spartan, he initiates new ideas with a few papers in which he gives by far its most elegant formulation and the complete, basic, and generally most succinct solution of the key problem. In other words, he finds the core kernel of the problem and brings it out for everyone to see. After that, rather than exploring all the manifold effects of his own key discovery, he addresses a different problem. He is clearly the great originator, rather than the best follower or expositor, of new ideas.

An examination of the broader scope of his work reveals, however, that his pursuit of knowledge lies deeper than that exposed by his individual studies. Thus, these apparently disparate perspectives are the characteristics of a subtle and more noble effort. The breadth of his work, and the way he mostly starts new tracks, is probably why only a few cognoscenti - those whose souls are really troubled at that moment by the specific problem that he solves with such clarity, grace, and perfection - can possibly recognize the greatness of his work. Each of us sees only a small part of this giant. The rest do not see him sweating long in the field, so they often do not even know that he has worked in the area. His seminal papers are quoted but often go unread and not really understood until considerable time has passed.

The deep common thread of his efforts is an attempt to resolve the fundamental conflict between descriptions of objects (*being*) and processes (*becoming*). There is a unifying theme in George's work: irreversible processes and transformations (including decay, measurements, etc) in strictly reversible dynamical quantum systems. The understanding of this contradiction is still the most fundamental unsolved problem. Through what can be categorized as his seven quests, George has made major contributions in trying to piece this puzzle together.

These contributions deserve far greater recognition than they have got so far. This may be because the unity in his contributions may not have been widely understood. In modern times, when much "work" in science is done by huge groups, and valued more by the quantity and the glitz of the wrapper than the quality of its core content, someone like George is sure to go unrecognized. This symposium is a step in correcting this natural oversight by recognizing a genius creator as distinct from prolific and important builders.

Ennackal Chandy George Sudarshan's efforts are ongoing. The Quest continues ...

Journal of Physics: Conference Series **196** (2009) 011002 doi:10.1088/1742-6596/196/1/011002

Welcome to "Sudarshan's 7 Science Quests"

Mary Ann Rankin

Rodger M Walser

Juan Sanchez

Sheldon Glashow

Journal of Physics: Conference Series **196** (2009) 011003 doi:10.1088/1742-6596/196/1/011003

Message for Sudarshan Symposium

Sheldon Glashow

Dept. of Physics, Boston University, Boston, MA 02215

E-mail: sig@bu.edu

Professor E.C.G. Sudarshan has made extraordinary contributions to four distinct areas of physics. I shall focus upon the one of them lying closest to my own expertise; his development on 1957 (with the late Robert Marshak) of the universal V-A theory of the weak interactions. Indeed, I am uniquely qualified to comment on this development because Sudarshan was a postdoc at Harvard in 1958 while I was a graduate student. During that year, we had frequent discussions of the weak interactions, our common scientific interest. Indeed, his 1958 Physical Review Letter on the V-A theory contains the first known scientific acknowledgement to me!

The mid 1950s were a time of ferment and confusion in particle physics. The newly discovered failure of the mirror symmetry reinvigorated the study of nuclear beta decay, while the decay properties of the elementary particles were just being observed. How were the new results to be incorporated into Fermi's decades-old framework? To make matters difficult, it seemed that no one model could agree with all the data. T.D. Lee concluded that "a universal Fermi interacion could not be realized," while my thesis advisor, Julian Schwinger, choosing the wrong data to ignore, published a deeply flawed version of the universal theory of weak interactions.

Marshak and Sudarshan, having pondered these matters for some months, met with Felix Boehm in July, 1957, where they learned that the latest data disfavored the VT possibility that Schwinger had adopted. They soon were able to complete and submit their ground-breaking paper, "The Naure of The Four-Fermion Interaction," of which Sudarshan (then a mere graduate student) was the first-named author. They presented a comprehensive analysis of the weak interaction data, which along with the imposition of an elegant symmetry principle, allowed them to deduce a unique form for the weak interactions, the so-called V-A theory. Their daring hypothesis was accompanied by a list of four experimental results that, they wrote, "cannot be reconciled with this hypothesis... All of these experiments should be redone... If any of the four experiments stand, it will be necessary to abandon the hypothesis." **This is theoretical physics at its zenith!** The experiments were redone with results that now confirmed their hypothesis. It was a stunning accomplishment, yet one which has never been recognized with a prize.

The Sudarshan-Marshak paper (SM) was submitted practically at the same time as the Feynmann-Gell-Mann paper (FG), "Theory of the Fermi Interaction." The two papers cover much the same ground. They offer somewhat different arguments leading to the same V-A theory, and both identify the same conflicting experiments. However, several reasons underlie my belief that Sudarshan and Marshak deserve priority in this matter:

(1) The FG paper includes the footnote: "A universal V, A interaction has also been proposed by Sudarshan and Marshak (to be published)." This surely suggests that FG had access to the SM work. In contrast, the SM paper submitted on the very same day, includes the polite acknowledgement, "We are grateful to Prof. Gell-Mann....for valuable discussions". These facts are consonant with

Sudarshan's later claim "that Gell-Mann was informed of our work on the V-A theory [by] the first week of July, 1957, at which time our (SM) paper was completed."

(2) Richard Feynman, himself a protagonist in the priority issue, made the following public and published remark much later, in 1974: "We have a conventional theory of weak interactions [the V-A theory] invented by Marshak and Sudarshan [sic], published by Feynmann and Gell-Mann, and completed by Cabibbo". [N.B. In 1962, Cabibbo showed how to extend the universality of the V-A theory to strange particles.] So much, I would say, for the issue of priority.

In my view, Sudarshan's seminal contribution to weak-interaction theory, representing only a small portion of his oevre, would itself justify the award of a major prize in Physics.

G. Bhamathi, G. Sudarshan, and S. Glashow (1991)

Robert Marshak and George Sudarshan (1991)

Journal of Physics: Conference Series **196** (2009) 011004 doi:10.1088/1742-6596/196/1/011004

Welcome to Sudarshan's Seven Science Quests Symposium

Mary Ann Rankin

Dean, College of Natural Sciences, The University of Texas at Austin

E-mail: rankin@mail.utexas.edu

I want to welcome you all to Austin and thank you for helping us celebrate George's 75th year, and I also want to say again - Happy Birthday. We are very happy to celebrate this occasion.

George Sudarshan came to UT Austin in 1969, so he has spent a fair amount of those 75 years here with us, and for that we are very grateful. As you know, he has had a remarkable career. I will just mention a few of his accomplishments - and there are an amazing number. George has Honorary degrees from University of Wisconsin, University of Delhi, Chalmers University of Technology in Sweden, University of Madras, University of Braswan, University of Calcutta. He has been awarded the Padma Vibhushan (Supreme Order of the Lotus) by the President of India, the Bose Medal of Indian Academy of Sciences, and the Sarbadhikari Medal by the University of Calcutta. He has supervised 30 Ph.D. students here, and he still has many more. He has, of course, given hundreds of invited lectures, authored or co-authored nine books - some of which explain science to the novice. He has written over 500 papers, including some that were cited in the Royal Swedish Academy of Sciences this year in the description of the work that led to the 2005 Nobel Prize in Physics. Indeed many feel that he should have been the winner of that prize.

On behalf of the College of Natural Sciences, I just want to say that we recognize very much the value and quality and impact of Dr. Sudarshan's work and we have been honored by our association with this fine man for these many years. In addition to the Universal Theory of Weak Interactions, he has also given the actual first formulation of the Quantum Mechanical Theory of Optical Coherence, which led to the Nobel Prize in 2005 which we feel he should have won. He has also given us the Theory of Instablity, Decay, and Quantum Zeno Effect, Quantum Mechanics of Open Systems and Stochastic Maps, and the Theory of Tachyons - all of which will have impacts in the future. He has been nominated six times for the Nobel Prize, including for the Theory of Weak Interactions and Quantum Optics.

What an extraordinary career and how fortunate for us to share it with him! We recognize it here in the college and we are honored to be here today to celebrate your 75th birthday.

Journal of Physics: Conference Series **196** (2009) 011005 doi:10.1088/1742-6596/196/1/011005

Welcome to The University of Texas at Austin

Juan Sanchez

Vice President for Research, The University of Texas at Austin

E-mail: jsanchez@mail.utexas.edu

I am very happy to be here to honor our very own George Sudarshan on his 75th birthday. I welcome all of you who are here to honor his remarkable scientific contribution. We are very proud to count George among our colleagues here at University of Texas, and we thank you for helping us celebrate this occasion. Happy birthday, George!

The program of this symposium is built around the many and wide contributions that George has made to science. I am sure each of you has a special affection for different parts of his work. I myself am fond of his fundamental contribution to weak interactions - it unveils a central part of the great design of nature. Of course you know that eventually led to the unification of the weak and electromagnetic forces by Prof. Weinberg. I cannot help thinking that if Newton, Maxwell, and Einstein were alive, they would have gladly come to Austin today to honor George Sudarshan.

We have a very exciting program planned for this meeting for the next two days. Despite what Rodger Walser told me, most of this is way over my head, but I will enjoy listening to a few selected talks. Thank you, and have a wonderful time honoring a remarkable scientist and a remarkable man.

Introduction to “V-A”: the first quest

Jan Nilsson

Rector (former), Chalmers Technical University, Göteberg, Sweden

E-mail: kaw@kaw.se

I have known George since 1958, so I experienced first hand the excitement of his important paper that we will hear about later. I have wondered what it takes for great discoveries to happen. I think there are a number of things that are at least very helpful. It needs a brilliant mind, a proper timing, and also a substantial amount of courage. And I think Sudarshan’s discovery of the nature of universal weak interactions is an example of that.

You all recall that after parity violation was discovered, it opened up a number of new possibilities for experimentalists - in data that was already there but not looked at very carefully. And there were a number of new experiments carried out - a confusing situation really. That was the time when George carried out a very careful analysis of the experimental situation and he came to two very important conclusions. First of all that the experimental situation was inconsistent - there were experiments that could not all be right. And the other one was that if there was a universal Fermi interaction it had to be of the V-A type. He also indicated which experiments had to be redone because they did not conform with V-A. That was very brave, because some of those experiments were done by giants in the experimental community. And for a young theorist to point out that they should redo their experiments because they were wrong, I think it takes a good deal of courage. Well, you know that the experiments were redone, and the idea of V-A interactions actually was proven correct. How profound the impact of that was on physics, we will hear some of it at this meeting. I will not get into that because we have the best possible speakers here to tell us about that.

It is my pleasure and privilege to introduce Prof. Steven Weinberg to give the first talk. He has chosen the title: “V-A was the key.”

George Sudarshan

Robert Marshak

$$H_{int} = g\overline{\Psi}_1\gamma_\mu(1+\gamma_5)\Psi_2\overline{\Psi}_4\gamma_\mu(1+\gamma_5)\Psi_3$$

Speakers at V-A: Sudarshan's first quest

Jan Nilsson

Steven Weinberg

Susumu Okubo

Ashok Das

Sandeep Pakvasa

$$H_{int} = g\overline{\Psi}_1\gamma_\mu(1+\gamma_5)\Psi_2\overline{\Psi}_4\gamma_\mu(1+\gamma_5)\Psi_3$$

Journal of Physics: Conference Series **196** (2009) 012002 doi:10.1088/1742-6596/196/1/012002

V-A was The Key

Steven Weinberg

Department of Physics, The University of Texas at Austin

weinberg@physics.utexas.edu

My scientific interaction with George Sudarshan goes back almost half a century to 1958. I was a new post-doc at Columbia University, when I got a call from Bob Marshak, who was the head of the physics department at the University of Rochester, inviting me to come up to Rochester for a week or two to work with him and George Sudarshan and two other people. One of these two I am glad to see here today – Susumu Okubo. Okubo is famous, but the other one, Werner Teutsch, sadly died before he made a name for himself. The invitation was to come up to Rochester and work with them on the currents of weak interactions. I was absolutely thrilled by this invitation for two reasons. One reason was that I was a brand new post-doc, and this was the first time that anyone had ever invited me to go anywhere and pay my plane fare. The other reason for being excited was that, at that time, the whole world of physics was excited by the work that Marshak and Sudarshan had done. They were trailing clouds of glory, because of their development of the $V-A$ theory of weak interactions. This was a great opportunity to work with them.

We wrote two papers together, and I don't think I am going to hurt George's feelings if I say that these were not the most important papers that either of us wrote. They were of some importance, but not great. To me the important thing was getting to know George and Bob Marshak and Okubo, and participating in the intellectual excitement over the breakthrough in the theory of weak interactions.

At that time I did not understand how important that breakthrough would prove to be. It opened up the door to a line of theoretical work, which in less than a decade led to a thorough understanding of things that had earlier, when I was a graduate student, seemed like a mélange of miscellaneous facts that had no coherence and certainly no beautiful theoretical interpretations. What I'd like to do is first go backward in time from 1958 and remind you of some of the prehistory of this and then from 1958 go forward and describe the developments that followed from the $V-A$ theory.

I suppose it really starts from Enrico Fermi's great paper on the theory of beta decay in 1933, a paper, incidentally, that was rejected by *Nature* magazine when it was submitted on the grounds that it was purely mathematical and not really physical. What Fermi did was to initiate the theory of weak interactions. Weak interactions, and in particular the weak interaction of nuclear beta decay, were supposed to occur because of a term in the Hamiltonian of the world involving the product of four fields - electron, neutrino, proton, and neutron. Another way one can think of this is that an electron and neutrino together would interact with a current operator that, unlike the electromagnetic current, does not leave the charge of the nucleus unchanged, but actually converts a neutron into a proton or vice versa. Because of the possibility of an analogy with electromagnetism, Fermi originally thought of this as a vector current, like the electric current. It soon became clear on experimental grounds that this could not be the whole story. Fermi's theory beautifully accounted for the spectrum shape of beta decay (that is the distribution of electron energies between the maximum and minimum) in the fastest beta decays, the so-called allowed beta decays; beta decays that we later understood were decays in which the electron and neutrino receive no orbital angular momentum from the decay process. These allowed beta decays have just the spectrum shape that Fermi's theory predicted, but Fermi's theory also led to selection rules that simply were violated in nature. With a purely vector current, in allowed beta decay – where there's no orbital angular momentum given to the electron and neutrino – the nucleus cannot change its spin. This created no problem at all with, say, the decay of a neutron into a proton, both of which have spin 1/2. That process was allowed. But there were other known beta decays, which Fermi's theory would forbid. One example is the decay of the ground state of Boron-12 or Nitrogen-12 into the ground state of Carbon-12. While the ground states of Boron-12 and Nitrogen-12 both have spin 1, the ground state of Carbon-12 has spin 0, so that decay would be forbidden with just a vector current. And yet, those were discovered to be perfectly good allowed beta decays, going as fast as you would expect of decays in which no orbital angular momentum was exchanged. So it was clear that something more than vector currents was needed.

I think it was Belinfante who first showed that the most general possibility, again, assuming that no orbital angular momentum is transferred in the fundamental decay process, was a mixture of five types of current: scalar, vector, tensor, axial vector, and pseudoscalar. This may be unfamiliar to today's graduate students in physics, because the work of Marshak and Sudarshan made this kind of theoretical classification obsolete. We now know what the right answer is, so who studies these other possibilities anymore? But, I find history fascinating, so allow me here to look back on this bit of history.

It was clear that you had to have scalar or vector, because the other currents would not produce transitions between nuclear states of 0 spin, and there are transitions like the decay of Oxygen-14 into Nitrogen-14 where both nuclei have spin 0. That is allowed by the scalar and vector current interactions, but would not be allowed by the other interactions. Then there are transitions like the one I mentioned when Boron-12 and Nitrogen-12 decay into the ground state of Carbon-12, which require the presence of tensor or axial vector currents. So you have scalar and/or vector plus tensor and/or axial vector currents - in some mixture that no one knew. (The pseudoscalar interaction doesn't produce allowed beta decay, so it is a somewhat tangential possibility.)

And this is where beta decay theory stayed for a long time. There were experiments on forbidden beta decays where angular momentum is transferred, not in the fundamental decay process but through more complicated nuclear physics – these are highly suppressed beta decays – which show that you couldn't have scalar and axial vector, and you couldn't have tensor and vector. And then there were experiments that showed you couldn't have scalar and vector, or tensor and axial vector. After a long history, it appeared that the interaction had to be scalar and tensor or vector and axial vector. The experimental evidence was for a while overwhelming in favor of scalar and tensor, which would quite rule out any analogy with electromagnetism, whose current is a vector current.

As I recall, there were a number of experiments that indicated scalar and tensor, and every once in a while one of those experiments was shown to be wrong. But then another experiment was done which restored the status quo, and again showed scalar and tensor. (I'm talking from my memory, rather than from serious historical research.) By the late 1950s, this picture was beginning to fray at the edges and, in particular, was inconsistent with the idea of universality. I think the thing that most stood out was the problem of the decay of the pi-meson. You know, the main decay of the pi-meson is into a muon and a neutrino. (I'm not being careful in this talk to distinguish neutrinos from anti-neutrinos, but you all know which it is in these decays.) But, a pion can occasionally decay instead into an electron and a neutrino. Now the ratio of those two processes – if you assume the universal interaction, that is if you assume that electrons and muons, and their associated neutrinos (in those days we didn't know that there were associated neutrinos - we thought there was only one type of neutrino - but today we can say 'and their associated neutrinos') interact in exactly the same way, as they seem to be in all respects identical except for their mass, then as Ruderman and Finkelstein showed, you can very neatly predict the ratio of the decay rates. There are only two possibilities for the currents in this decay that are allowed by Lorentz invariance: axial vector and pseudoscalar. Since, at that time most people thought the fundamental interaction was scalar and tensor, and with pseudoscalar being allowed but unconstrained by nuclear beta decay, the natural assumption was that in pion decay the interaction is pseudoscalar rather than axial vector, since there didn't seem to be any axial vector in beta decay. Well, that would predict that the pion would decay overwhelmingly into electrons rather than into muons, because since electrons are lighter, there's more energy available for the decay. But this was in complete contradiction with the observed fact that the pion almost always decays into a muon, not an electron, plus a neutrino.

On the other hand, if the current was axial vector, then because of the peculiar fact that in that case the decay amplitude would be proportional to the mass of the muon or the electron, and since the muon is 210 times heavier than the electron, the decay rate would be almost 210 squared times greater for pion decay into muons than into electrons, which was what was observed.

The whole issue was clouded by the fact that for a long time experimenters couldn't find the electron mode at all. Since the electron made had a rate that apparently was even below the theoretical expectation for axial vector decay, many physicists assumed that for some reason the rate must be zero. (I lost a dollar to Norman Kroll betting on that.) But it was also clearly a possibility that the experimenters had just missed the electron mode, and that in fact the current in pion decay was pure axial vector. But that contradicted everything that was known from beta decays.

So, as Jan Nilsson said here, it took tremendous courage for Marshak and Sudarshan to propose, in order to save universality, and on the grounds of other experiments who perhaps other speakers will discuss, that in fact the weak interaction was vector and axial vector, and not, as had been thought, scalar and tensor. And, of course, subsequent experiment showed that they were right.

This may all sound like a technicality: scalar, vector, tensor, axial vector, pseudoscalar -- who really cares? Fermi didn't even think it necessary to explore these other possibilities, and then experimentalists had beaten away on this problem for decades, and now we had vector and axial vector. So what?

Well, it turned out that this really was the key to the future. The analogy with electromagnetism was restored in the first place, the analogy that originally motivated Fermi. It wasn't long before Feynman and Gell-Mann suggested that to a good approximation the vector current in beta decay is actually conserved, in the same sense that the electromagnetic current was conserved. That is that the rate of change of its time component plus the divergence of its space component equals zero. Gell-Mann discussed various experiments that would test this, and very soon these experiments showed that the idea was right. Because of their suggestion of the conservation of current, Feynman and Gell-Mann (who had independently suggested the $V - A$ theory) I think received more than their share of credit for the $V - A$ theory. I believe the first public statement of the $V - A$ theory was in a paper by Sudarshan and Marshak presented by George in a conference at Padua. (If my memory of that is wrong, doubtless others will correct me). But, perhaps because of their suggestion of a conserved vector current, or perhaps because Feynman and Gell-Mann were already justly famous for other work, and George, after all, was just beginning his career, people tended to refer to this as the Feynman and Gell-Mann Theory. They don't now. I think that with the passage of time, the history has gotten sorted out and it's now generally realized that the earliest statement of the $V - A$ theory was due to Marshak and Sudarshan. Marshak was very upset about this, and I remember a series of international conferences where some speaker would stupidly refer to the Feynman Gell-Mann Theory and Marshak would get up and complain. George was much more of a gentleman and would never publicly complain, and left it to the judgment of history, which has finally been rendered with fairness on this point.

Now, the vector current was known to be conserved. The axial vector current rather obviously couldn't be conserved. In fact, there were obvious objections to that idea. Axial vector current conservation would imply effects in beta decay that were grossly in contradiction to observation. Nevertheless, Yoichiro Nambu saw through this problem and realized that if the pion mass were zero, then the exchange of a pion between the initial and final nucleus and the leptons - the muon or electron and the neutrino - would provide a canceling current which would allow the axial vector current to be conserved without any violation of experimental knowledge.

So this we had a kind of peculiar situation, that in a mythical world with a zero pion mass, the axial vector current would be conserved. This led to a number of interesting predictions. Right away, this allowed an explanation of the success of the calculation of Goldberger and Treiman of the rate of pion decay, which had seemed mysterious because the approximations made by Goldberger and Treiman were difficult to understand. But, with Nambu's idea of an approximately conserved axial vector current, which would be exactly conserved if the pion were massless, you could actually derive the Goldberger-Treiman formula as an exact formula for the pion decay amplitude in the limit when the pion mass is zero. The pion mass, of course, is not zero, but the pion is by far the lightest of all the strongly interacting particles. So, as an approximation, this didn't seem outrageous.

But this was still a peculiar situation. After all, the electromagnetic current was exactly conserved, and this continues to be true to this day. No one challenges it. But the vector current was not exactly conserved, as Feynman and Gell-Mann had recognized, because neutrons and protons don't have exactly the same mass. And the axial vector current is even less exactly conserved. It couldn't be conserved if the pion mass were not zero, and it isn't.

So what's going on here? There was a lot of concentration at this time on the nature of these currents, as my discussion shows. But it was gradually realized that instead of concentrating so much on the properties of the currents themselves, we should concentrate instead on the symmetries of the strong interactions. After all, electromagnetic current conservation is just a manifestation of the fact that there is a symmetry governing electromagnetic phenomena known as electromagnetic gauge invariance. The conserved vector current of Feynman and Gell-Mann is approximately conserved because of the approximate symmetry of strong interactions known as the isotopic spin symmetry. The approximately conserved axial vector current is conserved -- in that weird sense that Nambu suggested -- because of an approximate symmetry of the strong interaction known as chiral symmetry. But this symmetry is not apparent in formulas for ratios of masses or decay rates. If it were, then for instance the nucleon mass would have to be zero. Instead chiral symmetry is spontaneously broken. That is, though it is a symmetry of the underlying equations, it is not a symmetry of the solution of these equations, the particles and their interactions. Goldstone first suggested that whenever you have a spontaneously broken symmetry, you get a massless particle – a so called Goldstone boson, sometimes called a Nambu-Goldstone boson – of which the pion is an archetype. (This was proved as an apparently quite general theorem by Goldstone, Salam, and me.) The pion would be massless if the axial vector symmetry were exact.

There was a lot of confusion at the time about the relation between approximate symmetry and broken symmetry, because many people thought they had something to do with each other. I remember as late as 1975, Heisenberg was suggesting that isotopic symmetry was really an exact symmetry, and only appeared to be an approximate symmetry because it was spontaneously broken. This is really nonsense, because in general when a symmetry is broken, it does not appear as an approximate symmetry – it is completely gone and is manifested not in symmetry relationships like ratios of amplitudes, but instead in theorems about interactions of the Goldstone bosons. It took a while for this understanding to sort itself out.

My own work in the mid 1960's was on the theory of spontaneously broken symmetry – developing it not from the point of view of weak interaction currents but from the point of view of spontaneously breaking an approximate symmetry of strong interactions – the chiral symmetry. But of course, that work, which led to all sorts of soft pion theorems and so on, could not have been done without the realization that there was a partially conserved axial vector current, which could not have been done without the first realization that beta decay proceeds through vector and axial vector currents.

For a while the theorem of Goldstone, Salam, and me – that every time you have a spontaneously broken symmetry, you get a massless Goldstone boson – was regarded as an obstacle to the further development of spontaneous symmetry breaking in particle physics. Because you don't need a giant accelerator to produce massless particles. Especially if the particles have strong interactions, they should be pouring out of experiments at just about any accelerator. Since this doesn't happen, one could easily rule out the existence of new massless particles. In fact, when Salam, Goldstone, and I wrote our paper on this, I contributed an epigraph, a line from King Lear: "nothing can come from nothing," expressing our sense of disappointment with the failure of the idea that approximate symmetries could be explained through the spontaneous symmetry breaking of exact symmetries, or in other words, by the lack of symmetry in the vacuum.

Well, it turned out that spontaneous symmetry breaking did have a future ahead of it, and not only in the theory of low-energy pions. There was an exception to the theorem of Goldstone, Salam, and me, found about the same time by Higgs, by Brout and Englert, and by Guralnik, Hagen, and Kibble, That exception was (as already realized by Anderson in the theory of superconductivity) that if the symmetry is a local symmetry (a gauge symmetry) then instead of the Goldstone boson appearing as a massless particle, it instead appears as a zero helicity state of a massive vector particle, so you would not have any of these unwanted massless Goldstone bosons. Salam and I independently both seized on this as something that would provide an opportunity for a theory of weak and electromagnetic interactions based on an exact but spontaneously broken symmetry. Instead of Goldstone bosons, which should have been observed, you have massive vector particles which transmit the weak and electromagnetic interactions.

That idea of massive vector particles was by no means new. Indeed this is another way in which the V-A theory was the key. As long as you had scalar and tensor currents in the weak interactions, there would be no possibility of weak interactions being transmitted by anything like a photon – a vector field. As soon as the V-A theory came along, many theorists – I remember papers by T. D. Lee and others --- Schwinger, Glashow --- started exploring the possibility that the weak interactions were transmitted by a field which interacts both in a parity conserving and a parity non-conserving way to produce both a vector and an axial vector coupling of beta decay. This idea of intermediate vector bosons was ubiquitous . Everyone knew that an intermediate vector boson had to be heavy because it had not been seen. (Generally speaking, when people said "heavy," they meant ten times the proton mass or something like that.) Experimenters started searching for intermediate vector bosons in cosmic rays and accelerators.

Several theorists showed that having a massive vector boson would lead to mathematical problems, precisely because of their helicity zero component. Spin-1 massless particles have only helicity ±1 states but a spin-1 massive particle has helicity ±1 and 0 states. That zero helicity state of the spin-1 massive particle leads to terrible problems - in particular ultraviolet divergences that could not be removed by renormalization theory. When Salam and I, in the late 60's, proposed our theory of weak and electromagnetic interactions based on the exact but spontaneously broken symmetry, we hoped – and both of us expressed it in our respective papers – that because the local symmetry in the theory was exact, the theory would avoid the problems that were present with a massive vector particle. I attempted unsuccessfully to prove it. The proof was finally presented in 1971 by a young graduate student, Gerardus ‘t Hooft. That finally convinced at least the theoretical community of particle physicists that there was something to this theory.

Those were great days and I am enjoying reminiscing about them. But none of this – not the development of spontaneous symmetry breaking ideas, not the idea of a conserved vector current, and not the unified theory of weak and electromagnetic interactions which is based on spontaneous symmetry breaking ideas – none of that could have occurred without the key breakthrough of the realization that there really is a profound analogy between weak and electromagnetic interactions. It is not that the weak interactions are produced only by a vector current, as Fermi had thought; it certainly is not that weak interactions are produced by scalar and tensor interactions which had dominated the thinking of nuclear physicists for so many years; rather, weak interactions are produced by both vector and axial vector currents, as with great courage and physical insight and élan, Marshak and Sudarshan were the first to propose in 1957.

Comments on Works of Sudarshan in the 1950's at Rochester

Susumu Okubo
Dept. of Physics and Astronomy, University of Rochester, Rochester, NY 14627
E-mail: okubo@pas.rochester.edu

Abstract. I shall discuss some works done by Sudarshan in the 1950's at Rochester. Especially, the topic of magnetic moments of Σ-baryons as well as their electromagnetic mass differences was explained with some historical background. Also, works on weak interaction theory after the discovery of the V-A theory are discussed.

This paper is dedicated to the 75th birthday of Professor E.C. George Sudarshan with whom I coauthored 7 joint papers [1-7] in the 1950's at the University of Rochester, and I will review some of these papers. Although they are now perhaps only of historical interest, it may offer some lesson of interest especially for the younger generation of physicists who are less familiar with the history of the subject. Except for the first paper [1], the rest of them [2-7] deal with further development of the pivotal V-A theory of Sudarshan and Marshak [8] in 1957 and subsequently by Feynman and Gell-Mann [9]. First, I will discuss the earliest joint paper [1] on electromagnetic mass differences among Σ-baryons.

1. Magnetic Moment of Baryons and Electromagnetic Mass Differences

1.1. Historical Background

In the early 1930's, Heisenberg [10] for the nucleon and then Kemmer [11] for pions have proposed the notion of the isotopic spin invariance based upon SU(2) group, which has been supposed to be a exact symmetry group for strong interactions. The experimentally observed small mass differences between the proton, p and the neutron, n for the nucleon, N as well as between π^+, π^0, and π^- for pions have been attributed to be due to effects of electromagnetic self-energies, although the lowest order perturbation theory will lead to a incorrect sign for the mass difference, $M(n) - M(p) \simeq 1.29$ MeV. Feynman and Speisman [12] have, nevertheless, observed that we could explain the correct mass difference, if we take into account the anomalous magnetic moments of the nucleon with very sharp electromagnetic form factors. Although the latter assumption on the form factors was found later to be incorrect from studies of e-p elastic scatterings in the late 1960's, it was not yet known in 1950. Soon after the discovery of the Σ baryon, $\Sigma = (\Sigma^+, \Sigma^0, \Sigma^-)$ which belongs to the isotopic spin 1 realization of the SU(2) group, it has been noted that they have also small mass differences among them. The current experimental values [13] are

$$M(\Sigma^-) - M(\Sigma^+) \simeq (8.08 \pm 0.08) \text{ MeV},$$

$$M\left(\Sigma^{-}\right)-M\left(\Sigma^{0}\right) \simeq (4.807 \pm 0.035)\ \mathrm{MeV}. \tag{1}$$

Sudarshan and Marshak [14] in 1956 have proposed that these mass differences may similarly be explained by anomalous magnetic moments of these baryons, and George decided to compute them.

1.2. Magnetic Moment Sum Rule

Although Sakata [15] has proposed the so-called Sakata model in 1956 in which all hadrons are assumed to be bound states of three fundamental objects of p, n, and Λ, the idea was not so well-known, until Gell-Mann [16] and Zweig [17] in 1964 transformed it by introducing the quark model based upon three fractionally charged quark, u, d, and s instead. (See reference 18 for some brief historical account on this subject.) At any rate, the most particle theorists around the 1950's and early 1960's thought that all hadrons such as N, Λ, Σ, Ξ, π, and K are elementary particles with Yukawa interactions of form:

$$\begin{aligned} H_{\pi} &= i\Big\{g_{N\pi\pi}\left(\overline{N}\gamma_5\underline{\tau}N\right)+g_{\Sigma\Lambda\pi}\left(\overline{\underline{\Sigma}}\gamma_5\underline{\tau}\Lambda+\text{ h.c.}\right) \\ &\quad + g_{\Sigma\Sigma\pi}\left(\overline{\underline{\Sigma}}\times\gamma_5\underline{\Sigma}\right)+g_{\Xi\Xi\pi}\left(\overline{\Xi}\gamma_5\underline{\tau}\Xi\right)\Big\}\underline{\pi} \end{aligned} \tag{2}$$

for pion and

$$\begin{aligned} H_K &= i\Big\{g_{NK\Lambda}\left(\overline{N}\gamma_5\Lambda\right)K+g_{NK\Sigma}\left(\overline{N}\gamma_5\underline{\tau}\cdot\underline{\Sigma}\right)K \\ &\quad + g_{\Xi\Lambda K}\left(\overline{\Lambda}\gamma_5\Xi\right)i\tau_2 K+g_{\Xi\Sigma K}\left(\overline{\underline{\Sigma}}\cdot\underline{\tau}\gamma_5\Xi\right)i\tau_2 K\Big\}+\text{ h.c.} \end{aligned} \tag{3}$$

for kaon interactions.

In the lowest order perturbation calculations for these coupling constants $g_{\Sigma\Sigma\pi}$, $g_{NK\Lambda}$ etc., George (Sudarshan) found a interesting sum rule

$$\mu\left(\Sigma^{+}\right)+\mu\left(\Sigma^{-}\right)=2\mu\left(\Sigma^{0}\right) \tag{4}$$

among magnetic moments $\mu(\Sigma)$'s.

Because the relation does not appear to be accidental and seems to be of more general validity, Professor Marshak organized special research study seminars on the problem. He himself conjured up a Bohr-like model of the Σ-hyperon in which the Σ's are regarded as bound states of core Λ or N with circulating π or $\overline{K}$. On the basis of such a model, he could derive the sum rule of Eq. (4). However, we could find the following more satisfactory explanation. In 1954, Gell-Mann [19] and Nakano-Nishijima [20] have proposed the formula for electric charge Q of hadrons in terms of the baryon number B and the strangeness quantum number S by

$$Q=I_3+\frac{1}{2}(B+S) \tag{5}$$

where I_3 is the 3rd component of the iso-spin $I=(I_1,I_2,I_3)$. This formula was very successful to explain the semi-stability (against strong interactions) of these newly discovered hadrons. Corresponding to Eq. (5), the electromagnetic current operator $j_{\mu}^{e\ell m}(x)$ and hence the quantum mechanical magnetic moment operator μ must have the same iso-spin structure, i.e.,

$$\mu=V_3^{(I=1)}+V_0^{(I=0)} \tag{6}$$

where $V_3^{(I=1)}$ is the 3rd component of some iso-vector operator $\underline{V}^{(I=1)}$ and $V_0^{(I=0)}$ stands for an iso-scalar one. Then, the experimentally observed magnetic moment $\mu(I_3)$ of a baryon with isotopic spin I can be computed by

$$\mu(I_3) = < I, I_3|\mu|I, I_3 > = aI_3 + b \tag{7}$$

according to the well known angular momentum theory [21]. If $I \geq 1$, then Eq. (7) will give a sum rule

$$\mu(I_3 = +1) + \mu(I_3 = -1) = 2\mu(I_3 = 0) \tag{8}$$

since we calculate

$$\mu(I_3 = \pm 1) = \pm a + b, \quad \mu(I_3 = 0) = b,$$

noting that constants a and b are independent of I_3. Applying this formula for the Σ's with $I = 1$, it reproduces the sum rule Eq. (4) which holds valid independently of any dynamical details, as long as the strong interaction is invariant under the SU(2) group.

Remark 1

In 1959, Weinberg and Treiman [22] have applied the same technique to prove the validity of

$$M(\Delta^{++}) - M(\Delta^{-}) = 2\left\{M(\Delta^{+}) - M\left(\Delta^{0}\right)\right\}$$

for electromagnetic mass differences among the Δ resonances with $I = \frac{3}{2}$.

Remark 2

Let us consider the flavor SU(3) symmetry based upon three quarks u, d, and s. Then, the major mass difference among them (ignoring the small mass difference between $M(u)$ and $M(d)$) behaves as a T_3^3 component of a octet tensor $T_\nu^\mu(\mu, \nu = 1, 2, 3)$ under the SU(3) group. This leads to the SU(3) mass formula [18] for hadrons composed of these quarks. However, we can proceed also as in the following way by using the concept of the U-spin [23]. In the U-spin basis, the mass operator M with the mass splitting operator T_3^3 can now be expressed as

$$M = M_3^{(U=1)} + M_0^{(U=0)} \tag{9}$$

analogously to Eq. (6). Then just as Eq. (7), we calculate

$$M(U_3) = < U, U_3|M|U, U_3 > = a'U_3 + b' \tag{10}$$

for the masses of hadron multiplets with the U-spin U. Especially for $U = \frac{3}{2}$, this yields

$$\begin{aligned} M\left(U_3 = \frac{3}{2}\right) - M\left(U_3 = \frac{1}{2}\right) &= M\left(U_3 = \frac{1}{2}\right) - M\left(U_3 = -\frac{1}{2}\right) \\ &= M\left(U_3 = -\frac{1}{2}\right) - M\left(U_3 = -\frac{3}{2}\right). \end{aligned}$$

Applying this to the baryon decouplet of $(\Omega^-, \Xi^*, Y, \Delta)$, we find the well known mass formula

$$M(\Omega^-) - M(\Xi^{*-}) = M(\Xi^{*-}) - M(Y^-) = M(Y^-) - M(\Delta^-)$$

which established the SU(3) theory into a firmer ground.

Remark 3 (Global Symmetry)

Although this is a unsuccessful attempt, it was perhaps one of the most popular models of strong interactions before the SU(3) theory. In this model, we assume

$$g = g_{NN\pi} = g_{\Sigma\Lambda\pi} = g_{\Sigma\Sigma\pi} = g_{\Xi\Xi\pi} \tag{11}$$

among coupling constants of H_π in Eq. (2). If we set

$$\begin{aligned} N_1 &= \begin{pmatrix} p \\ n \end{pmatrix}, \quad N_2 = \begin{pmatrix} \Sigma^+ \\ \frac{1}{\sqrt{2}}(\Lambda - \Sigma^0) \end{pmatrix}, \\ N_3 &= \begin{pmatrix} \frac{1}{\sqrt{2}}(\Lambda + \Sigma^0) \\ \Sigma^- \end{pmatrix}, \quad N_4 = \begin{pmatrix} \Xi^0 \\ \Xi^- \end{pmatrix}, \end{aligned} \tag{12}$$

then H_π can be rewritten (see e.g. ref. 24 and earlier references quoted therein) as

$$H_\pi = ig\left\{\overline{N}_1\gamma_5\underline{\tau}N_1 + \overline{N}_2\gamma_5\underline{\tau}N_2 + \overline{N}_3\gamma_5\underline{\tau}N_3 + \overline{N}_4\gamma_5\underline{\tau}N_4\right\}\underline{\pi}, \tag{13}$$

which has a very large symmetry group of SU(2) $\otimes$ U(4), where U(4) stands for the four-dimensional unitary group among N_1, N_2, N_3, and N_4. Although the kaon Hamiltonian H_K of Eq. (3) does not possess such a nice symmetry, it was believed at that time that H_K is small in comparison to H_π and can be ignored in the first order approximation. As we shall show shortly, this model in this approximation will then predict

$$\begin{aligned} M(\Sigma^+) - M(\Sigma^-) &= 0, \\ M(\Xi^-) - M\left(\Xi^0\right) &= M(p) - M(n) \end{aligned} \tag{14}$$

for electromagnetic mass differences [25], as well as

$$\begin{aligned} \mu(p) &= \mu(\Sigma^+) = -\mu(\Sigma^-) = -\mu(\Xi^-), \\ \mu(n) &= -\mu\left(\Xi^0\right), \quad \mu\left(\Sigma^0\right) = \mu(\Lambda) = 0 \end{aligned} \tag{15}$$

for magnetic moments. To demonstrate the validity of these relations, we utilize the concept of the Weyl symmetry group, which is the 4-dimensional symmetric group S_4 of 4-objects for the present problem. We note first that the electromagnetic interaction is also invariant under the Weyl symmetry of $N_1 \leftrightarrow N_2$ and also of $N_3 \leftrightarrow N_4$, this immediately proves the validity of $\mu(p) = \mu(\Sigma^+)$, and $\mu(\Sigma^-) = \mu(\Xi^-)$. In order to prove the rest of the relations, we consider antiparticle states N_3' and N_4' of N_3 and N_4 with

$$\begin{aligned} N_3' &= i\tau_2 C\overline{N}_3, \\ N_4' &= i\tau_2 C\overline{N}_4 \end{aligned}$$

where C is the particle-antiparticle (or charge conjugate) operator. Then H_π is rewritten as

$$H_\pi = ig\left\{\overline{N}_1\gamma_5\underline{\tau}N_1 + \overline{N}_2\gamma_5\underline{\tau}N_2 - \overline{N}'_3\gamma_5\underline{\tau}N'_3 - \overline{N}'_4\gamma_5\underline{\tau}N'_4\right\}\underline{\pi}.$$

Although its invariance group is no longer U(4) but the non-compact one of U(2,2), both H_π and electromagnetic interaction are invariant under the modified Weyl reflection of

$$N_1 \leftrightarrow N'_3, \quad N_2 \leftrightarrow N'_4, \quad \underline{\pi} \rightarrow -\underline{\pi}.$$

Since the magnetic moment of antiparticle differs in sign in comparison to that of the particle, this gives the rest of the relations in Eq. (15). Such a method also explains the validity of Eq. (14) for electromagnetic mass differences [25].

We note that relations, Eqs. (14) and (15) are very badly satisfied in comparison to the experimental values. For example, Eq. (14) implies 6.48 MeV $= -1.29$ MeV, which is wrong in sign. These are one reason why we abandoned the global symmetry model in favor of the SU(3) theory. Note that the Coleman-Glashow relation [26] based upon the SU(3) symmetry predicts

$$\mu(p) = \mu\left(\Sigma^+\right), \quad \mu\left(\Sigma^-\right) = \mu\left(\Xi^-\right), \quad \mu\left(\Xi^0\right) = \mu(n) = 2\mu\left(\Lambda\right), \tag{16}$$

in contrast to Eq. (15), which are reasonably consistent with the present experimental values [13]. Also, it predicts the mass relation of

$$M\left(\Xi^-\right) - M\left(\Xi^0\right) = M\left(\Sigma^-\right) - M\left(\Sigma^+\right) + M(p) - M(n), \tag{17}$$

i.e.,

$$6.48 \text{ MeV } = 8.08 \text{ MeV } - 1.29 \text{ MeV } (= 6.73 \text{ MeV})$$

again in a reasonable agreement with the experiment. We remark that these SU(3) relations can be more easily derived from Weyl symmetry group S_3 of the SU(3) group. The same method has been extensively used by Levison, Lipkin, Meshkov [27] and by Dullemond, McFarlane, and Sudarshan [28] for various other processes. However, these topics will be covered by the talk of Professor Meshkov.

Remark 4 (Quark Model)

In the 1960's, a significant paradigm change has occurred in particle physics. First, it was the quark model in which the SU(3) octet baryon can be realized as

$$p = (uud), \quad n = (ddu), \quad \Sigma^+ = (uus), \quad \Sigma^- = (dds), \quad \Xi^0 = (ssu), \quad \Xi^- = (ssd) \tag{18}$$

as bound states of three quarks u, d, and s. Moreover, the iso-spin group SU(2) may also be violated by a small mass difference between masses of u and d quarks. In the naive quark model, the mass difference among hadrons are simply calculated accordingly as

$$\begin{aligned} M(n) - M(p) &= M(d) - M(u), \\ M\left(\Sigma^-\right) - M\left(\Sigma^+\right) &= 2\{M(d) - M(u)\} \\ M\left(\Xi^-\right) - M\left(\Xi^0\right) &= M(d) - M(u) \end{aligned}$$

from Eq. (18). Note also that $M(d) = 3 \sim 7$ MeV, $M(u) = 1.5 \sim 3.0$ MeV, $M(s) \simeq 95 \pm 25$ MeV with $M(u)/M(d) \simeq 0.3 \sim 0.6$. Although these predict the correct sign for mass differences, it leads to a very bad relation of

$$M\left(\Xi^-\right) - M\left(\Xi^0\right) = M(n) - M(p) = \frac{1}{2}\left\{M\left(\Sigma^-\right) - M\left(\Sigma^+\right)\right\}.$$

This fact shows the deficiency of the naive quark model. Moreover, a better Coleman-Glashov relation of Eq. (17) indicates the necessity of taking into account the electromagnetic interactions.

1.3. *Concluding Remark*

Presently, no particle theorist will doubt that these electromagnetic mass differences as well as magnetic moments of baryons can be explained from calculations based upon the lattice QCD theory. However the computation is notoriously complicated and this has not yet been achieved. Hence, the problem posed by George Sudarshan in 1956 has not yet been answered even after 50 years. Perhaps we have to wait at least another 10 years for the final answer.

2. ΔI $=\frac{1}{2}$ Selection Rule for Semi-leptonic Hadron Decays

This section deals with two papers in references [2] and [4]. We first note:

2.1. *Historical Background*

In the present standard model of the weak interaction, the weak Hamiltonian responsible for semi-leptonic hadronic decays involving the first family of particles; u, d, s, μ, e, ν_μ and ν_e is given by

$$H_W = \frac{1}{\sqrt{2}}G\left\{\cos\theta J_\mu^{(\Delta S=0)} + \sin\theta J_\mu^{(\Delta S=1)}\right\}\left\{\overline{\mu}\gamma_\mu\left(1+\gamma_5\right)\nu_\mu + \overline{e}\gamma_\mu\left(1+\gamma_5\right)\nu_e\right\} \tag{19}$$

where the strangeness-conserving current $J_\mu^{(\Delta S=0)}$ and the strangeness violating current $J_\mu^{(\Delta S=1)}$ are expressed as

$$\begin{aligned} J_\mu^{(\Delta S=0)} &= \overline{u}\gamma_\mu\left(1+\gamma_5\right)d, \\ J_\mu^{(\Delta S=1)} &= \overline{u}\gamma_\mu\left(1+\gamma_5\right)s. \end{aligned} \tag{20}$$

Note that they have particular isotopic spin structures of $I = 1$ and $I = \frac{1}{2}$, respectively. However, even after the pivotal discovery of the V-A theory in 1957, by Sudarshan and Marshak, such a simple isotopic spin behavior for $J_\mu^{(\Delta S=0)}$ and $J_\mu^{(\Delta S=1)}$ were completely unknown. This is mostly due to the fact that the quark model did not appear until 1964. As a matter of fact we simply expressed $J_\mu^{(\Delta S=1)}$ for example to be

$$\begin{aligned} J_\mu^{(\Delta S=1)} &= g_1\overline{p}Q_\mu\Lambda + g_2\overline{p}Q_\mu\Sigma^0 + g_3\overline{n}Q_\mu\Sigma^- + \\ &+ g_4\overline{\Sigma}^+Q_\mu\Xi^0 + g_5\overline{\Sigma}^0Q_\mu\Xi^- + g_6\overline{\Lambda}Q\Xi^- + \ldots \end{aligned} \tag{21}$$

with $Q_\mu = \gamma_\mu\left(1+\gamma_5\right)$ in terms of supposedly elementary objects p, n, Σ^+, Σ^0, Σ^-, Λ, Ξ^0, and Ξ^- etc. just as Eqs. (2) and (3) for strong interaction Hamiltonians with so many unknown coupling constants g_1 g_2, ...

In the summer of 1958, Professor R.E. Marshak had organized a discussion seminar on this problem at Rochester by inviting very young promising theorists, Dr. S. Weinberg from Columbia University and Dr. E.C.G. Sudarshan from Harvard University. Note that Sudarshan has finished his Ph.D thesis at Rochester under Marshak in 1957 and went to Harvard as a postdoctoral fellow. As for me, I was still a graduate student at Rochester because of a problem with my thesis advisor. At any rate, after several discussions, we produced two joint papers (see ref. 2 and 3). The idea was simply that $J_\mu^{(\Delta S=1)}$ has the isotopic spin property of $I = \frac{1}{2}$. Although this idea is now so trivial, it was regarded as quite a novel approach at that time. In fact, we could derive the so called $\Delta Q = \Delta S$ rule for $J_\mu^{(\Delta S=1)}$ from our hypothesis as follows. From the Gell-Mann-Nakano-Nishijima relation, Eq. (5), we observe

$$\Delta Q = \Delta I_3 + \frac{1}{2}(\Delta B + \Delta S) = \Delta I_3 + \frac{1}{2}\Delta S \tag{22}$$

since $J_\mu^{(\Delta S=1)}$ preserves the baryonic number so that $\Delta B = 0$. However, we must have $\Delta Q = \pm 1$, $\Delta S = \pm 1$ for $J_\mu^{(\Delta S=1)}$ with $\Delta I_3 = \pm\frac{1}{2}$. There are only two solutions consistent with Eq. (22) of $\Delta Q = \Delta S = 1$ and $\Delta I_3 = \frac{1}{2}$ or $\Delta Q = \Delta S = -1$ and $\Delta I_3 = -\frac{1}{2}$, both of which satisfies $\Delta Q = \Delta S$. Again, this rule is now so obvious from the structure of $J_\mu^{(\Delta S=1)}$ given by Eq. (20b). However, it was again a very new discovery at that time.

2.2. $\Delta I = \frac{1}{2}$ *Selection Rule*

Since the isotopic spin group SU(2) is expected to be a good symmetry for strong interactions, any consequence from the isospin $\frac{1}{2}$ structure of $J_\mu^{(\Delta S=1)}$ will constitute a good test for the standard model. Using the Wigner-Eckart theorem, we can then readily find the following relations

(i)

$$\begin{aligned} M\left(K^+ \to \pi^0 \bar{\ell}\nu\right) &= \frac{1}{\sqrt{2}} M\left(K^0 \to \pi^- \bar{\ell}\nu\right), \\ M\left(\overline{K}^0 \to \pi^- \bar{\ell}\nu\right) &= 0, \end{aligned} \tag{23}$$

for decay rates of these semi-leptonic modes. Especially, this implies

$$M\left(K^+ \to \pi^0 \bar{\ell}\nu\right) = M\left(K_2^0 \to \pi^- \bar{\ell}\nu\right) = M\left(K_1^0 \to \pi^- \bar{\ell}\nu\right). \tag{24}$$

Experimentally, this yields

$$\frac{B\left(K_L^0 \to \pi^\pm e^\mp \nu\right)}{\tau\left(K_L^0\right)} = \frac{B\left(K_S^0 \to \pi^\pm e^\mp \nu\right)}{\tau\left(K_S^0\right)} = \frac{2B\left(K^+ \to \pi^0 e^+ \nu\right)}{\tau\left(K^+\right)}$$

for branching rates B and decay lifetime τ. Experimentally, it is quite well satisfied to be given by

$$7.94 \times 10^6\,\mathrm{sec}^{-1} = 7.85 \times 10^6\,\mathrm{sec}^{-1} = 7.90 \times 10^6\,\mathrm{sec}^{-1}.$$

(ii) Similarly, for muon decays

$$\frac{B\left(K_L^0 \to \pi^\pm \mu^\mp \nu\right)}{\tau\left(K_L^0\right)} = \frac{2B\left(K^+ \to \pi^0 \mu^+ \nu\right)}{\tau\left(K^+\right)},$$

which should be compared experimentally

$$5.29 \times 10^6 \,\mathrm{sec}^{-1} = 5.32 \times 10^6 \,\mathrm{sec}\,.$$

(iii) We would get

$$\begin{aligned} M\left(\Sigma^+ \to n\bar{\ell}\nu\right) &= 0, \\ M\left(\Xi^0 \to \Sigma^-\bar{\ell}\nu\right) &= 0. \end{aligned} \tag{25}$$

The current experimental limit for decay rates are

$$\frac{R\,(\Sigma^+ \to ne^+\nu)}{R\,(\Sigma^- \to ne\bar{\nu})} < 0.9 \times 10^{-2}.$$

(iv) We also find

$$M\left(\Xi^0 \to \Sigma^+\ell\bar{\nu}\right) = \sqrt{2}M\left(\Xi^- \to \Sigma^0\ell\bar{\nu}\right) \tag{26}$$

which gives

$$\frac{B\,(\Xi^0 \to \Sigma^+ e\bar{\nu})}{\tau\,(\Xi^0)} = \frac{2B\,(\Xi^- \to \Sigma^0 e\bar{\nu})}{\tau\,(\Xi^-)}$$

or

$$1.0 \times 10^6 \,\mathrm{sec}^{-1} = 0.93 \times 10^6 \,\mathrm{sec}^{-1}$$

(v) S. Oneda has noted the validity of

$$\begin{aligned} M\,(K^+ \to \pi^+\pi^+ e\bar{\nu}) &= 0, \\ M\left(K^0 \to \pi^+\pi^0 e\bar{\nu}\right) &= 0, \end{aligned} \tag{27}$$

as well as a sum rule

$$\begin{aligned} \sqrt{2}M\left(K_2^0 \to \pi^0\pi^- e^+\nu\right) &= \sqrt{2}M\left(K_1^0 \to \pi^0\pi^- e^+\nu\right) \\ &= \frac{1}{\sqrt{2}}M\left(K^+ \to \pi^0\pi^0 e^+\nu\right) + M\,(K^+ \to \pi^+\pi^- e^+\nu) \end{aligned} \tag{28}$$

on the basis of the $\Delta I = \frac{1}{2}$. The latter can give a triangle inequality of

$$\left| R\,(K^+ \to \pi^+\pi^- e^+\nu) - \sqrt{\frac{1}{2}R\,(K^+ \to \pi^0\pi^0 e^+\nu)} \right|$$

$$\leq \sqrt{R\,(K_L^0 \to \pi^0\pi^\pm e^\mp\nu)} \leq \sqrt{R\,(K^+ \to \pi^+\pi^- e^+\nu)} + \sqrt{\frac{1}{2}R\,(K^+ \to \pi^0\pi^0 e^+\nu)}$$

for the decay rates. This should be compared to the experimental values

$$36.6 - 29.3 \leq 57.1 \leq 36.6 + 29.3 \; (= 65.9).$$

2.3. Conclusion

The $I = \frac{1}{2}$ current hypothesis for $J_\mu^{(\Delta S=1)}$ is now just a simple consequence of the standard model of Weinberg-Salam-Glashow theory of the weak interaction. Nevertheless, it provides good experimental tests of the theory in good agreements with currently available data. In this sense, our 1958 paper has stood a test of 50 years of time.

3. Conserved Current Hypothesis for Vector Part of $\mathbf{J}_\mu^{(\Delta=1)}$

This is based upon two papers of ref. 3 and 7.

3.1. Historical Background

Now, we know that $\Delta S = 0$ vector current $V_\mu^{(\Delta S=0)} = \overline{u}\gamma_\mu d$ will satisfy the conservation law

$$\partial^\mu V_\mu^{(\Delta S=0)} = 0$$

in the SU(2) limit in which we ignore the small mass difference of $M(d) - M(u)$. This leads to no renormalization theorem for various $\Delta S = 0$ semi-leptonic decays. We wondered in 1958, whether a similar situation exists for the vector part of $\Delta S = 1$ current, i.e.,

$$\partial^\mu V_\mu^{(\Delta S=1)} = \partial^\mu \left(\overline{u}\gamma_\mu s\right) = 0$$

in the current terminology. We now know that this is possible for the SU(3) limit of $M(s) = M(u)$, which is far worse than that of the SU(2) limit. However, in 1958, the situation was moot, and we computed its possible consequence for semi-leptonic decay modes $K \to \pi \ell \overline{\nu}$.

3.2. Conclusion

As is expected, the hypothesis yields a pole in the variable q^2 (q is the momentum transfer between K and π mesons) for one of the decay form factors. Now we know that this is a unrealistic consequence of the conservation law. The correct approach is to compute these decay form factors directly from the lattice QCD and we have nothing more to say about the subject.

4. Asymmetry Parameters of Ξ and Λ Decays

The paper of ref. 6 was concerned with the asymmetry parameters of non-leptonic decays of $\Xi \to \Lambda\pi$ followed by $\Lambda \to N\pi$. If the parity is violated in these hadronic decays, the decay matrix element of $\Lambda \to p\pi^-$ for example will be given by

$$M\left(\Lambda \to p\pi^-\right) = A + B\underline{\sigma}\cdot\underline{n} \tag{29}$$

where $\underline{n} = \frac{\underline{p}}{|\underline{p}|}$ for the momentum $\underline{p}$ of the decaying pion in the rest frame of Λ. The asymmetry parameters α, β, γ of the decay are defined by

$$\begin{aligned} \alpha &= \frac{2\mathrm{Re}(A^*B)}{|A|^2+|B|^2}, \quad \beta = \frac{2\mathrm{Im}(A^*B)}{|A|^2+|B|^2}, \\ \gamma &= \frac{|A|^2-|B|^2}{|A|^2+|B|^2} \end{aligned} \tag{30}$$

with $\alpha^2 + \beta^2 + \gamma^2 = 1$. The parity violation implies $AB \neq 0$. Especially, if $\alpha \neq 0$ or $\beta \neq 0$, the parity is violated in the decay.

Now consider the successive decay mode of $\Xi^- \rightarrow \Lambda\pi^-$ followed by $\Lambda \rightarrow p\pi^-$. Then, we discovered [6] that the angular correlation of two decaying pions has a simple form of

$$P(\theta) = 1 + \alpha_\Lambda \alpha_\Xi \cos\theta \tag{31}$$

where

$$\cos\theta = \frac{\underline{p_1} \cdot \underline{p_2}}{|\underline{p_1}| \cdot |\underline{p_2}|}$$

is the cosine of the angle θ for two pion moments $\underline{p_1}$ and $\underline{p_2}$. Here the second pion momentum $\underline{p_2}$ must be measured in the rest frame of Λ. The important point is that the formula $P(\theta)$ does not depend upon polarization of the initial Ξ. Hence, by measuring $P(\theta)$, we can experimentally determine $\alpha_\Lambda \alpha_\Xi$. The currently available experimental values [13] are

$$\begin{aligned} \alpha_{\Lambda\rightarrow p\pi^-} &= 0.64, \quad \alpha_{\Lambda\rightarrow n\pi^0} = 0.65, \\ \alpha_{\Xi^-\rightarrow\Lambda\pi^-} &= -0.46, \quad \alpha_{\Xi^0\rightarrow\Lambda\pi^0} = -0.41 \end{aligned}$$

which are non-zero, indicating the parity violations for all these hadronic decays.

5. V-A Theory and Λ-Decay

In ref. 5, we attempted to compute the decay rate and the asymmetry parameter α_Λ for $\Lambda \rightarrow p\pi^-$ decay from the first principle based upon the V-A theory. We used a dispersion theoretical calculation to estimate the effect of the final state interaction. But the attempt was definitely premature. We have yet to wait for the correct calculation based upon the lattice QCD.

5.1. Acknowledgement

This paper is a tribute to the 75th birthday of Professor E.C. George Sudarshan. It was a pleasure to have seen him at Austin, to be healthy and in good spirits. Also, this note is supported in part by the U.S. Department of Energy Grant no. DE-FG02-91ER40685.

6. References

[1] Marshak R E, Okubo S and Sudarshan E C G 1957 *Phys. Rev.* **106** 599
[2] Weinberg S, Marshak R E, Okubo S, Sudarshan E C G and Teutsch W 1958 *Phys. Rev. Lett.* **1** 25
[3] Okubo S, Marshak R E, Sudarshan E C G, Teutsch W and Weinberg S 1958 *Phys. Rev.* **112** 665
[4] Okubo S, Marshak R E and Sudarshan E C G, 1959 *Phys. Rev. Lett.* **2** 12
[5] Okubo S, Marshak R E and Sudarshan E C G 1959 *Phys. Rev.* **113** 944
[6] Teusch W, Okubo S and Sudarshan E C G 1959 *Phys. Rev.* **114** 1148
[7] Hatsukade S, Marshak R E, Okubo S and Sudarshan E C G 1960 *Nuovo Cimento* **16** 332
[8] Sudarshan E C G and Marshak R E 1957 *Proc. of the Conf. on Mesons and Recently Discovered Particles*
[9] Feynman R and Gell-Mann M 1958 *Phys. Rev.* **109** 193
[10] Heisenberg W 1932 *Z. Phys.* **80** 587
[11] Kemmer N 1937 *Phys. Rev.* **52** 906
[12] Feynman R P and Speisman G 1954 *Phys. Rev.* **94** 500
[13] Particle Data Group 2006 *J. Physics, Nuclear and Particle Physics* **30**
[14] Sudarshan E C G and Marshak R E 1956 *Phys. Rev.* **104** 267
[15] Sakata S 1956 *Prog. Theor. Phys.* **16** 686
[16] Gell-Mann M 1964 *Phys. Lett.* **8** 214
[17] Zweig G 1964 *CERN report. no. 8182/TH401*
[18] Okubo S 2006 *Mod. Phys. Lett.* A**21** 373

[19] Gell-Mann M 1953 *Phys. Rev.* **92** 833
[20] Nakano T and Nishijima K 1953 *Prog. Theor. Phys.* **10** 581
[21] Cohen-Tannoudji C, Din B and Laloë F 1977 *Quantum Mechanics* (Wiley) Vol. 2
[22] Weinberg S and Treiman S B 1959 *Phys. Rev.* **116** 465
[23] Lipkin H 1965 *Lie Groups for Pedestrians* (North-Holland)
[24] Feinberg G and Gürsey F 1959 *Phys. Rev.* **114** 1153
[25] Sakurai J J 1959 *Phys. Rev.* **114** 1152
[26] Coleman S and Glashow S L 1961 *Phys. Rev. Lett.* **6** 423
[27] Levinson C A, Lipkin H I and Meshkov S 1962 *Nuovo Cimento* **23** 236
[28] Dullemond C, McFarlane A J and Sudarshan E C G 1963 *Phys. Rev. Lett.* **10** 423

Journal of Physics: Conference Series **196** (2009) 012004 doi:10.1088/1742-6596/196/1/012004

V-A theory: a view from the outside

Ashok Das

Department of Physics and Astronomy, University of Rochester,
Rochester, NY 14627, USA
and
Saha Institute of Nuclear Physics,1/AF Bidhannagar,Calcutta 700064, INDIA

E-mail: `das@pas.rochester.edu`

Abstract. In this talk I will review the V-A theory within the context of the prevalent experimental results at the time.

It is a great pleasure for me to be able to participate in this celebration in honor of Professor George Sudarshan. Let me say right away that unlike the other speakers in this session, I haven't had the opportunity to collaborate with George yet. On the other hand, George is still young and who knows!

Once upon a time there was .....

Much of what is being discussed in this session happened many, many years ago. I am, of course, not going to tell you how old I am. Let me simply say that all of this happened long before my time, which is the reason for the title of my talk. The only reason I agreed to speak in this session is because George is a dear friend, a former student of Rochester and I felt that I had access to all the resources of Rochester. This was before I knew who else were going to be speaking in this session and that the organizers had kindly deprived me of part of my resources, thank you very much!

Second Rochester Conference, January 11-12, 1952

So, I had to work very hard, go back to the old papers and the archives and reconstruct events. What I learnt is that the story of *V-A* theory is a fascinating one. You can even think of this as one of the stories of *Sheherazade* in the *Arabian Nights*. As we all know these are very intricately woven stories that naturally lead from one story to the next. In fact, if only *Sheherazade* were aware of the story of *V-A*, you can hear her starting the story as, "A long time ago, there was....". The difference is that the "long time" in this case can be quantified. You have already heard that it all happened 50 years ago, but let me simply say that "it all happened that many years ago".

The story of *V-A* has two aspects to it. There is, of course, the physics aspect that I will come to. But, more important is the sociological aspect. Many younger people in the audience may not know that the Rochester conference which started in Rochester in 1951 was one of the first successful international conferences in high energy physics in this country. Over the years, the character of this conference has, of course, changed

Journal of Physics: Conference Series **196** (2009) 012004 doi:10.1088/1742-6596/196/1/012004

enormously. The first Rochester conference, for example, had only 50 participants and lasted for just one day.

The discussions in these conferences were friendly but intense and this was the venue for announcing new results. So, it was an exciting place to be in. If you think arranging visas for foreign visitors is difficult now, you should think again. Things were not at all easy at that time.

The other sociological aspect is that it is around this time that Bob Marshak started recruiting smart foreign graduate students into the university. Even the university was not fully supportive, but in his characteristic way, Marshak prevailed. It is in this way that both George Sudarshan and Susumu Okubo arrived into the middle of all the exciting things that were happening in Rochester around the Rochester conferences.

Young graduate students in the middle of all the excitement.

The physics aspect can be summarized by saying that it was a very exciting period, I guess because everything was so confusing. To me the most impressive thing about the V-A theory is that it was formulated in the face of experimental results that did not support its predictions. In physics experiments are expected to give guidance to theorists. Sometimes an experiment may itself be wrong, but then it is the responsibility of other experiments to "weed out" the wrong experiment. On the other hand, when repeated experiments stand by a result, it is generally foolish to propose a theory that contradicts the accepted experimental results. However, this is exactly what Sudarshan and Marshak did in proposing the V-A theory simply because they had a desire to have a universal theory of weak interactions and the predictions of their theory were subsequently vindicated by more careful experiments. Basically, there was a combination of wrong experiments that had completely dominated the physics scene before the V-A theory. Let me explain this in some detail.

The story of weak interactions started with the observations of nuclear β decays and the developments in the field can be divided into two phases: a phase prior to the discovery of parity violation and a phase after this discovery. In the earlier phase, with more and more studies on nuclear decays, it was established that there are three kinds of nuclear beta decays [1, 2]:

(i) Fermi transitions: $\Delta J = 0, \quad 0 \to 0. \quad (O^{14} \to N^{14} + e^+ + \nu)$

(ii) Gammow-Teller transitions: $\Delta J = 1. \quad (He^6 \to Li^6 + e + \bar{\nu})$

(iii) Mixed transitions: $\Delta J = 0, \quad 0 \not\to 0. \quad (n \to p + e + \bar{\nu})$

In addition to the nuclear β decays, it was also observed (1948) that the *muon* decays through weak interactions as

$$\mu^- \to e + \nu + \bar{\nu}.$$

Tiomno and Wheeler [3] were the first to carry out a systematic study of the strengths of all these decays which suggested a universal nature of the weak force governing all of these decays (Puppi's triangle):

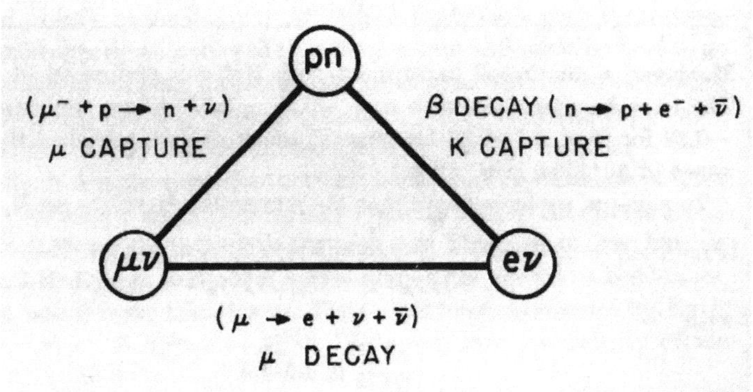

As a result, the Fermi theory of β decay [1] involving a current-current interaction of vector currents (in analogy with QED) was generalized to an interaction of the form

$$H_{\text{int}} = \sum_i C_i J_i J^{i\,\dagger}, \quad J_i = J_i^{\text{hadron}} + J_i^{\text{lepton}},$$

where the generalized "current" had the generic form

$$J_i = \bar{\psi}_1 \Gamma_i \psi_2, \quad i = S, V, T, A, P.$$

Namely, the generalized "current" involved all the five bilinear covariants of Dirac matrices. On the other hand, since the nucleons inside the nucleus were nonrelativistic, one could immediately deduce:

(i) Only S, V terms contribute to Fermi transitions (since they do not change spin).

(ii) Only T, A terms can contribute to Gamow-Teller transitions (since these can change spin).

The pseudoscalar term vanishes in the nonrelativistic limit and, therefore, nothing could be deduced about this coupling. Furthermore, since Fermi and Gammow-Teller transitions were observed, it meant that one cannot have vanishing couplings for both S, V or both T, A terms in the interaction. The form of the interaction Hamiltonian can, of course, be constrained further by experimental studies.

(i) Fermi transitions: $\Delta J = 0, \quad 0 \to 0. \quad (O^{14} \to N^{14} + e^+ + \nu)$

(ii) Gammow-Teller transitions: $\Delta J = 1. \quad (He^6 \to Li^6 + e + \bar{\nu})$

(iii) Mixed transitions: $\Delta J = 0, \quad 0 \not\to 0. \quad (n \to p + e + \bar{\nu})$

Journal of Physics: Conference Series **196** (2009) 012004 doi:10.1088/1742-6596/196/1/012004

In addition to the nuclear β decays, it was also observed (1948) that the *muon* decays through weak interactions as

$$\mu^- \rightarrow e + \nu + \bar{\nu}.$$

Tiomno and Wheeler [3] were the first to carry out a systematic study of the strengths of all these decays which suggested a universal nature of the weak force governing all of these decays (Puppi's triangle):

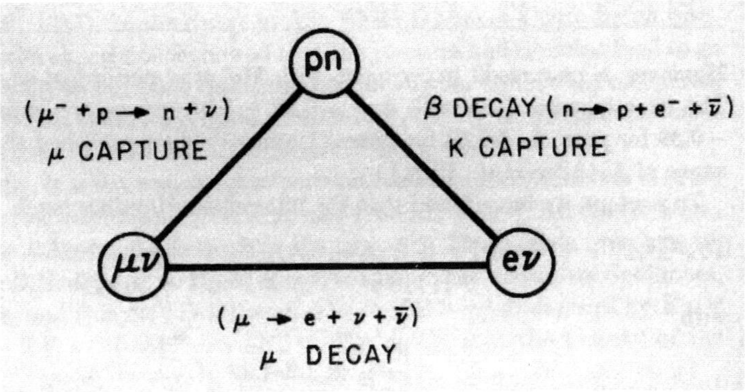

As a result, the Fermi theory of β decay [1] involving a current-current interaction of vector currents (in analogy with QED) was generalized to an interaction of the form

$$H_{\text{int}} = \sum_i C_i J_i J^{i\,\dagger}, \quad J_i = J_i^{\text{hadron}} + J_i^{\text{lepton}},$$

where the generalized "current" had the generic form

$$J_i = \bar{\psi}_1 \Gamma_i \psi_2, \quad i = S, V, T, A, P.$$

Namely, the generalized "current" involved all the five bilinear covariants of Dirac matrices. On the other hand, since the nucleons inside the nucleus were nonrelativistic, one could immediately deduce:

(i) Only S, V terms contribute to Fermi transitions (since they do not change spin).

(ii) Only T, A terms can contribute to Gamow-Teller transitions (since these can change spin).

The pseudoscalar term vanishes in the nonrelativistic limit and, therefore, nothing could be deduced about this coupling. Furthermore, since Fermi and Gammow-Teller transitions were observed, it meant that one cannot have vanishing couplings for both S, V or both T, A terms in the interaction. The form of the interaction Hamiltonian can, of course, be constrained further by experimental studies.

However, before I describe the experimental results, let me point out a theoretical result that will play a crucial role in the later developments. It was already known from cosmic ray studies that the *pion* decays through weak interactions through the channel

$$\pi^- \rightarrow \mu^- + \bar{\nu}.$$

However, the decay involving the electron

$$\pi^- \to e^- + \bar{\nu},$$

had never been seen. That was a bit puzzling at the time since e, μ had similar properties except for their mass difference. Based on the universal interaction hypothesis of Tiomno and Wheeler, a theoretical calculation was carried out by Ruderman and Finkelstein [4] for the ratio of the decay rates for the two modes

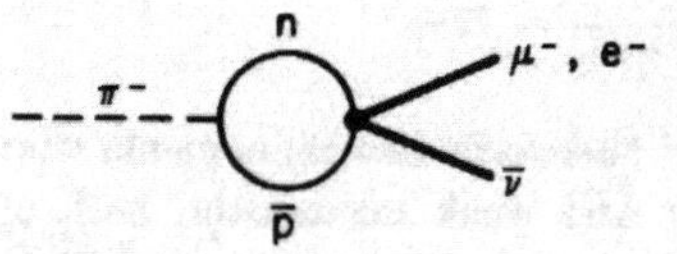

and their results are expressed in the table (f denotes forbidden)

TABLE I. Ratio of $\pi \to (e, \nu)$ to $\pi \to (\mu, \nu)$-decay for couplings (1) and (7).

Meson	Type of β-decay: Scalar	P-scalar	Vector	P-vector	Tensor
Scalar	5.1	f	f	f	f
P-scalar	f	5.1	f	1.0×10^{-4}	f
Vector	f	f	4.0	f	2.4
P-vector	f	f	f	4.0	f

I have highlighted the relevant numbers for the appropriate couplings. Since the only allowed channels through which the *pion* can decay to electrons are the P and the A channels and since this decay is so rare, theoretically, the P (P-scalar) channel was ruled out (This is the channel on which we had no information in the nonrelativistic limit). Therefore, experimentally one had to look at only the S, V, T, A channels.

The way experiments looked for these terms in the decay process is by looking at the end spectrum of the electron energy in the decay or by looking at the angular correlation between the electron and the neutrino which can be determined from the recoil of the nucleus. For example, the angular correlation can be parameterized as

$$f(\theta) \sim 1 + \lambda\, \frac{v}{c} \cos\theta,$$

where v denotes the speed of the electron. (In writing this I have neglected a term known as the *Fierz interference* term which was not observed experimentally.) Theoretically, the parameter λ can be calculated for the different channels and takes the values

	S	V	T	A
λ	-1	$+1$	$+1/3$	$-1/3$

So, for example in the decay of He^6 (which is a Gammow-Teller decay), λ was determined to have the value $\lambda = \frac{1}{3}$ so that one would conclude the coupling for Gamow-Teller transitions to be primarily of T type [5].

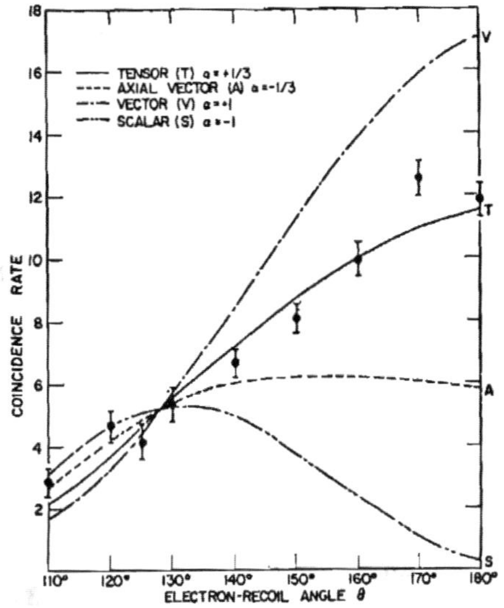

One can summarize the results of many of these experiments at the time as [6]

Nuclear decay	Method	λ	Conclusion
Prior to parity non-conservation			
^{6}He	Recoil Spectrum		T (not A)
n	Recoil Spectrum	$+0{\cdot}09 \pm 0{\cdot}11$	S and T or V and A
		$-0{\cdot}21 \pm 0{\cdot}08$	
^{19}Ne	Recoil Spectrum	$+0{\cdot}14 \pm 0{\cdot}2$ $-0{\cdot}15 \pm 0{\cdot}2$	S and T or V and A
^{6}He	Angular Correlation	$+0{\cdot}34 \pm 0{\cdot}12$	T (not A)

This suggested that the β decays were governed by S, T couplings. On top of that the leading search for the $\pi \to e + \nu$ led to the result [7]

$$\frac{\Gamma(\pi \to e + \nu)}{\Gamma(\pi \to \mu + \nu)} = (-3 \pm 9) \times 10^{-5}.$$

This further ruled against an interaction of A type and supported the S, T couplings. Thus, the dominant sentiment in the theoretical physics community at the time was that the generalized Fermi theory with S, T couplings was responsible for the weak decays.

Then came the proposal by Lee and Yang [8] that parity is violated in weak interactions which was soon [9] established in the Gamow-Teller decay of Co^{60} which also showed that parity is violated maximally in these decays. Violation of parity necessitated that the interaction Hamiltonian should be further generalized to

$$H_{\text{int}} = \sum_i \bar{\psi}_1 \Gamma_i \psi_2 \, \bar{\psi}_3 \Gamma^i \left(C_i + C_i' \gamma_5\right) \psi_4 + \text{hermitian conjugate}, \quad i = S, V, T, A.$$

As a result, experiments needed to be analyzed afresh. Surprisingly, however, the large volume of experimental data in β decays were unaffected by violation of parity since they involved measuring parity invariant quantities. However, there were two new experiments that are worth talking about. First, there was the experiment [10] which studied the transitions:

$$\mathrm{Ar}^{35} \rightarrow \mathrm{Cl}^{35} + e^{+} + \nu, \quad (\mathrm{Fermi}), \quad \mathrm{Ne}^{19} \rightarrow \mathrm{F}^{19} + e^{+} + \nu \quad (\mathrm{Mixed}).$$

The angular correlation measurements in these decays gave a value of $\lambda \approx 0.9$ supporting a strong V coupling in the interactions.

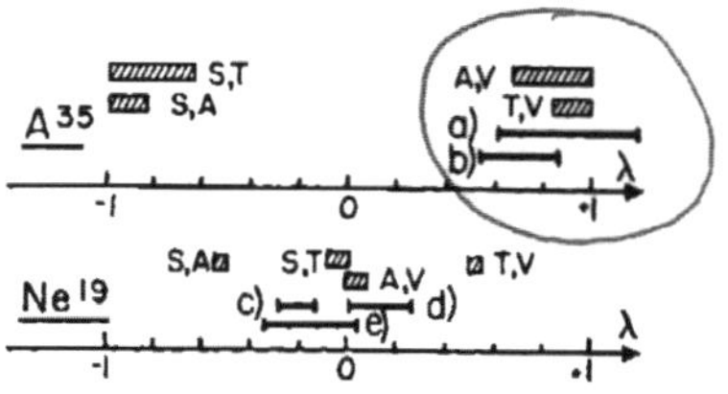

However, these results were in contradiction with most other results including one of their own earlier experiments. (Let me remark here parenthetically that IUPAC changed the symbol for Argon from A^{35} to Ar^{35} in 1957.) The second was a better experiment [11] designed to search for $\pi \rightarrow e + \nu$ decays and it gave even a lower limit on the ratio to be

$$\frac{\Gamma(\pi \rightarrow e + \nu)}{\Gamma(\pi \rightarrow \mu + \nu)} = (-0.4 \pm 9) \times 10^{-6},$$

and, therefore, in spite of the new component of parity violation in the theory, the general sentiment continued to be that the interactions were of S, T type.

However, at this point, there was a new twist from the theoretical side. Observing that all the parity violating experiments involved a neutrino, it was argued that it is the neutrino that is responsible for the violation of parity and this resurrected the two component theory of neutrinos. Salam [12] was the one who introduced the concept of γ_5 invariance into the neutrino equations. "Chirality" or handedness crept into this discussion through the works of Watanabe [13] and it was clear that unlike other particles, neutrinos were either left-handed or right-handed. Various β asymmetry measurements showed that if the neutrino was left-handed, the interaction will have a (V, A) form while if it was right-handed (S, T) form would be favored. (Handedness of neutrino will be measured only a year later in a beautiful experiment by Goldhaber et al. [14]) The *muon* decay experiments, on the other hand, seemed to support a (V, A) form of the interactions. The idea of a universal Fermi theory was, therefore, in jeopardy and the confusion that prevailed is best summarized by the remarks of T. D. Lee in the seventh Rochester conference (1957)

"... *We turn to the universal Fermi interaction, which is an attempt to gain a more unified understanding of certain of the weak interactions. We draw the famous triangle representing the interactions of interest. Beta decay information tells us that the interaction between* (p, n) *and* (e, ν) *is scalar and tensor, while the two component*

Journal of Physics: Conference Series **196** (2009) 012004 doi:10.1088/1742-6596/196/1/012004

theory plus the law of conservation of leptons implies that the coupling between (e,ν) and (μ,ν) is vector. This means that the universal Fermi interaction cannot be realized in the way we have expressed it $\cdots$"

This basically summarizes the climate in which Sudarshan and Marshak proposed the universal *V-A* theory [15]. Their desire was to have a universal Fermi theory and was not based primarily on symmetry principles which came only after the fact. Let me try to reconstruct here their reasoning for proposing such a structure for the theory. First, since neutrinos have a definite handedness, they satisfy

$$(1 \pm \gamma_5)\,\psi^{(\mathrm{R,L})} = 0 = \bar{\psi}^{\,(\mathrm{R,L})}\,(1 \mp \gamma_5)\,,$$

depending on their handedness. Since the longitudinal polarization of the electron in the Co^{60} decay is negative (predominantly left-handed), this then determined that the current in the lepton sector would involve (S,T) couplings if the neutrino is right handed and (V,A) couplings if the neutrino is left handed. From the β decay experiments, therefore, it would correspond to choosing between the He^6 or the Ar^{35} results respectively. I want to emphasize here that the He^6 results were the commonly accepted ones at the time and the Ar^{35} results were not generally in favor. The muon decay, on the other hand, favored the (V,A) coupling which can be seen in the following way. The current involving the neutrino and the antineutrino (with well defined handedness)

$$\bar{\psi}_\nu\,(1 \mp \gamma_5)\,\Gamma_i\,(1 \pm \gamma_5)\,\psi_\nu = 0, \quad \text{for } i = S, T, P,$$

and is nontrivial only for $i = V, A$. Therefore, if one wants a universal theory for all the interactions, one must choose the results of the Ar^{35} experiment over the commonly accepted He^6 results. That is exactly what Sudarshan and Marshak did. Such a choice then leads to the universal *V-A* theory of the form

$$H_{\text{int}} = G\bar{\psi}_1\gamma_\mu(1+\gamma_5)\psi_2\bar{\psi}_3\gamma^\mu(1+\gamma_5)\psi_4 + \text{hermitian conjugate}, \quad C_i = C_i',\ i = V, A;\ C_V = C_A,$$
$$C_i = C_i' = 0,\ i = S, T.$$

On the other hand, such an interaction would predict the *pion* to decay into an electron. Subsequently, several new experiments measured the ratio for this decay leading to the value [16]

$$\frac{\Gamma(\pi \to e + \nu)}{\Gamma(\pi \to \mu + \nu)} = (1.03 \pm 0.20) \times 10^{-4},$$

which vindicated the *V-A* hypothesis. Many of the β decay experiments were soon redone and corrected for the errors in the older results.

Once a *V-A* structure for the interaction was proposed, its symmetry properties, which are important for later developments, followed. For example, it was realized that the interaction Hamiltonian is invariant under a γ_5 transformation. It was also observed that the *V-A* structure is invariant under the *Fierz rearrangement*. In fact, the combination S-$T + P$ is also invariant, but as I have tried to emphasize, the P interactions do not enter the weak interaction Hamiltonian. However, I do not have time to get into these.

You can almost hear it now. At the conclusion of this triumphant story, the curious emperor, *Sheheriyar*, asks *Sheherazade*, "What happened next?" Like a good story teller, politely and respectfully, *Sheherazade* reminds the emperor that it is almost dawn and that it may be better to rest a little and continue the next day, to which the emperor agrees. After a sumptuous dinner the following evening, *Sheherazade* meets the emperor at the appointed hour and resumes her story. "You see, your excellency, ...", she continues and as we have learnt from Professor Weinberg's talk this morning, the story of V-A leads naturally to the story of the Standard Model. The story (of *Sheherazade* and of scientific discoveries), of course, continues. However, unlike *Sheherazade*, I have only half an hour and, therefore, let me thank you all for your attention and stop here.

Acknowledgment:
This work was supported in part by US DOE Grant number DE-FG 02-91ER40685.

References

[1] Fermi E 1934 Z. Physik **88**, 161
[2] Gamow G and Teller E 1936 Phys. Rev. **49**, 895
[3] Tiomno J and Wheeler J A 1949 Rev. Mod. Phys. **21**, 153. See also Klein O 1948 Nature **161**, 897; Puppi G 1948, Nuovo Cim. **5**, 587
[4] Ruderman M and Finkelstein R 1949 Phys. Rev. **76**, 1458
[5] Rustad B M and Ruby S L 1955 Phys. Rev. **97**, 991
[6] Wu C S 1960 in *Theoretical Physics in the Twentieth Century*, ed. M. Fierz and V. F. Weisskopf (Interscience Publishers Inc, New York)
[7] Lokanathan S and Steinberger J 1955 Suppl. Nuovo Cim. **1**, 151
[8] Lee T D and Yang C N 1956 Phys. Rev. **104**, 254
[9] Wu C S *et al* 1957 Phys. Rev. **105**, 1413
[10] Herrmannsfeldt W B , Maxson D R, Stäheln P, and Allen J S 1957 Phys. Rev. **107**, 641
[11] Anderson H L and Lattes C M G 1957 Nuovo Cim. **6**, 1356
[12] Salam A 1957 Nuovo Cim. **5**, 229; Landau L 1957 Nucl. Phys. **3**, 127; Lee T D and Yang C N 1957 Phys. Rev. **105**, 1671
[13] Watanabe S 1957 Phys. Rev. **106**, 1306
[14] Goldhaber M, Grodzins L, and Sunyar A W 1958 Phys. Rev. **109**, 1015
[15] Sudarshan E C G and Marshak R E 1957 *Proceedings of Padua-Venice Conference on Mesons and Newly Discovered Particles*; *ibid* 1958, Phys. Rev. **109**, 1860
[16] Anderson H L, Fuji T, Miller R H, and Tau L 1960 Phys. Rev. **119**

50 Years of V-A (Living Dangerously)

Sandip Pakvasa

Department of Physics & Astronomy, University of Hawaii, 2505 Correa Road, Honolulu, HI 96822

E-mail: pakvasa@phys.hawaii.edu

Abstract. How V-A has survived for so long despite many a threat to its survival is described with a somewhat idiosyncretic choice of topics.

In 1957, George Sudarshan undertook an analysis of all the data on beta decay, μ-decay, μ-capture and π-decay. The conclusion he reached in collaboration with Bob Marshak was as follows: A universal weak interaction of the form $H_w \sim J^\dagger J$ with J as the weak current is only possible if the space time structure of the current J is V-A. This was presented by Marshak at the April 1957 Padua-Vernice conference and published (eventually) in the proceedings[1]. The importance of V-A to the subsequent development of the standard model of electro-weak interactions as well as of the eventual theory of strong interactions has been stressed in the talk of Prof. Weinberg ("V-A was the key"). Here I will concentrate on phenomenological and experimental aspects.

At the time, their proposal was already in conflict with four experimental results, two published and two preliminary (unpublished). The published results were the $e-\nu$ correlation in He^6 beta decay[2] and the $\pi \to e\nu$ branching ratio[3]; the unpublished preliminary results were the e-polarization in μ decay [4]. and the e-asymmetry in polarized neutron decay[5]. Sudarshan and Marshak urged that these four experiments be repeated (to quote from their paper: "all of these experiments should be re-done...").

In $He^6 \to Li^6 + e^- + \bar{\nu}_e$, which is a pure Gamaw-Teller (G.T.) transition with $\Delta J = 1$, the angular correlation between e and $\bar{\nu}_e$ is given by

$$P(\theta) = 1 + \lambda v/c \ \cos\theta_{e\nu} \tag{1}$$

where $\lambda = -1/3$ for Axial Vector and + 1/3 for Tensor interaction. The result from Ruby-Rustad[2] experiment was $\lambda = +0.34 \pm 0.09$ clearly favoring tensor coupling. A few months after the (V-A) was proposed, Wu and Schwarschild[6] presented a critique of the Ruby-Rustad experiment at an APS meeting (April 21, 1958). They pointed out that the detection efficiency had a strong angular dependence, the He^6 gas pressure gradient had to be known to great accuracy, and there were back-scattering effects. They concluded that when all these uncertainties were included, the errors were large enough so that λ could be just as consistent with -1/3 as with + 1/3 in the Ruby-Rustad experiment. Ruby and Rustad themselves presented a similar critique[7]. Eventually a measurement[8] of the energy spectrum of the recoiling nucleus which is given by

$$d\Gamma \sim \{E(E_0 - E) + \lambda\,[A + E_0(E_0 - E)....]\} \tag{2}$$

yielded $\lambda = -033 \pm 0.003$, ruling out tensor coupling.

The ratio $R = \Gamma(\pi \to ev)/\Gamma(\pi \to \mu\nu)$ has a strong dependence on the space time structure of the decay interaction. R is given by

$$R(V-A) = \frac{m_e^2(m_\pi^2 - m_e^2)^2}{m_\mu^2(m_\pi^2 - m_\mu^2)^2} \approx 1.2.10^{-4}$$

$$R(S-P) = \frac{(m_\pi^2 - m_e^2)^2}{(m_\pi^2 - m_\mu^2)^2} \cong 5.5 \tag{3}$$

The negative search[3] had yielded a bound on R of $R < 10^{-5}$. A new search after the V-A proposal found positive evidence for $\pi \to e\nu$ at the expected rate[9] of $1.2.10^{-4}$.

The electron energy spectrum in polarized μ-decay is proportional to

$$d\Gamma \sim [x(1-x) + \rho f(x) + \eta x_0(1-x) \pm \; P_\mu \cos\theta_e \{\xi g(x)\}] \tag{4}$$

where $x = E_e/m_\mu, x_0 = m_e/E_{\max}$, P_μ is the muon polarisation, $\pm$ stands for the charge of the muon f and g are known functions of x, and θ_e is the angle between muon polarization and electron momentum. For V-A, $\rho = 3/4$ and $\xi = 1$. At the time of the V-A proposal, ρ had been measured to be about 0.7 and a preliminary value for ξ was reported to be $\xi << 1$. However, the final published value was $\xi = 0.9 \pm 0.1$[10].

In the decay of polarized neutron, the coefficient B of the term

$$B\{\underline{S}_{\,n} \cdot \underline{P}_{\,e}\,(W - E_e)/E_e\} \tag{5}$$

is predicted to be 1 for V-A. Again, a preliminary value of $B << 1$ was reported[5]; but the final published value was $B = 0.88 \pm 0.3$. The measurement of neutrino helicity in the elegant experiment of Goldhaber, Grodzins and Sunyar[12] was an additional confirmation of the V-A nature of the weak current.

The universality had to be modified to take into account the low rate for the beta decay of Λ[14]. After the early suggestion of Gell-Mann and Levy[13], the final form as proposed by Cabibbo[15] still survives today:

$$J_\mu = \cos\theta_c \; J_\mu^{(0)} \; (\Delta I = 1) + \sin\theta_c \; J_\mu^{(1)} \; (\Delta I = 1/2) \tag{6}$$

where θ_c is the Cabibbo angle, the superscript stands for change of strangeness (ΔS).

The main features of the Cabibbo current are: (i) J_μ the weak charged current transforms as 8 in flavor SU(3), (ii) Conserved vector current hypothesis makes the vector current transform as the anti-symmetric 8_F and the Axial vector current is a mixture of D (symmetric) and F (antisymmetric) couplings. There are two parameters, θ_c and D/F ((D+F) is fixed by g_A/g_V in neutron beta decay). A recent detailed fit is given in [16].

The agreement of the hyperon semi-leptonic decay rates with the data is impressive. The most interesting case is of $\Sigma \to ne^-\bar{\nu}_e$; which has a predicted matrix element of $V + 0.33A$ and was confirmed by the data only recently.

The final step in establishing the low energy effective weak interaction was taken when the deviation of g_A/g_V from one was accounted for, using the proposal of Gell-Mann's current algebra[17] with the closed form for the equal time commutator of [A,A] to close on V. This non-linearity in the current commutators allowed Adler and Weisberger[18] to calculate g_A/g_V.

With the structure of the weak charged current in hand, the non-leptonic weak decays (especially of hyperons which had a rich structure) were thought ripe for attack. Before turning to that, let us first summarize the main features of the complex of non-leptonic hyperon decays. There are seven decay modes $\Lambda \to p\pi^-, \Lambda \to n\pi^0, \Sigma^+ \to p\pi^0, \Sigma^+ \to n\pi^+, \Sigma^- \to n\pi^-$, $\Xi \to \Lambda\pi^-$ and $\Xi^0 \to \Lambda\pi^0$; and each has s-wave (parity violating) and p-wave (parity conserving) amplitudes. The main properties of these decays can be summarized as: (i) approximate $\Delta I = 1/2$ rule (the ratio of the amplitudes $\Delta I = 3/2/\Delta I = 1/2$ being small, i.e. less than 5%):

$$\begin{aligned} \Delta(\Lambda) &= \Lambda^0_- + \sqrt{2}\,\Lambda^0_0 \cong 0 \\ \Delta(\Xi) &= \Xi^-_- - \sqrt{2}\,\Xi^0_0 \cong 0 \\ \Delta(\Sigma) &= \Sigma^-_- - \Sigma^+_+ - \sqrt{2}\,\Sigma^+_0 \cong 0 \end{aligned} \tag{7}$$

where Λ^0_- stands for amplitude for $\Lambda \to p + \pi^-$ etc; and (ii) the (B.W.) Lee- H. Sugawara sum rule: $\Delta(L.S.) = \Lambda^0_- + 2\Xi^-_- - \sqrt{3}\Sigma^+_0 \cong 0$, also at a 5% level. The above regularities are found for both s-waves and p-waves[19].

In a 1966 paper [20] the most general current-current interaction with the Cabibbo form for weak currents was considered and shown to lead to the following sum rules for the s-wave (parity violating) amplitudes:

$$\begin{aligned} \Delta(\Lambda) + \Delta(\Xi) = 0 \\ 2\Delta(L.S.) = \sqrt{\frac{3}{2}}\,\Delta(\Xi) + \Delta(\Sigma) \end{aligned} \tag{8}$$

It was shown soon after that the same sum rules follow for p-wave amplitudes with a dynamical assumption of pole-dominance[21]. With the use of current algebra (or chiral symmetry) the vanishing of the Σ^+_+ s-wave amplitude was tied to the vanishing of $\Delta I = 3/2$ amplitude[22]. Hence, if it could be shown that the $\Delta I = 1/2$ amplitudes dominated over $\Delta I = 3/2$ ones, almost all the observed properties of non-leptonic hyperon decays could be accounted for. With the application of QCD renormalization group to the amplitudes a modest enhancement of O(5-6) could be obtained but not the factor of O(20) needed[23].

By this time the Standard Model (SM) of electro-weak interactions[24] was already in the picture. In the Glashow-Weinberg-Salam SM, described by the spontaneously broken SU(2) x U(1) gauge interaction, the charged weak current is pure V-A and the neutral weak current J_n is of the form[24]

$$J_n \sim (V - A)_{I_3} + \sin^2\,\theta\ J_{e.m.} \tag{9}$$

where θ is the weak mixing angle. Hence the neutral weak current is V-A with an admixture of a known vector current. There were several proposals to modify the charged current by addition of V+A currents. One was to add a small admixture of (V+A) to account for the observed CP violation[25]. A more radical proposal was the so-called "Vector Model" with equal strengths for Left-handed (V-A) and Right-handed (V+A) charged currents with a resulting pure vector neutral current[26]. This was motivated by the experimental indications that $\sigma^{N.C.}(\nu) \cong \sigma^{N.C.}(\bar{\nu})$ and hence the hint that the neutral current may be pure vector: There was also the added attraction that the addition of a R-H-C for a heavy quark current may enhance the $\Delta I = 1/2$ amplitude for non-leptonic hyperon decay by supplying an extra ingredient; e.g. via an effective coupling such as

$$H_{NL} \sim (\bar{d}_R\ \gamma_\mu c_R)\ (\bar{c}_L\ \gamma_\mu\ s_L) \tag{10}$$

It was soon shown that this was illusory due to constraints from soft pion theorems[27]. Furthermore, the data on high energy neutrino neutral current scattering settled down to a value $\sigma(\nu)/\sigma(\bar{\nu}) \neq 1$; thus ruling out a pure vector neutral current. There was a valiant attempt wherein the effective neutral current interaction remained parity conserving (viz

$H_{NC} \sim VV + AA$) and the result $\sigma_v/\sigma_{\bar{\nu}} \neq 1$ was achieved by the handedness of ν and $\bar{\nu}$[28], but this too was soon ruled out by data.

In 1978 two measurements changed the scene and settled in favor of the standard model $SU(2)_L x U(1)$ and stopped the fury of model building. One was the SLAC measurement[29] of left-right asymmetry A, where $A = (\sigma_L - \sigma_R)/(\sigma_L + \sigma_R)$ with $\sigma_L(\sigma_R)$ being the cross-section for left (right) polarized electrons scattering on deuterons. In the Standard Model, A is given by

$$A_{SM} = -\frac{G_F Q^2}{2\sqrt{2}\pi\alpha}(9/16)\left[\left(1 - \frac{20}{9}\sin^2\theta\right) + (1 - 4\sin^2\theta)K(y)\right] \tag{11}$$

where the first term is from $A_e V_q$ coupling and the second is from $V_e A_q$ coupling and

$$K(y) = \frac{\{1 - (1-y)^2\}}{\{1 + (1-y)^2\}} \tag{12}$$

for $y = E'/E$ and Q^2 is the momentum transfer. The experimental value(measured at $Q^2 \approx 1.6(GeV/c)^2$) was found to be

$$A_{exp} = -(9.5 \pm 16)10^{-5}(Q^2/GeV^2) \tag{13}$$

This corresponds to a value for $\sin^2\theta$ of about 0.224 and was consistent with the value for $\sin^2\theta$ deduced from neutrino neutral current data.

The second result was the first observation of parity violation in atoms, which is dominantly due to the coupling $A_e V_q$. The effective coupling is

$$H_{eff} = \frac{G_F}{4\sqrt{2}}\left(\frac{Q_W}{m_e}\right)\underline{S}_e \cdot \left[\delta^3(x)\underline{p}_e + \underline{p}_e\,\delta^3(x)\right] \tag{14}$$

The experiment in Novosibirsk by Barkov and Zolotorev[30] measured the rotation of plane of polarization of light by Bismuth. The rotation is given by $\delta\phi = \frac{1}{2}Re(n_+ - n_-)k\ell$ where n_+ and n_- are the refractive indeces for the two polarizations, k is the wave number and ℓ the distance in the medium $(n_+ - n_-)$ in turn is proportional to $(f_+ - f_-)$ where f is the scattering amplitude and $(f_+ - f_-)$ in turn is proportional to Q_W the effective weak charge of the atom.

For Bismuth,

$$Q_W(Bi) = -\left[Z(4\sin^2\theta - 1) + N\right] \approx -130 \tag{15}$$

for $\sin^2\theta$ as given above. The observations led to

$$Q_W(Bi)_{exp} \approx -140 \pm 40 \tag{16}$$

in good agreement with the expectation.

These two results showed that the weak neutral current for neutrinos, electrons and quarks is given correctly by the Standard Model as in Eq. (8) and is described by a single unique value of the mixing angle θ.

As we now know, the Standard Model with V-A interaction has been unchallenged from the scale of eV to several hundred GeV. No hint for any deviation or any small admixture of V+A or S,P,T currents has yet been seen. A recent very accurate measurement of the parameters in muon decay was carried out by the TWIST experimental collaboration at TRIUMPH, and no deviation from V-A values was found[31]. Severe constraints on many models including V+A interactions were placed by these and many other results.

Problems such as a complete understanding of the structure of the non-leptonic hyperon decay amplitudes or understanding of the weak radiative decays (especially the large asymmetry in

$\Sigma^+ \to p\gamma$ decay mode) most probably have to do with the low energy soft QCD rather than the fundamental structure of weak interaction. Someday QCD calculations on the lattice should be able to account for these phenomena in detail.

The V-A structure in the Standard Model is robust and is tested from a q^2 of eV^2 to $(TeV)^2$ over some 12 orders of magnitude. It will continue to be tested!

Acknowledgments
I thank Tom Browder for a careful reading of the paper. This work is supported in part by USDoE Grant #DE-FG02-04ER41291.

References

[1] Sudarshan E C G and Marshak R E 1957 The Nature of the Four-Fermion Interaction, in Proc. Conf. on Mesons and Newly Discovered Particles, Padua-Venice (Conference held in Sept 1957), ibid. 1958 ed. Zonichelli N, Bologna, p. v-14. By then the Feynman-Gell-Mann paper on V-A was also published: Feynman R P and Gell-Mann M 1958 Phys. Rev. 109, 193
[2] Rustad B M and Ruby S L 1953 Phys. Rev. 89, 880
[3] Anderson H L, and Lattes C M G 1957 Nuovo Cim. 6, 1356; Lokanathan S and Steinberger J 1955 Nuov. Cim, 2, 151 (1955)
[4] Coffin T, Berley D, Garwin R, Lederman L and Weinrich M, presented by Lederman L 1957 Proc. of VII International Conference on High Energy Physics, Rochester ed., VII-31
[5] Novey T B and Telegdi V L (unpublished), cited as Ref. 13 in [1]
[6] Wu C S and Schwarschild A 1958 Columbia Report CV-173, (unpublished). See Wu C S, Proc. of the Wingspread Conference , 50 Years of Weak Interactions, ed. Cline D and Riedasch G, University of Wisconsin, p. 208
[7] Raby S L and Rustad B M 1958 Post Deadline paper at the annual New York APS meeting
[8] Hermmansfeldt W B *et. al.* 1958 Phys. Rev. Lett. 1, 61
[9] Impeduglia G *et. al.* 1958 Phys. Rev. Lett. 1, 249
[10] Maeg P C, Crowe K M, and Haddack R P 1958 Phys. Rev. 112, 2061
[11] Allen J *et. al.* 1958 Phys. Rev. 116, 134
[12] Goldhaber M, Goodzins L, and Sunyar A W 1957 Phys. Rev. 109, 1015
[13] Gell-Mann M and Levy M 1950 Nuov. Cim. 16, 705
[14] Ely R P *et. al.* 1962 Proc. I.C.H.E.P. CERN, Geneva ed. Prentki J, p.445
[15] Cabibbo N 1963 Phys. Rev. Lett. 10, 531
[16] Cabibbo N, Swallow E C and Winston R 2003 Ann. Rev. Nucl. Part. Sci., 63, 69
[17] Gell-Mann M 1962 Phys. Rev. 126, 1067
[18] Adler S L1965 Phys. Rev. Lett. 14, 1051; Weisberger W I, Phys. Rev. Lett. 14, 1047
[19] Sugawara H 1964 Phys. Theo. Phys, 31, 213; Lee B W, Phys. Rev. Lett. 12, 83
[20] Rosen S P, Pakvasa S and Sudarshan E C G 1966 Phys. Rev. 146, 118 ; Suzuki M, Phys. Rev. 140, B1405
[21] Pakvasa S, Graham R, and Rosen S P 1966 Phys. Rev. 149, 1200
[22] Sugawara H 1965 Phys. Rev. Lett. 15, 870,997(E); Suzuki M 1965, ibid., 15, 986
[23] Altarelli G and Maiani L 1974 Phys. Lett. B52, 351; Gaillard M K and Lee B W 1974 Phys. Rev. Lett 33, 108
[24] Sakurai J J 1979 Proc. of the Eight Topical Conference in Particle Physics, UH Press (Honolulu) ed. Peterson V Z and Pakvasa S, p. 375
[25] Mohapatra R N 1972 Phys. Rev. D6, 2023
[26] Fritsch H, Gell-Mann M and Minkowski P1975 Phys. Lett. B59, 256; Pakvasa S, Simmons W A and Tuan S F 1975 Phys. Rev. Lett. 351 702; de Rujula A, Georgi H, and Glashow S L 1975 Phys. Rev. D12, 3588; Wilczek F A *et. al.* 1975 Phys. Rev. D12, 2768; Mohapatra R N 1975 Phys. Rev. D13, 680
[27] Golowich E and Holstein B1975 Phys. Rev. Lett., 35, 831
[28] Ma F 1977 Nucl. Phys. B121, 421; de Rujula A, Georgi H, and Glashow S L 1977 Annals. Phys. 109, 242; Pati J C, Rajpoot S, and Salam A 1978 Phys. Rev. D17, 131
[29] Prescott C Y *et. al.* 1978 Phys. Lett. 77B, 347
[30] Barkov L M and Zolotorev M S 1978 JETP Lett. 27, 357
[31] TWIST Collaboration, Musser J R *et. al.* 2005 Phys. Rev. Lett. 94, 101805 ; Jamieson B *et. al.* 2006 Phys. Rev. D74, 072007

Journal of Physics: Conference Series **196** (2009) 011007 doi:10.1088/1742-6596/196/1/011007

Introduction to ``Symmetry'': the second quest

Richard Arnowitt

Dept. of Physics, Texas A&M University, College Station, Texas

E-mail: arnowitt@physics.tamu.edu

Symmetries played a basic role in the discovery and elucidation of particles in high energy physics. It began in atomic physics, then was extended to the study of nuclear physics. In the sixties, there was an explosion of group theoretical methods applied to exploit the symmetry properties of physical systems. Symmetries existed in classical physics, but in quantum mechanics, with the superposition principle, one could use them to find the selection rules.

George Sudarshan played an important role in these advances, as you will hear from Sydney Meshkov and speakers following him in this session.

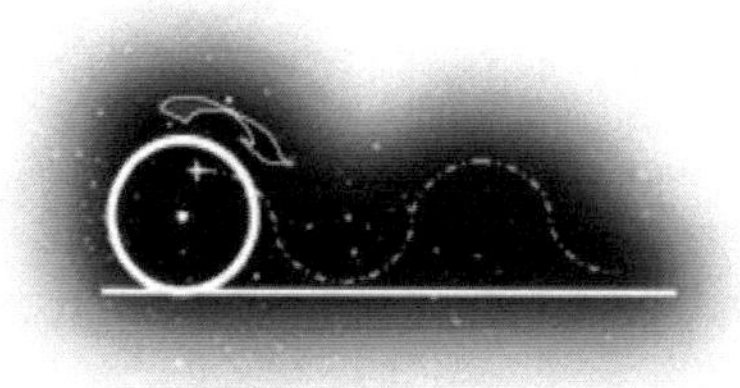

E. C. G. Sudarshan with two sons, daughter-in-law and two of his grandchildren.

Journal of Physics: Conference Series **196** (2009) 012006 doi:10.1088/1742-6596/196/1/012006

The Deterministic set of Operators, Quantum Interference Phenomena, and Quantum Reality

Jeff Tollaksen and Yakir Aharonov

Chapman University, Schmid College of Science

Department of Physics, Computational Science, and Engineering, 1 University Dr, Orange, CA 92866

E-mail: tollaksen@chapman.edu

Abstract.
We develop an approach to quantum interference phenomenon which makes the characteristics of the problem more conspicuous and brings out more physical concepts. We do this by discussing the idea of deterministic quantum experiments using the Heisenberg picture. The variables that are relevant for describing interference are modular variables. These variables obey non-local equations of motion and can be used to follow the time evolution of a system without disturbing it. We present a physical explanation for the different behaviors of a single particle when the distant slit is open or closed: instead of having a quantum wave that passes through all slits, we have a localized particle with non-local interactions with the other slit(s). While the Heisenberg picture and the Schrodinger pictures are equivalent formulations of quantum mechanics, nevertheless, the results discussed here support a new approach to quantum mechanics which has led to new insights, new intuitions, new experiments, and even the possibility of new devices that were missed from the old perspective.

Written to celebrate the 75th birthday of E.C.G. Sudarshan: deep and subtle

1. Introduction

The two-slit experiment is the quintessential example of the dual character of quantum mechanics. The initial incoming particle seems to behave as a wave when falling on the (left and right) slits, but when recorded on the screen, its wavefunction "collapses" into that of a localized particle. By repeating the experiment for an ensemble of many particles, the interference pattern manifests through the density of hits along the screen (aligned with, say the x direction): $dn(x)/dx \sim |\psi_{\rm L}(x) + e^{i\alpha}\psi_{\rm R}(x)|^2$ with $\psi_{\rm L}(x)$ coming from the left slit, $\psi_{\rm R}(x)$ from the right (located a distance D away), and α the relative phase between the left and right parts of the wavefunction.

There are two ways to think about such phenomenon:

The first accepts the Schrödinger description as given, with wavepackets evolving in time. Indeed the Schrödinger description has been extremely useful, having served, for example, as the starting point for the Feynman path integral. The apparent analogy between Schrödinger wave interference and classical wave interference (arising from the use of identical calculations), presents a conceptually simple interpretation of quantum phenomena in terms of our classical picture. One is often advised to apply this consistent formalism for statistical predictions (providing, in this case, probability distributions for the positions of *many* particles) *without* asking questions about it's interpretation. In fact, the belief that the Schrödinger picture is the *only* way by which the interference and relative phase can be inferred, played a central role in the development of the probability amplitude interpretation in the quantum formalism.

The second way of thinking maintains that this is not the end of the story and advocates further inquiry. For example, Feynman [32] stated that such phenomena "..have in it the heart of quantum mechanics. In reality, it contains the only *mystery*." Such proponents often seek to obtain as close a correspondence as possible between theory and measurement. As a consequence, they try to weed out "classical" notions when they have been mis-applied to the quantum realm. For example, classical waves involve many degrees of freedom (e.g. field phenomenon such as sound and electromagnetic waves) and their phase *can* of course be measured by local experiments. But the meaning of a quantum phase is very different. Multiplying the wavefunction $\psi_{\rm L}(x) + e^{i\alpha}\psi_{\rm R}(x)$ by an overall phase ϕ does not change the relative phase α and thus does not yield a different state. Furthermore, it seems that the relative phase α cannot be measured directly on a single particle since it cannot be represented by a Hermitian operator. That is, $\psi_{\rm L}(x) + e^{i\alpha}\psi_{\rm R}(x)$ and $\psi_{\rm L}(x) + e^{i\beta}\psi_{\rm R}(x)$ are not generally orthogonal and thus cannot be eigenstates belonging to different eigenvalues of a Hermitian operator. In further contrast to the classical phase, a change in the relative quantum phase - say from $\psi_{\rm L}(x) + \psi_{\rm R}(x)$ to $\psi_{\rm L}(x) - \psi_{\rm R}(x)$ - would not result in a measureable change in any local properties. The change only shows up in certain non-local properties or much later when the two separate components $\psi_{\rm L}(x)$ and $\psi_{\rm R}(x)$ eventually overlap and interfere. It seems that the relative phase cannot be thought of simply as the difference between a local phase at $\psi_{\rm L}(x)$ and another local phase at $\psi_{\rm R}(x)$.

Another aspect of this second way of thinking is the realization that the Schrödinger wave only has a measureable meaning for an ensemble of particles, not for a single particle (generally speaking). This therefore leaves important questions unanswered concerning the *physics* of interference from the perspective of a single particle: if physics obeys local dynamics, then how does the localized particle passing through the right slit sense whether or not the distant left slit is open (closed), causing it to scatter (or not scatter) into a region of destructive interference? Interference experiments have been performed with electron/photon beams whose intensity is sufficiently small such that only one electron/photon traverse the interference apparatus at a time. The interference pattern with light and dark bands is nevertheless built up successively, mark by mark, with each individual "particle-like" electron/photon, [37]. One is then confronted with the fact that a single degree of freedom created the interference pattern. This mystery led Feynman to declare: "Nobody knows how it can be like that." [32]
We follow the second way of thinking and offer a fresh approach to this time honored problem [13] [12] [26] [29]. To motivate the first step, involving a fundamental shift in the types of observables utilized, we make several observations:

First, most discussions of this problem are based on measurements which disturb the interfering particle. This is one of the main reasons that quantum interference is generally considered to be intimately associated with the problems that stem from the statistical character of the quantal description.

Second, the observables studied to date have been simple functions of position and momentum. These observables, however, are **not** sensitive to the relative phase between different "lumps" of the wavefunction (centered around each slit). Nevertheless, the subsequent interference pattern of course is entirely determined by the relative phase between these "lumps," suggesting that simple moments of position and momentum are not the most appropriate dynamical variables to describe quantum interference phenomena.

Third, operators that ***are*** sensitive to the relative phase are exponentials of the position and momentum.

We address the first observation with non-disturbing measurements. To date, several non-disturbing measurements, such as weak measurements and protective measurements, have stimulated lively debates and have proven useful in separating various aspects of quantum theory from the probabilistic aspects [14]. The underlying framework for the approach to interference presented in this paper is based on another kind of non-disturbing measurement on the "set of deterministic operators" or "deterministic experiments" [12] [29]. This set ($\hat{A}_\psi$) involves measurement of only those variables for which the state of the system under investigation is an eigenstate. This question is dual to the more familiar question "what are the eigenstates of a given operator?" Measurement of these operators $\hat{A}_\psi$ does not collapse the wavefunction, since the wavefunction is initially an eigenstate of the operator being measured. The essential point emphasized here is the relevance of deterministic experiments for a *single* particle since they can be performed without causing a disturbance. As an example, consider a spin-1/2 particle placed in a magnetic field $\vec{B}$ pointing in the x-direction and with $\sigma_z = +1$ initially. If at some time t later we measure σ_z again, the result is not predictable; we get fluctuations and the experiment is not a deterministic one. However, if instead we rotate our apparatus and measure the spin in the yz-plane in a direction $\alpha = \omega t$, where ω is the Larmor frequency, then the result is predictable and we have a deterministic experiment. As a second example, consider a free particle of mass m which at $t = 0$ is in the state $\psi = C\exp\{-\frac{x^2}{\Delta x^2}\}$. This state is an eigenstate of the Hermitian operator $\hat{A}(0) = \hat{x}^2 + \hat{p}^2$. At time t we can measure the operator $\hat{A}(t) = \left\{\hat{x} - \frac{\hat{p}}{m}t\right\}^2 + \hat{p}^2$ with a predictable outcome, and thus we have again a deterministic experiment.

We address the second and third observations by performing yet another kind of non-disturbing measurement, namely weak measurements, on the observables that *are* sensitive to the relative phase. These observables that are senstive to the relative phase are functions of ***modular variables***. For the special case of interference in space, as considered here, the relevant modular variable is modular momentum, not ordinary momentum. These observables are also members of the "deterministic set of operators" and are relevant for an individual particle. We then see that in the context of interference phenomenon, the Heisenberg equations of motion for these modular variables are *non-local*. The nonlocality of these observables is quite intuitive: the operators sensitive to the relative phase simply translate the different "lumps" of the wavefunction. The appropriate translation may cause one lump to overlap with another lump or to overlap simply with the region where the distant slit is either open (or closed). This provides a **physical** explanation for the different behavior of a *single* particle when the distant slit is open or closed. It therefore provides the under-pinnings for a new ontology based on localized particles with non-local interactions, rather than an unphysical Schrödinger "wave of probability" traveling throughout all of space.

This kind of non-locality which is revealed in the equations of motion, is *dynamical* non-locality, to distinguish it from kinematic non-locality [36] [16] implicit in quantum correlations. These two kinds of non-locality are fundamentally different: kinematic non-locality arises from the structure of Hilbert space and does not create any change in probability distributions, causes and effects cannot be distinguished and therefore "action-at-a-distance" cannot manifest. Kinematic non-locality has been extremely useful, having catalyzed, e.g., much of the progress in quantum information science. On the other hand, dynamical non-locality, arises from the structure of the equations of motion and *does* create explicit changes in probability, though in a "causality-preserving" manner. This approach was first introduced by Aharonov, Pendelton and Petersen (APP) [12] in order to explain the nonlocality of topological phenomena such as the Aharonov-Bohm (AB) effect [4] [38]. The AB effect conclusively proved that a magnetic (or electric) field inside a confined region can have a measureable impact on a charged particle which never traveled inside the region. In order to represent the closest correspondence between measurement and theory, APP introduced nonlocal interactions between the particle and field.

This was in contrast to the prevailing approach of reifying local interactions with (unphysical) non-gauge invariant quantities outside the confined region, such as the vector (and/or scalar) potential.

Both dynamic and kinematic non-locality are generic and can be found in almost every type of quantum phenomenon. [33] Prior to APP, dynamical nonlocality was avoided due to the possibility that it could violate causality. However, in a beautiful theorem, APP proved that the dynamical nonlocality they introduced could never violate causality. They considered the general set of conditions necessary to see the non-local exchange of modular variables, for example when the left slit is either monitored or closed and the particle is localized around the right slit. APP proved that these are precisely the same conditions which make the non-local exchange completely uncertain and therefore "un-observable."

While it was beautiful that quantum mechanics allowed "action-at-a-distance" to "peacefully-coexist" with causality, this theorem nevertheless proved to be somewhat anti-climatic: if we cannot actually observe the nonlocal exchange of modular variable, then have we not violated the dictum of maintaining the closest correspondence between measurement and theory by claiming the existence of a new kind of nonlocal - yet un-observable - effect?

It was recently pointed out [28], that these non-local interactions ***can*** be observed. To measure this nonlocality (when it occurs) in a causality-preserving manner, various tools must be utilized such as pre-selection, post-selection, and weak measurements.

This development thereby underscores a fundamental difference between classical mechanics and quantum mechanics that is easily missed from the perspective of the Schrödinger picture: the equations of motion for observables relevant to quantum mechanical interference phenomenon can be **non-local** in a peculiar way that preserves causality. These novel results motivate a new approach to quantum mechanics starting from the Heisenberg picture and involving the set of deterministic operators. While the new framework and associated language are, in principle, equivalent to the Schrödinger formulation, it has led to new insights, new intuitions, new experiments, and even the possibility of new devices that were missed from the old perspective. These types of developments are signatures of a successful re-formulation.

We briefly mention one important conceptual shift evident in this new approach: when quantum mechanics is compared to classical mechanics, often the uncertainty or indeterminism of quantum mechanics is emphasized and the profound, fundamental differences in the dynamics is ignored. This is perhaps a result of the similarity between the classical dynamical description (Poisson bracket) and the quantum dynamical description (commutator) for simple functions of momentum or position. Furthermore, uncertainty is viewed in a kind of "negative" light: as a result of the uncertainty in quantum mechanicics, we have lost the ability that we had in classical mechanics to predict the future. Not only is nature "capricious," but it seems that we do not even gain anything from the uncertainty.

The new approach allows us to change this perspective by *deriving* uncertainty from principles that we argue are more fundamental, namely from non-locality and causality. This changes the meaning of uncertainty from one with a "negative" connotation to one with a "positive" connotation. Something similar happened with special relativity when the axioms of relativity were discovered. This inspired a modification of the old language: e.g. that light has the same velocity in all reference frames is certainly highly unusual, but everything works in a self consistent way due to the axiomatic framework, and because of this, special relativity is rather easy to understand.

Similarly, we are convinced that the new approach arising from this paper will lead to a deeper understanding of the nature of quantum mechanics.

2. The Deterministic Set of Operators (DSOs)

The approach to the issue of the quantum reality for a single particle is based on the notion of "deterministic" experiments. We call an experiment a deterministic experiment when we measure only variables for which the state of the system under investigation is an eigenstate. In other words, for any state ψ it is possible to ask the following question: "What is the set of Hermitian operators $\hat{A}_\psi$ for which ψ is an eigenstate?" That is, which satisfy:

$$\hat{A}_\psi = \{\hat{A}_i \textit{ such that } \hat{A}_i|\psi(t)\rangle = a_i|\psi(t)\rangle \,, a_i \in \Re\} \tag{2.1}$$

where $\Re$ is the set of real numbers. This question is dual to the more familiar question which asks "What are the eigenstates of a given operator?" A measurement of such operators $\hat{A}_\psi$ would not lead to collapse of the wave function, since the wave function is initially an eigenstate of the operator being measured. A succession of such experiments can be performed in time and still, no statistics need be associated with the experimental results. The results are completely predictable, i.e. the experiments are deterministic. In deterministic experiments, it is possible to give a detailed physical description of the evolution of the system.

The set $\hat{A}_\psi$ is closed in the sense that if $\hat{A}_i, \hat{A}_j \in \hat{A}_\psi$ then:

(i) $[\hat{A}_i, \hat{A}_j] = \hat{A}_k \in \hat{A}_\psi$ since $\hat{A}_k\psi = 0$. $\hat{A}_k$ is an operator proportional to a projector perpendicular to ψ.
(ii) $\hat{A}_i + \hat{A}_j$ and $\hat{A}_i\hat{A}_j$ are in $\hat{A}_\psi$

The set $\hat{A}_\psi$ is also unique in the sense that it defines a single state $|\psi(t)\rangle$ which satifies the set membership criteria of eq. (2.1)[1]. This suggests that we can equivalently specify either the state $|\psi(t)\rangle$ or the set $A_\psi(t)$ of all time dependent eigenoperators of $|\psi(t)\rangle$. It follows that, since we know H and $A_\psi(0)$, that we know $A_\psi(t)$. The logic is straightforward: knowing the set $A_\psi(0)$ gives us uniquely $|\psi(0)\rangle$. $|\psi(0)\rangle$ together with H gives us $|\psi(t)\rangle$, which has a unique set $A_\psi(t)$ of eigenoperators. The proof of the uniqueness is trivial in a finite (and denumerably infinite) dimensional Hilbert space.

If we have a Hilbert space of dimension n then we can choose a basis such that a state vector $|\psi\rangle$ can be represented by an $n \times 1$ column vector (a unitary transformation can always bring us to this basis) given by 2.2.a and orthogonal vectors of the form given by 2.2.b:

$$\begin{pmatrix} 1 \\ 0 \\ 0 \\ \vdots \\ 0 \end{pmatrix} \; a) \qquad \begin{pmatrix} 0 \\ a_2 \\ a_3 \\ \vdots \\ a_n \end{pmatrix} \; b) \tag{2.2}$$

The Hermitian operators that operate on this space, represented (in that same basis) by an $n \times n$ matrix are in turn an n^2 dimension space. Operators of the form eq. 2.3 can be the DSO (where $a_{ij} = a^*_{ij}$) with dimensionality $(n-1)^2 + 1$.

The extension of these results to denumerably infinite spaces is straightforward: any matrix can be written as a sum over a complete set of vectors $|\Psi_n\rangle\langle\Psi_m|$ i.e. $\sum a_{nm}|\Psi_n\rangle\langle\Psi_m|$. An infinite matrix representing DSO is then given by:

$$a_{11}|\Psi_1\rangle\langle\Psi_1| + \sum_{n,m\neq 1} a_{nm}|\Psi_n\rangle\langle\Psi_m| \tag{2.4}$$

[1] This is so because any other statevector will be disturbed by at least one operator in $\hat{A}_\psi$. In addition, any other operator not in $\hat{A}_\psi$ will disturb $|\psi(t)\rangle$ (see §2.1)

$$\overbrace{\quad}^{dimension\ 1}$$

$$\begin{pmatrix} a_{11} & 0 & 0 & \dots & 0 \\ 0 & a_{22} & a_{23} & \dots & a_{2n} \\ 0 & a_{32} & a_{33} & \dots & a_{3n} \\ \vdots & \vdots & \ddots & & \vdots \\ 0 & a_{n2} & a_{n3} & \dots & a_{nn} \end{pmatrix} \tag{2.3}$$

$$dim(n-1)\ form\ \sum_{n,m\neq 1} a_{nm}|\Psi_n\rangle\langle\Psi_m|$$

Figure 1. Deterministic subset of operators with dimension $(n-1)^2 + 1^2$

There is a single unique eigenvector associated with this that is given by eq. (2.2.a) with either finite n or infinite $n \to \infty$. This is the only vector that is not disturbed by operators of the form $\sum_{n,m\neq 1} a_{nm}|\Psi_n\rangle\langle\Psi_m|$ in eq. (2.4). If the only operators being used or measured are those for which eq. (2.1) is satisfied, then the wavefunction evolves in a deterministic way, i.e. if we define ψ at one time, then we know ψ uniquely at some later time. Hence, if we define DSO at one time, it must evolve in a completely deterministic fashion to the DSO at a later time.

2.1. The Completely Uncertain Operators

In addition to the set $\hat{A}_\psi$ given by 2.1, there is also a set B_ψ of operators whose results are completely uncertain. If we define $\hat{\Pi}_\psi$ such that $\hat{\Pi}_\psi|\psi\rangle = |\psi\rangle$ where $\hat{\Pi}_\psi^2 = \hat{\Pi}_\psi$ and $\hat{\Pi}_\psi|\psi'\rangle = 0$ if and only if $\langle\psi'|\psi\rangle = 0$, then this operator is the projector on $|\psi\rangle$, $\hat{\Pi}_\psi = |\psi\rangle\langle\psi|$. We can also decompose any operator $\hat{C}(x,p)$ as follows:

$$\hat{C}(x,p) \equiv \hat{A}_\psi^{(C)} + \hat{B}_\psi^{(C)} \tag{2.5}$$

where $\hat{A}_\psi^{(C)} = \hat{\Pi}_\psi\hat{C}\hat{\Pi}_\psi + (1-\hat{\Pi}_\psi)\hat{C}(1-\hat{\Pi}_\psi)$ and $\hat{B}_\psi^{(C)} = (1-\hat{\Pi}_\psi)\hat{C}\hat{\Pi}_\psi + \hat{\Pi}_\psi\hat{C}(1-\hat{\Pi}_\psi)$.
In the above basis, the completely uncertain operators for 2.2.a have the matrix form:

$$\underbrace{\begin{pmatrix} 0 & a_{12} & a_{13} & \dots & a_{2n} \\ a_{21} & 0 & 0 & \dots & 0 \\ a_{31} & 0 & 0 & \dots & 0 \\ \vdots & \vdots & \ddots & & \vdots \\ a_{n1} & 0 & 0 & \dots & 0 \end{pmatrix}}_{completely\ uncertain\ operators} \begin{pmatrix} 1 \\ 0 \\ 0 \\ \vdots \\ 0 \end{pmatrix} = \begin{pmatrix} 0 \\ a_{21} \\ a_{31} \\ \vdots \\ a_{n1} \end{pmatrix} \tag{2.6}$$

Note that the uncertain operators have non-zero entries a_{12}, a_{13}, ...a_{2n} and a_{21}, a_{31}...a_{n1} which take the state given by 2.2.a to orthogonal states 2.2.b.

2.2. Geometric interpretation

Traditionally, it was believed that if a measurement interaction is weakened so that there is no disturbance on the system, then no information will be obtained. However, it has been shown [21] that information can be obtained even though not a single particle (in an ensemble) was disturbed. To begin to introduce this point, let us consider a general theorem for any vector (state) in Hilbert space:

Theorem I: $\hat{A}|\psi\rangle = \langle\hat{A}\rangle|\psi\rangle + \Delta A|\psi_\perp\rangle$ where $\langle\hat{A}\rangle = \langle\psi|\hat{A}|\psi\rangle$, $|\psi\rangle$ is any vector in Hilbert space, $\Delta A^2 = \langle\psi|(\hat{A} - \langle\hat{A}\rangle)^2|\psi\rangle$, and $|\psi_\perp\rangle$ is a vector (state) in the perpendicular Hilbert space such that $\langle\psi|\psi_\perp\rangle = 0$ (see fig. 2.a). Proof: left multiplication by $\langle\psi|$ yields the first term; evaluating $|(A - \langle A\rangle)|\psi\rangle|^2 = \Delta A^2$ yields the second.

Using this theorem, it is easy to see that in the basis of eq. 2.2, the diagonal elements of 2.3 correspond to the averages of the observables, e.g. $a_{11} = \langle \hat{A} \rangle$ for state 2.2.a, which again, can be measured without uncertainty. We can also see how the relation $\hat{A}|\Psi\rangle = \langle \hat{A} \rangle |\Psi\rangle + \Delta A |\Psi_\perp\rangle$ (Theorem I) works in the DSO picture. As a result of the Hermiticity and the closed algebra, we can think of these operators as vectors in a real vector space. We can also consider a product, $\hat{A}_1 \odot \hat{A}_2 = trace(\hat{A}_1 \cdot \hat{A}_2)$ which defines the length of the vectors. The normalization of 2.6 then is $N \equiv |a_{21}|^2 + |a_{31}|^2 + \cdots + |a_{n1}|^2$. Using $a_{ij} = a^*_{ij}$, we see that this is the same as $\sqrt{N}$, which is the uncertainty in the operator, ΔA. We can therefore construct the geometric interpretation for any operator 2.5 illustrated in figure 2.b: the length of the projection onto the DSO is given by the average of the operator, and the length of the projection onto the completely uncertain operators is given by the uncertainty in the operator. Subtracting the dimensionality of the whole set of Hermitian operators, n^2, from the dimensionality of the DSO, $(n-1)^2 + 1$, gives us $O(n)$ the dimensionality of the completely uncertain operators. Therefore, as $n \to \infty$, most of the parameters belong to the DSO.

Figure 2. a) geometry of $\hat{A}|\Psi\rangle = \langle \hat{A} \rangle |\Psi\rangle + \Delta A |\Psi_\perp\rangle$, b) geometry of $\hat{C}(x,p) \equiv \hat{A}^{(C)}_\psi + \hat{B}^{(C)}_\psi$, i.e. any operator can be decomposed into it's projection (proportional to $\langle \hat{C} \rangle$) onto the DSO $\hat{A}_\psi$ and its projection (proportional to ΔC) onto the completely uncertain set B_ψ

3. Interference phenomena from the traditional perspective

To show the utility of DSOs in interference situations, we review past attempts to analyze the disappearance of interference whenever it is possible to detect through which slit the particle passes. The original debate was famously conducted by Einstein and Bohr. Einstein attempted to challenge the consistency of quantum mechanics by arguing that a Which Way Measurement (WWM) could be performed without destroying the interference pattern by measuring the transverse recoil (i.e. the transverse momentum kick) of the double-slit screen after the particle passed through. Bohr maintained that the consistency of quantum mechanics depended on the destruction of the interference pattern when WWM information is obtained. He showed that the measurement-induced uncertainty created in the transverse position of the screen by an accurate measurement of the transverse momentum was sufficient to destroy the interference pattern.

This reasoning leads to a paradox which helps to motivate our approach. It has been argued (borrowing from the discussion of the "Heisenberg microscope") that if the particle were "observed" at the right slit, then the photon involved in this observation should have a wavelength $\lambda \leq D/2$ and a corresponding momentum uncertainty $\Delta p > 2\hbar/D$. This momentum uncertainty is imparted to the particle making its wave number $k = p/\hbar$ uncertain, thereby destroying the interference pattern.

This argument is incorrect. To see this, assume that a sensitive detector, placed at the left slit, failed to detect any particle. We then know that all particles passed through the right slit. The interference pattern will then be completely destroyed despite the fact that there was no interaction with the detector! [12] [34] One might suppose that since the action of opening/closing the left slit never caused an interaction with the particle at the right slit, then nothing associated with the particle should change. But, it was first pointed out by APP [12] that in this scenario when a WWM is performed without actually interacting with the interfering particle, then the probability distribution of the momenta *does* change, although none of the moments of the momenta change.

To best resolve this paradox, we need to take a step back. We note that the effect of a generic interaction or collision between any two quantum systems can be characterized by a change in the probability distribution of the momentum i.e. going from an initial probability distribution, $\rho_i(p)$, to a final distribution, $\rho_f(p)$. We can analyze this change in two ways[2]:

(i) Look at moments such as $\langle p^n\rangle = \int \rho(p)p^n dp$ and calculate $\delta\langle p^n\rangle = \langle p^n\rangle_f - \langle p^n\rangle_i$, and thus ask how the interaction affected these averages. This is the usual approach.

(ii) Or, we may look at the fourier transform of the probability distribution $\int \rho(p)e^{\frac{i}{\hbar}pD}dp$. (We will later see that these functions, $\langle e^{\frac{i}{\hbar}pD}\rangle$, are precisely the observables that are sensitive to the relative phase.) To analyze the effect of the interaction, we calculate $\langle e^{\frac{i}{\hbar}pD}\rangle_f - \langle e^{\frac{i}{\hbar}pD}\rangle_i$ and ask how the interaction affected these averages.

In principle, one can discuss the effect of interactions using (1) or (2), since knowing (2) for all D is equivalent to knowing (1) for all n.

The first approach has been used universally to analyze interference phenomena [34]. However, there are inherent limitations to any approach based on analyzing changes in the probability for momenta through changes in the moments. For example, while momentum is of course conserved, there is no definite connection between the probability of an individual momentum before and after an exchange between the interfering particle and the slit. Furthermore, the analysis in terms of moments does not offer any intuition as to ***how*** or ***why*** the probability of momentum changes.

When compared to the first (traditional) approach based on the moments, the second approach focusing on the fourier transform of the probability distribution has many advantages, both mathematical and physical. In this section, we briefly review some of the mathematical advantages, leaving much of the physical advantages to the rest of the article.

The first "moments" approach to interference derived from intuitions developed with wavefunctions consisting of just one "lump." In these cases, the averages of x (or of p) evolve according to *local* classical equations of motion. Also the uncertainties $(\Delta x)^2 \equiv \overline{(\hat{x})}^2 - \overline{\hat{x}}^2$ and $(\Delta p)^2 \equiv \overline{(\hat{p})}^2 - \overline{\hat{p}}^2$, describing the spread in these variables, have properties similar to those of the spread of variables in a classical situation with unsharply defined initial conditions and which evolve according to diffusion-like rules.

This drastically changes when we have two or more separate "lumps" of the wavefunction. Indeed, the wavefunction, after passing through the symmetric two-slits, consists of a superposition of two identical, but physically disjoint "lumps," $\psi_{\rm L}$ and $\psi_{\rm R}$:

$$|\Psi_\alpha\rangle = \frac{1}{\sqrt{2}}\{|\psi_{\rm L}\rangle + e^{i\alpha}|\psi_{\rm R}\rangle\} \tag{3.7}$$

Collapsing it to just $\psi_{\rm R}(x) \equiv \langle x|\psi_{\rm R}\rangle$ does *not* change Δp nor the expectation values of any finite order polynomial in p, as none of these local operators have a non-vanishing matrix element between the disjoint "lumps" of the wavefunction. In fact, the superposition of any countable number of such wavepackets (so that Δx becomes arbitrarily large) does not require the uncertainty in momentum, Δp, to increase. To see this, consider a simple example: let $\psi_o(x)$ be a wave packet centered at $x = 0$ that is exactly equal to zero for $x \in [-\frac{L_o}{2}, \frac{L_o}{2}]$. For this wave function we have $\Delta p \approx \frac{\hbar}{\Delta x}$. We now define $\psi_n(x) = \psi_o(x + nL)$ where $n = 1, 2, ...$ and $L > L_o$.

[2] We consider momentum here, but our comments apply to any conserved quantity.

We now show that for the class of states $\Psi(x) = \sum_n a_n \psi_n(x)$ with $\sum_n |a_n|^2 = 1$, the uncertainty in momentum is independent of a_n. We have $\Delta p = (\langle p^2\rangle - \langle p\rangle^2)^{1/2}$. For $\langle p\rangle$ we have:

$$\langle p\rangle = \int_{-\infty}^{+\infty} dx \Psi^*(x)(-i\hbar)\frac{\partial}{\partial x}\Psi(x) = \int_{-\infty}^{+\infty} dx \sum_{n,m} a_n^* a_m \Psi_n^*(x)(-i\hbar)\frac{\partial}{\partial x}\psi_m(x) \tag{3.8}$$

But, by definition ψ_m and ψ_n do not overlap, so that ψ_m and its derivatives are different from zero only where $\psi_n = 0$.[3] Thus

$$\langle p\rangle = \sum_{n,m} a_n^* a_m \frac{\hbar}{i}\int_{-\infty}^{+\infty} dx \psi_n^*(x)\frac{\partial}{\partial x}\psi_n(x) \tag{3.9}$$

Further, since $\psi_n(x) = \psi_o(x + nL)$, we have

$$\int_{-\infty}^{+\infty} dx \psi_n^*(x)\frac{\partial}{\partial x}\psi_n(x) = \int_{-\infty}^{+\infty} dx \psi_o^*(x+nL)\frac{\partial}{\partial x}\psi_o(x+nL) = \int_{-\infty}^{+\infty} dx \psi_o^*(x)\frac{\partial}{\partial x}\psi_o(x) \tag{3.10}$$

Therefore,

$$\langle p\rangle = \sum_n |a_n|^2 \frac{\hbar}{i}\int_{-\infty}^{+\infty} dx \psi_o^*(x)\frac{\partial}{\partial x}\psi_o(x) = \frac{\hbar}{i}\int_{-\infty}^{+\infty} dx \psi_o^*(x)\frac{\partial}{\partial x}\psi_o(x) \tag{3.11}$$

I.e. $\langle p\rangle$ is independent of the choice of a_n. In a similar way, one can show that

$$\langle p^2\rangle = \sum_n |a_n|^2 \frac{\hbar}{i}\int_{-\infty}^{+\infty} dx \psi_o^*(x)\left\{\frac{\partial}{\partial x}\right\}^2 \psi_o(x) = \frac{\hbar}{i}\int_{-\infty}^{+\infty} dx \psi_o^*(x)\left\{\frac{\partial}{\partial x}\right\}^2 \psi_o(x) \tag{3.12}$$

I.e. $\langle p^2\rangle$ is independent of a_n and thus $(\Delta p)^2 = (\langle p^2\rangle - \langle p\rangle^2)$ is independent of the choice of a_n.

From this surprising result we learn, for example, that if in a typical quantum interference experiment we want to measure through which slit the particle passes, then we do not have to increase the uncertainty in momentum. In fact, what our intended experiment does is to change the situation from one in which $a_1 = a_2 = \frac{1}{\sqrt{2}}$ to one in which, e.g., $a_1 = 1$, $a_2 = 0$. In both situations Δp is the same. Therefore, it is misleading to say that in measuring which slit the particle passes through we destroy the interference pattern by increasing the uncertainty in momentum.

Up until now we have focused on the disappearance of interference upon WWM. But the other fundamental mystery highlighted by Feynman remains: namely, how does a particle localized at the right slit "know" whether the left slit is open or closed? The first approach based on moments tell us nothing about this mystery. The decisive importance of the second "fourier transform" approach for this mystery is best illustrated through a basic theorem which characterizes all interference phenomenon: all **moments** of both position and momentum are *independent* of the relative phase parameter α (until the wavepackets overlap):

Theorem II: Let $\Psi_\alpha = \psi_{\rm L}(x,t) + e^{i\alpha}\psi_{\rm R}(x,t)$ such that there is no overlap of $\psi_{\rm L}(x,0)$ and $\psi_{\rm R}(x,0)$. If n and m are integers, then for all values of t, and choices of α, β:

$$\int [\Psi_\alpha^*(x,t)\Psi_\alpha(x,t) - \Psi_\beta^*(x,t)\Psi_\beta(x,t)] x^m p^n dx = 0 \tag{3.13}$$

For the particular double-slit wavefunction, it is easy to see that if there is no overlap between $\psi_{\rm L}$ and $\psi_{\rm R}$ then nothing of the form $\int_{-\infty}^{\infty} \Psi^* x^m p^n \Psi dx$ will depend on α for any value of m and n.

[3] $\psi^*(x)\frac{\partial}{\partial x}\psi(x+L) = 0$ for all x if $\psi(x)$ vanishes outside the interval, then so does its derivatives.

Furthermore, expanding $\int\{\psi_L+e^{-i\alpha}\psi_R\}^* x^m p^n\{\psi_L+e^{i\alpha}\psi_R\}dx$, we see that only the cross terms, i.e. $\langle\psi_L|x^m p^n|e^{i\alpha}\psi_R\rangle$, have the *possibility* of depending on α; but operators of the form $x^m p^n$ cannot change the fact that ψ_R and ψ_L do not overlap.It is also easy to show that $\langle x^n p^m\rangle$ at $t=0$ is independent of α by using the Heisenberg representation $\langle x^n(t)\rangle = \int \Psi^*_\alpha(x,0)x^n(t)\Psi_\alpha(x,0)dx$ and noting that $x(t) = x(0) + p(0)\frac{t}{m}$ and $p(t) = p(0)$ in this representation, we must have $\langle x^n(t)\rangle = \int \Psi^*_\alpha(x,0)[x(0)+p(0)\frac{t}{m}]^n(t)\Psi_\alpha(x,0)dx$. This is independent of α, since term by term it is independent of α. Eq. (3.13) then follows, and holds for p^n, as long as we retain the proper $\Psi^*_\alpha p^n \Psi_\alpha$ order.

This suggests that these dynamical variables (e.g. $\langle x\rangle$, $\langle p\rangle$, Δx, Δp) are not the most appropriate to describe quantum interference phenomena. What observables, then, are sensitive to this interference information which appears to be stored in a subtle fashion? To fully capture the physics of these scenarios with wavefunctions composed of multiple lumps, non-polynomial and *non-local* operators, connecting the disjoint parts are required. For many, equi-distant slits, these are the discrete translation by $\pm D$, namely $\exp\{\pm\frac{i}{\hbar}\hat{p}D\}$, effecting $\exp\{-\frac{i}{\hbar}\hat{p}D\}\psi_R(x) \to \psi_R(x-D)$ which overlaps with $\psi_L(x)$. The expectation value of the translation operator $\exp\{\frac{i}{\hbar}\hat{p}D\}$ ***does*** depend on α: $\langle\Psi_\alpha \mid \exp\{i\hat{p}D/h\} \mid \Psi_\alpha\rangle = e^{-i\alpha}/2$.

This provides the basis for a mechanism to explain ***how*** the particle at the right "knows" what is happening at the left slit. As we will see, the second "fourier transform" approach even provides us with the parameters relevant for this question (namely the distance between the slits), while the first "moments" approach remains silent.

Before proceeding in the next section to the *physics* of interference for single particles, we briefly mention three additional mathematical advantages concerning the second "fourier transform" approach.

First, all the moments $\langle p^n\rangle$ are averages of unbounded quantities, while $\langle\exp\{\frac{i}{\hbar}\hat{p}D\}\rangle$ are averages of bounded quantities. There are problems with unbounded quantities (as pointed out by Mir et al). Infinitesimal changes in $\rho(p)$ can cause very large changes in the moments $\langle p^n\rangle$. To see this, consider a negligible change, $\delta\rho(p)$, in $\rho(p)$. By negligible, we mean there is only a small change in the probability distribution. If we calculate $\delta\langle p^n\rangle = \int\delta\rho(p)p^n dp$, we could get a finite change if $\delta\rho(p)$ differs from zero at a sufficiently large p. In the limit, we could in fact consider $p\to\infty$ and $\delta\rho(p)\to 0$, in such a fashion that Δp^n is finite. Then clearly $\delta\langle p^{n+1}\rangle$ diverges as do all higher moments. The second "fourier transform" approach never has these kinds of problems and is always finite.

The second significant "mathematical" difference concerns the utility of conservation laws. While conservation of momenta is certainly maintained for the averages of moments, there is no definite connection between an individual momentum before and after an exchange in this general kind of setup. As we shall see below, the second "fourier transform" approach uncovers an exchange of a new conserved quantity. The conservation law for these quantities can be expressed in a "product-form" rather than a sum (as occurs for ordinary momentum). This product-form conservation law is more relevant for many situations such as a change in relative phase.

Finally, the Fourier transform method provides us with the parameters relevant to physical problem (e.g. the distance D between the slits), while the first "moments" approach remains silent. (We note that, in effect, looking at the modular variables is asking how the Fourier transform of the momentum distribution changes.) In addition, there are different conservation laws involved with the second method which are more relevant and useful. One of the basic notions used in the analysis of conserved quantities in any interaction is that as the probability of one conserved quantity changes ($prob(A)$), then the probability of another should also change ($prob(B)$), such that the probability of the sum ($prob(A+B)$) does not change. As we pointed out, there are situations where the probability of one variable does not change ($prob(B)$), while the probability of the other does change ($prob(A)$). This, per Theorem II, can only happen if that variable (e.g. B) is completely uncertain. That is, this can only happen if the fourier transform of $prob(B)$ below some value remains unaffected while the fourier transform of $prob(B)$ above some value is affected. This means that there is a whole range of modular variables that are being exchanged nonlocally and a large number of conservation laws which can be utilized.

4. Interference phenomenon from the Heisenberg perspective: modular variables

As we have argued previously, the basic gauge symmetry would be violated if any quantum experiment could measure the local phase in $|\Psi_\alpha\rangle$ and therefore there is no *locally accessible phase* information in $|\Psi_\alpha\rangle$. The relative phase is a truly *non-local* feature of quantum mechanics. This point is often missed when the Schrödinger picture is taught and classical intuitions are applied to interference. For this and other reasons, we maintain that the non-local aspect of interference is clearer in the Heisenberg picture.

4.1. Modular variables are the observables that are sensitive to the relative phase

We previously pointed to the significance of the Heisenberg translation operator, $\exp\{\pm\frac{i}{\hbar}\hat{p}D\}$, effecting $\exp\{-\frac{i}{\hbar}\hat{p}D\}\psi_{\rm R}(x) \to \psi_{\rm R}(x-D)$ overlapping with $\psi_{\rm L}(x)$. Therefore, the expectation value of the translation operator $\exp\{\frac{i}{\hbar}\hat{p}D\}$ ***does*** depend on α: $\langle\Psi_\alpha \mid \exp\{i\hat{p}D/h\} \mid \Psi_\alpha\rangle = e^{-i\alpha}/2$. More generally:

$$\langle e^{i\frac{pD}{\hbar}}\rangle = \int_{-\infty}^{+\infty} dx \sum_m a_m^* \psi_m^* e^{i\frac{pD}{\hbar}} \sum_n a_n^* \psi_n^* = \int_{-\infty}^{+\infty} dx \sum_m a_m^* \psi_m^* \sum_n a_n^* \psi_{n+1}^* = \sum_n a_{n+1}^* a_n \quad (4.14)$$

Similarly for $\langle e^{-i\frac{pD}{\hbar}}\rangle = \sum_n a_{n-1}^* a_n$ and thus:

$$\langle\cos\frac{pD}{\hbar}\rangle = \frac{1}{2}\sum_n \{a_{n+1}^* a_n + a_{n-1}^* a_n\} = \frac{1}{2}\sum_n \{a_{n+1}^* + a_{n-1}^*\}\, a_n \quad (4.15)$$

which obviously depends on the choice of a_n. In particular, if all the a_n are equal, then $\langle\cos\frac{pD}{\hbar}\rangle = 1$, while if only one a_n is different from zero, then $\langle\cos\frac{pD}{\hbar}\rangle = 0$. Thus, a measurement to find through which slit the particle passes in an interferometer will change the value of $\langle\cos\frac{pD}{\hbar}\rangle$.

But, exactly *what* information about α does $\langle\exp\{\pm\frac{i}{\hbar}\hat{p}D\}\rangle$ reveal? It is easy to see that if we replace p with $p - \frac{nh}{D}$, then $e^{\frac{i}{\hbar}\hat{p}D}$ changes by $e^{\frac{iD}{\hbar}\frac{nh}{D}} = e^{in2\pi} = 1$, i.e. nothing changes. Furthermore, suppose n is the largest integer such that $n\frac{h}{D} < p$ (i.e. satisfying $0 \le \hat{p} - n\frac{h}{D} \le \frac{h}{D}$). This means that $e^{\frac{i}{\hbar}\hat{p}D}$ gives us information about the remainder after this integer number of $\frac{h}{D}$ is subtracted from p. This is otherwise known of as the modular momentum $p_{\rm mod} \equiv \hat{p}$ modulo $\frac{h}{D}$ (see fig. 3) defined by: $\hat{p}$ modulo $\frac{h}{D} \equiv \hat{p} - n\frac{h}{D}$.

It is clear that p *mod* $\frac{h}{D}$ has the topology of a circle, as would any periodic function. Every point on the circle is another possible value for $p_{\rm mod}$. We deal with modular quantities every time we look at a wristwatch which displays the time modulo 12.

We can get back to ordinary momentum through the relation: $p = N_p\frac{h}{D} + p_{\rm mod}$. We can see this (fig 3) if we stack an integer number (N_p) of $\frac{h}{D}$ on top of the modular portion of p ($p_{\rm mod}$ is the lower portion of fig 3). Note that the eigenstates of the translation operator $\exp\{\frac{i}{\hbar}\hat{p}D\}$ are also eigenstates of the modular momentum $p_{\rm mod}$.

4.2. Modular variables satisfy non-local equations of motion

The key to our explanation of interference from the single particle perspective are the non-local equations of motion satisfied by these modular variables. Thus, using $H = \frac{p^2}{2m} + V(x)$ and

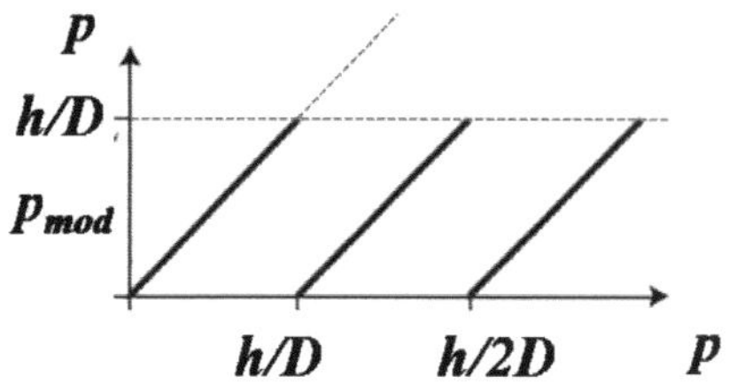

Figure 3. Stacking an integer number (N_p) of $\frac{h}{D}$ on top of the modular portion of p ($p_{\rm mod}$.

$e^{\frac{i}{\hbar}\hat{p}D}V(x)e^{-\frac{i}{\hbar}\hat{p}D} = V(x+D)$, we find **non-local** [13, 12] Heisenberg equations of motion for modular variables:

$$\frac{d}{dt}e^{\frac{i}{\hbar}\hat{p}D} = \frac{i}{\hbar}[H, e^{\frac{i}{\hbar}\hat{p}D}] = \frac{i}{\hbar}[V(x) - V(x+D)]e^{\frac{i}{\hbar}\hat{p}D} \tag{4.16}$$

with $e^{\frac{i}{\hbar}\hat{p}D}$ changing even when $\frac{\partial V}{\partial x} = 0$.

This, essentially quantum phenomenon, has no classical counterpart. The classical equations of motion for any function $f(p)$ derives from the Poisson bracket:

$$\frac{df(p)}{dt} = \{f(p), H\}_{PB} = -\frac{\partial f}{\partial p}\frac{\partial H}{\partial x} + \underbrace{\frac{\partial f}{\partial x}}_{=0}\frac{\partial H}{\partial p} = 0 \tag{4.17}$$

i.e. $f(p)$ changes only if $\frac{\partial V}{\partial x} \neq 0$ at the particle's location.

Unlike the Poisson bracket in classical mechanics, quantum mechanics has non-trivial and unique solutions to the commutator: $[f(p), g(x)] = 0$ if $f(p) = f(p+p_o)$, $g(x) = g(x+x_o)$ and $x_o p_o = h$. These are easy to miss in the Schrödinger picture.

However, in the Heisenberg picture, the non-local equations of motion that the modular variables satisfy show how the potential at the left slit *does* affect the evolution of the modular variable even when we consider a particle located at the right slit (and vice-versa). Modular variables obey non-local equations of motion independent of the specific state of the Schrödinger wavefunction, whether it is localized around one slit or in a superposition. Nevertheless, the modular momentum may change (non-locally) even if the wavefunction experiences no force. We can therefore see that the non-local effect of the open or closed slit is to produce a shift in the modular momentum of the particle while leaving the expectation values of moments of its momentum unaltered.

Let's consider a particular example with $f(p) \equiv \sin\frac{p}{p_o}$ (setting $\frac{\hbar}{D} = p_o$). Classically we have:

$$\frac{d}{dt}\sin\frac{p}{p_o} = \frac{dp}{dt}\frac{1}{p_o}\cos\frac{p}{p_o} = -\frac{\partial V}{\partial x}\frac{1}{p_o}\cos\frac{p}{p_o} \tag{4.18}$$

If the force vanishes ($\frac{\partial V}{\partial x} = 0$) at the position of the particle, then $\sin\frac{p}{p_o}$ will be a constant of the motion for every value of p_o. The quantum equations of motion for $\sin\frac{p}{p_o}$ is (using $\sin\frac{p}{p_o} = \frac{1}{2i}\left\{e^{i\frac{p}{p_o}} - e^{-i\frac{p}{p_o}}\right\}$):

$$\frac{d}{dt}e^{i\frac{p}{p_o}} = \frac{i}{\hbar}[H, e^{i\frac{p}{p_o}}] = \frac{i}{\hbar}\left\{V(x)e^{i\frac{p}{p_o}} - e^{-i\frac{p}{p_o}}V(x)\right\} = \frac{i}{\hbar}\left\{V(x) - V(x+D)\right\}e^{i\frac{p}{p_o}} \tag{4.19}$$

Similarly $\frac{d}{dt}e^{-i\frac{p}{p_o}} = \frac{i}{\hbar}\left\{V(x) - V(x-D)\right\}e^{i\frac{p}{p_o}}$ and therefore:

$$\frac{d}{dt}\sin\frac{p}{p_o} = \frac{1}{2i}\left\{\frac{d}{dt}e^{i\frac{p}{p_o}} - \frac{d}{dt}e^{-i\frac{p}{p_o}}\right\} = \frac{1}{2i}\frac{i}{\hbar}\left\{V(x)e^{i\frac{p}{p_o}} - V(x{+}D)e^{i\frac{p}{p_o}} - V(x)e^{-i\frac{p}{p_o}} + V(x{-}D)e^{-i\frac{p}{p_o}}\right\}$$

$$= \frac{1}{2\hbar}\{V(x-D)-V(x+D)\}\left\{\frac{e^{i\frac{p}{p_o}}+e^{-i\frac{p}{p_o}}}{2}\right\}+i\{2V(x)-V(x-D)-V(x+D)\}\left\{\frac{e^{i\frac{p}{p_o}}-e^{-i\frac{p}{p_o}}}{2}\right\}$$

$$= \frac{1}{2\hbar}\{V(x-D)-V(x+D)\}\cos\frac{p}{p_o}+i\{2V(x)-V(x-D)-V(x+D)\}\sin\frac{p}{p_o} \quad (4.20)$$

We see again that for variables like $\sin\frac{p}{p_o}$ there is a remarkable difference between the classical and the quantum equations of motion [4]. While the quantum equations of motion for the average of the momentum itself is identical to the classical equations of motion for the momentum, there is no such correspondence in the case of $\sin\frac{p}{p_o}$ [5] because quantum interference involves a non-local exchange of $\sin\frac{p}{p_o}$ between the particle and the screen with the slits.

4.3. Non-local exchange of modular variables in the double-slit setup

For the special case of two-slits, a set of spin-like observables can be identified as members of the set of deterministic operators. To obtain these operators, we take $\psi_{\rm L}$ and $\psi_{\rm R}$ as base states and consider the Hilbert space of all vectors:

$$\Psi(x) = a_1\psi_{\rm L}(x) + a_2\psi_{\rm R}(x) \quad (4.21)$$

with $|a_1|^2 + |a_2|^2 = 1$. Our purpose is to find the class of operators $\hat{A}_{a_1a_2}(x,p)$ which satisfy the condition:

$$\hat{A}_{a_1a_2}(x,p)\{a_1\psi_{\rm L}(x) + a_2\psi_{\rm R}(x)\} = a\{a_1\psi_{\rm L}(x) + a_2\psi_{\rm R}(x)\} \quad (4.22)$$

This will be the operators pertaining to the deterministic interference experiments. Since our Hilbert space is isomorphic to that of a spin-1/2 particle we expect that the $\hat{A}_\psi \equiv \hat{A}_{a_1a_2}(x,p)$ operators will have the same properties as the Pauli spin operators and therefore satisfy a spin algebra $[\tilde{\sigma}_i,\tilde{\sigma}_j]_- = i\epsilon_{ijk}\tilde{\sigma}_k$ and $[\tilde{\sigma}_i,\tilde{\sigma}_j]_+ = \delta_{ij}$. One of the three operators should give us information about which slit the particle went through:

$$\tilde{\sigma}_3 = \frac{\sin(\pi x/D)}{|\sin(\pi x/D)|} \quad (4.23)$$

E.g., if $\tilde{\sigma}_3 = 1$ is obtained, then this means the particle went through the right slit (i.e. $a_1 = 0$ and $a_2 = 1$) while $\tilde{\sigma}_3 = -1$ refers to the particle going through the left slit because $\sin(\pi x/D)$ in eq. (4.23) is positive for $\psi_{\rm R}$ and negative for $\psi_{\rm L}$. To derive the other two spin-like operators $\tilde{\sigma}_1$

[4] Consider also a potential $V(x)$, such that $V(x+D) = V(x)+V_o$. Suppose we have a single particle characterized by a wavefunction that is in a superposition of being isolated in region $(0,D)$ and $(2D,3D)$ (etc). Initially we take $\exp\frac{i}{\hbar}\{pD\} = 1$ and thus, initially, $p(t)mod\frac{h}{D} = 0$. Subsequently, $\frac{d}{dt}e^{\frac{i\hat{p}D}{\hbar}} = V_o e^{\frac{i\hat{p}D}{\hbar}}$ leading to $e^{\frac{i\hat{p}D}{\hbar}}(t) = \exp\frac{i}{\hbar}\{pD - V_o t\}$ and thus $p(t)mod\frac{h}{D} = V_o t\frac{h}{D}$. Therefore, irrespective of the system state and even though the potential vanishes and no forces are acting in the region $(0,D)$, the modular momentum [12, 26, 13, 29] is changing at a constant rate based on the particles' nonlocal interaction with the potential in other distant regions. We showed that these functions will depend on exponentials of the form $e^{\frac{i\hat{p}D}{\hbar}}$ and therefore obey non-local equations of motion.

[5] There was an attempt to give a counter-example to the non-local phenomenon (e.g. the Aharonov-Bohm effect) described by modular variables by considering a formulation of QM in terms of the density $\rho(x,p) = \psi^*(x,p)\psi(x,p)$ and the current $\vec{j} = \psi^*\nabla\psi - \psi\nabla\psi^*$. The equations of motion for ρ and $\vec{j}$ are non-linear and only involve the fields (rather than potentials and are thus gauge invariant) and are local. Consider again the superposed states with relative phase α between the two wavepackets. Because $\vec{j} = \rho\nabla\alpha$, the relative phase can be found by integrating $\int\frac{\vec{j}}{\rho}$ over all space. However, for wavefunctions that go to zero in between the wavepackets, this approach does not work. Nevertheless, this approach does reveal an interesting connection between non-locality and chaos. For example, there is an instability characterized by infinitesimal changes in ρ and/or $\vec{j}$ in the region between the wavepackets which will result in a large change later when the wavepackets overlap.

and $\tilde{\sigma}_2$, we will need an operator $\tilde{\sigma}_+$ such that $\tilde{\sigma}_+|\psi_L\rangle = |\psi_R\rangle$ and $\tilde{\sigma}_+|\psi_R\rangle = 0$, e.g. $\tilde{\sigma}_+ = \tilde{\sigma}_3 e^{-i\frac{pD}{\hbar}}$ and $\tilde{\sigma}_- = e^{i\frac{pD}{\hbar}}\tilde{\sigma}_3$. From this we construct $\tilde{\sigma}_1 = \tilde{\sigma}_+ + \tilde{\sigma}_-$ and $\tilde{\sigma}_2 = i(\tilde{\sigma}_+ - \tilde{\sigma}_-)$ and therefore:

$$\tilde{\sigma}_1 = \cos\frac{pD}{\hbar} - i\sin\frac{pD}{\hbar}\frac{\sin(\pi x/D)}{|\sin(\pi x/D)|} \tag{4.24}$$

$$\tilde{\sigma}_2 = \sin\frac{pD}{\hbar} + i\cos\frac{pD}{\hbar}\frac{\sin(\pi x/D)}{|\sin(\pi x/D)|} \tag{4.25}$$

Eq. 4.25 can be understood as translating the particle to the left when eq. 4.23 is positive (i.e. when the particle is located at the right slit then the operator is just $e^{\frac{ipD}{\hbar}}$). If $\tilde{\sigma}_1$ or $\tilde{\sigma}_2$ are measured, then information about the relative phase (i.e. the parameter α between the 2 wave-packets) is obtained, e.g. if we measure $\tilde{\sigma}_1$ and obtain +1, the state is $\frac{1}{\sqrt{2}}\{\psi_L + \psi_R\}$, i.e., $a_1 = 1 = a_2$; the result $\tilde{\sigma}_1 = -1$ yields $a_1 = 1$ and $a_2 = -1$. If we have the state: $\Psi = \frac{1}{\sqrt{2}}\left\{e^{i\frac{\alpha}{2}}\psi_L + e^{-i\frac{\alpha}{2}}\psi_R\right\}$ then the corresponding operator is $\tilde{\sigma}_1\cos\alpha + \tilde{\sigma}_2\sin\alpha = 1$. For any choice of a_1 and a_2 we can make a corresponding rotation in our state space, in full analogy to the ordinary spin-1/2 case.

For simplicity (without affecting the generality of our arguments), we can also express the relevant modular variable as the parity (exchange) operation $\hat{\mathsf{P}}$ (effecting $\hat{\mathsf{P}}|\psi_L\rangle = |\psi_R\rangle$ and $\hat{\mathsf{P}}|\psi_R\rangle = |\psi_L\rangle$). It **is** sensitive to the relative phase α between the disjoint lumps of eq. 3.7 [12] [13] [29]:

$$\begin{aligned}\langle\Psi_\alpha|\hat{\mathsf{P}}|\Psi_\alpha\rangle &= \frac{1}{2}\left\{\langle\psi_L| + e^{-i\alpha}\langle\psi_R|\right\}\hat{\mathsf{P}}\left\{|\psi_L\rangle + e^{i\alpha}|\psi_R\rangle\right\} \\ &= \frac{1}{2}\left\{e^{i\alpha} + e^{-i\alpha}\right\} = \langle\cos\alpha\rangle \end{aligned} \tag{4.26}$$

To simplify further, we will focus on the ±1 eigenstates of $\hat{\mathsf{P}}$: $\psi_L(x)+\psi_R(x)$ and $\psi_L(x)-\psi_R(x)$. A measurement of which slit the particle goes through (i.e. a WWM) will change the value of $\langle\hat{\mathsf{P}}\rangle$. For example if the initial state is $|\psi_L\rangle + |\psi_R\rangle$, then $\langle\cos\alpha\rangle = 1$, i.e. $\langle\hat{\mathsf{P}}\rangle = 1$. If we collapse the state to $|\psi_R\rangle$, then $\langle\cos\alpha\rangle = 0$ and $\langle\hat{\mathsf{P}}\rangle = 0$.

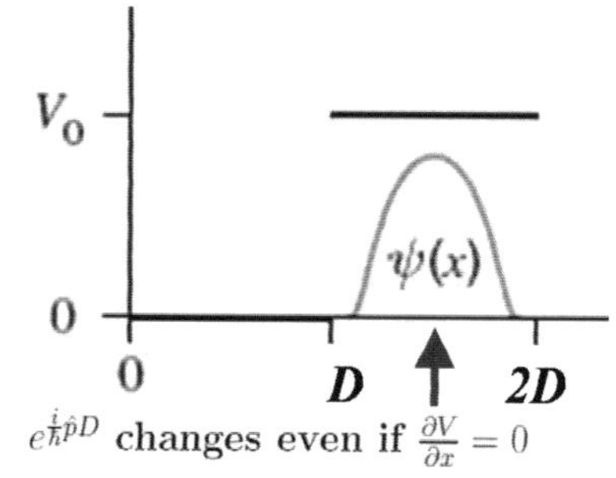

Figure 4. A potential with 2 values and a wave-packet with support only in a limited interval.

We can also see from eq. 5.27 that if the left slit is open, then $V(x) - V(x + D) = 0$, and therefore p_{mod} is conserved. However, if the left slit is closed, then $V(x) - V(x + D) \neq 0$ and p_{mod} is not conserved.

4.4. Why does the interference pattern disappear when the particle is localized?

When we obtain WWM information, we collapse the superposition from $|\Psi_\alpha\rangle$ to $|\psi_L\rangle$ or $|\psi_R\rangle$ (in the Schrödinger picture). In the Heisenberg picture, however, we cannot describe the collapse of a superposition. The wavefunction is still of course relevant as a boundary condition, but it does not evolve in time. Only the operators evolve in time according to the Heisenberg equation of motion: $\frac{dA_H}{dt} = \frac{i}{\hbar}[H, A_H] + U^{-1}(t)\frac{\partial A_s}{\partial t}U(t)$. But which operators become uncertain when WWM information is obtained?

Suppose again the particle travels through the right slit and we choose either to open or close the left slit. This action causes a non-local exchange of modular momentum between the potential at the left slit and the particle going through the right slit. Is this observable?

Up until this paper, it was believed that this could not be observabed. The reason is that modular momentum (unlike ordinary momentum) becomes, upon detecting (or failing to detect) the particle at a particular slit, *maximally* uncertain. In other words, the effect of introducing a potential at a distance D from the particle (i.e. of opening a slit) is equivalent to a rotation in the space of the modular variable - let's call it θ - that is exchanged nonlocally. Suppose the amount of nonlocal exchange is given by $\delta\theta$ (i.e. $\theta \to \theta+\delta\theta$). Now "maximal uncertainty" means that the probability to find a given value of θ is independent of θ, i.e. $P(\theta) = constant = \frac{1}{2\pi}$. Under these circumstances, the shift in θ to $\theta + \delta\theta$ will introduce no observable effect, since the probability to measure a given value of θ, say θ_1, will be the same before and after the shift, $P(\theta_1) = P(\theta_1 + \delta\theta_1)$. We shall call a variable that satisfies this condition a "completely uncertain variable". Using this, APP proved a stronger *qualitative* uncertainty principle for the modular momentum, instead of the usual quantitative statement of the uncertainty principle (e.g. $\Delta pD \geq \hbar$): if the nonlocal exchange of any modular variable θ came close to violating causality, then the probability distribution for all averages of that modular variable flattens out, i.e every value for θ became equally probable and change in θ becomes un-measureable:

Theorem III qualitative uncertainty principle for modular variables: if $\langle e^{in\theta}\rangle = 0$ for any integer $n \neq 0$ and if θ is a periodic function with period τ, then θ is completely uncertain if θ is uniformly distributed on the unit circle.
Proof: we expand the probability density $Prob(\theta)$ to a Fourier series $Prob(\theta) = \sum_{n=-\infty}^{+\infty} a_n e^{in\theta}$ (integer n is a requirement for the function to be periodic in θ), where $a_n = \int Prob(\theta)e^{in\theta}d\theta = \langle e^{in\theta}\rangle$ (since the average of any function is given the integral of the function with the probability). We see that $Prob(\theta) = const$ if and only if $a_n = 0$ for all $n \neq 0$, and therefore $\langle e^{in\theta}\rangle = 0$ for $n \neq 0$.

Consider how this works in the double slit setup. Let us start with a particular $|\Psi_\alpha\rangle$, namely the symmetric ($\alpha = 0$, $|\psi_L\rangle + |\psi_R\rangle$) or anti-symmetric ($\alpha = \pi$, $|\psi_L\rangle - |\psi_R\rangle$) states. The parity $\hat{\mathsf{P}}$ then has sharp eigenvalues ± 1. However, $\hat{\mathsf{P}}$ becomes maximally uncertain when the state is localized at one slit: by definition, $\overline{\hat{\mathsf{P}}^2} = 1$, however, $\langle\psi_L|\hat{\mathsf{P}}|\psi_L\rangle = \langle\psi_L|\psi_R\rangle = 0$ so $\bar{\hat{\mathsf{P}}} = 0$, and therefore $\Delta\hat{\mathsf{P}} \equiv \sqrt{\overline{(\hat{\mathsf{P}})}^2 - \bar{\hat{\mathsf{P}}}^2} = 1$, i.e. we have maximal uncertainty when the particle is localized at one slit. Stated differently, when the particle is at the right (or left) slit its' wave function is a superposition with *equal* weights of the two parity eigenstates $|\psi_L\rangle \pm |\psi_R\rangle$ with ± 1 eigenvalues which by definition is the state of maximal variance of the operator involved.

The vanishing of the expectation value of the modular momentum variable is the manifestation in our present picture of the loss of information on α and of the interference pattern, once we localize the particle at the left or right slit. [6]

[6] Although much of the discussion in this article focuses on the simplest interference example with 2-slits, our approach becomes clearer when it is applied to an infinite number of slits with (interfering) particles that are initially in an eigenstate of momentum. In this case, we can directly speak about modular momentum (instead of slightly more complicated functions for the double-slit setup). Also, both the non-local equation of motion for modular momentum is exact as is the conservation of modular momentum.

This brings us to what we believe to be a more physical answer (from the perspective of an individual particle) for the disappearance of interference: the momentum exchange with the left slit and resulting momentum uncertainty (destroying the interference pattern when the left slit is closed) is not that of ordinary momentum since as we noted Δp does not change. Rather, the closing of the left slit and localization of the particle at the right slit involves a non-local exchange of *modular* momentum. This phenomenon can also be demonstrated for any refinement of the double-slit. For example, any measurement at the left slit introduces an uncertain potential there. As a result of the non-local equations of motion, this introduces complete uncertainty in the modular variable. Thus, detecting which slit a particle passes through destroys all information about the modular momentum.

It therefore appears that no observable effect of one slit acting on the particle traveling through the other slit can be obtained via the nonlocal equations of motion of the modular variable and therefore, this non-locality "peacefully co-exists" with causality. Have we not violated the dictum of maintaining the closest correspondence between measurement and theory by claiming the existence of a new kind of nonlocal - yet un-observable - effect?

5. Gedanken-experiment to measure non-local equations of motion

The key novel observation we next make is that this non-locality *does have* an observable meaning. We discuss two different methods.

5.1. Method 1: von Neuman measurement after the non-local interaction

It is thus clear that if we want to design a deterministic interference experiment we must find a method to measure $\tilde{\sigma}$-type operators (which are diagonal in the given two-slit set-up). According to the von Neumann method [39] this can be done provided a potential proportional to $\tilde{\sigma}$ can be switched on. From the point of view of non-relativistic quantum theory we may in principle assume an arbitrary potential V which is a function of both $\hat{x}$ and $\hat{p}$, and thus a potential proportional to, for example, $\tilde{\sigma}_1$. However, in reality we are restricted to potentials which are local, i.e. functions of positions alone. The existence of potentials which are arbitrary functions of momentum would lead to non-local motion of a particle. To see this, consider a situation where the Hamiltonian contains a term proportional to $\sin \frac{pD}{\hbar}$. We then have

$$i\hbar \frac{\partial \Psi(x,t)}{\partial t} = \sin \frac{pD}{\hbar} \Psi(x,t) = \frac{1}{2i}\left\{e^{i\frac{pD}{\hbar}} - e^{-i\frac{pD}{\hbar}}\right\}\Psi(x,t) = \frac{1}{2i}\left\{\Psi(x+D) - \Psi(x-D)\right\} \quad (5.27)$$

By way of example, suppose that at $t = 0$ the wave function Ψ is different from zero only in a region around $x = 0$ which is small compared to D. Then immediately after application of eq. 5.27, Ψ will begin to grow around the positions $x = +D$ and $x = -D$. This violates the idea of locality or continuity of motion. Therefore, if we want to admit only deterministic experiments which do not violate locality we must find *local* potentials to measure the $\tilde{\sigma}$-type variables.

We do not have to look very far for such potentials: when the particle in the two-slit experiment has passed the slits, it moves again as a free particle. Thus, its position at time t after the passage is related to the position at the time of passage by

$$x(t) = x(0) + \frac{p}{m}t = x(0) + \frac{p(0)}{m}t \quad (5.28)$$

If a measurement of $\sin[x(t)]$ is equivalent to a measurement of $\sin[x(0) + \frac{p(0)}{m}t]$. But

$$\sin[x(0) + \frac{p(0)}{m}t] = \frac{1}{2i}\left\{\exp\left\{i[x(0) + \frac{p(0)}{m}t]\right\} - \exp\left\{-i[x(0) + \frac{p(0)}{m}t]\right\}\right\} \quad (5.29)$$

and using the identity $\exp(\hat{u}+\hat{v}) = \exp(\hat{u})\exp(\hat{v})\exp(-\frac{1}{2}[\hat{u},\hat{v}])$ (which holds for any pair of operators, u and v, that commute with $[u,v]$), it is easy to see that

$$\sin[x(0)+\frac{p(0)}{m}t] = \exp\left\{\frac{it\hbar}{2m}\right\}\left\{\sin[x(0)]\cos[\frac{p(0)}{m}t]+\cos[\frac{p(0)}{m}t]\sin[x(0)]\right\} \tag{5.30}$$

Thus, measurement of $\sin[x(t)]$ is related to measurement of sine and cosine of $x(0)$ and $p(0)$. This suggests that we can reach our goal by introducing a periodic potential in the x-direction. To exhibit the essential aspects of the procedure, consider a simple one-dimensional example of two disjoint wavepackets approaching each other, with a velocity $2\vec{v}$. We let $\Psi_2(x) = \Psi_1(x-2D)$, and at $t=0$ let the two wavepackets be

$$\Psi_1(x,0) = N\int_{-\infty}^{\infty} e^{-\frac{(k-k_o)^2}{\Delta k^2}} e^{ik(x+D)}dk \quad and \quad \Psi_2(x,0) = N\int_{-\infty}^{\infty} e^{-\frac{(k-k_o)^2}{\Delta k^2}} e^{ik(x-D)}dk \tag{5.31}$$

Then at time t we have

$$\Psi_1(x,t) = N\int_{-\infty}^{\infty} e^{-\frac{(k-k_o)^2}{\Delta k^2}} e^{ik(x+D)}e^{-i\frac{k^2t}{2m}}dk \quad and \quad \Psi_2(x,t) = N\int_{-\infty}^{\infty} e^{-\frac{(k-k_o)^2}{\Delta k^2}} e^{ik(x-D)}e^{-i\frac{k^2t}{2m}}dk \tag{5.32}$$

Again, these wavefunctions define an entire class of wavefunctions given by: $\Psi_\alpha = \Psi_1 + e^{i\alpha}\Psi_2$ While different α's yield different wavefunctions, the wavefunction $e^{-i\alpha}\Psi_\alpha$ is equivalent to Ψ_α (since absolute phase cannot be measured). This property, combined with the assumption that Ψ_1 and Ψ_2 do not overlap, means that no local experiment can measure α: i.e., if an apparatus only interacts with Ψ_2, it cannot determine α, unless it is later brought into contact with an apparatus which has interacted with Ψ_1. We could also wait a time $t > \frac{D}{c}$ for the packets to overlap without requiring 2 measuring devices (where $2D$ is the separation between the packets), thereby producing interference and the possibility of determining α. Let us further assume that we can neglect the spread of the wavepackets during the time it takes for the packets to become completely overlapping. Then, putting $k-k_o = \bar{k}$ and $t = T = \frac{Dm}{k_o}$, we have:

$$\Psi_1(x,t) = N\int_{-\infty}^{\infty} \exp\left\{-\frac{\bar{k}^2}{2(\Delta k)^2} - i\frac{\bar{k}^2T}{2m} + ik_o\frac{D}{2} + i\bar{k}x\right\} d\bar{k}e^{-ik_ox} = F(x)e^{ik_ox} \tag{5.33}$$

Similarly for Ψ_2. We consider now the two states

$$X_1 = \frac{1}{\sqrt{2}}\{\Psi_1(x,t)+\Psi_2(x,t)\} \quad and \quad X_2 = \frac{1}{\sqrt{2}}\{\Psi_1(x,t)-\Psi_2(x,t)\} \tag{5.34}$$

We have

$$X_1 = \frac{1}{\sqrt{2}}\left\{F(x)e^{ik_ox}+F(x)e^{-ik_ox}\right\} = \frac{2}{\sqrt{2}}F(x)\cos k_ox \tag{5.35}$$

Similarly, $X_2 = \frac{2i}{\sqrt{2}}F(x)\sin k_ox$. We want to find a method to distinguish between X_1 and X_2 that is analogous to the Stern-Gerlach experiment on a spin-1/2 particle, i.e. that makes X_1 move "up" and X_2 move "down".

The method that achieves this is the following: at the time when the complete overlap occurs we adiabatically switch on a weak potential of the form: $V(x) = A\cos k_ox$. This makes the state X_1 go into a new state X_1' which is an eigenstate of the Hamiltonian $H = \frac{p^2}{2m}+V(x)$. Similarly $X_2 \to X_2'$ which is a different eigenstate of H. The periodicity of the potential gives rise to the well known energy bands (see fig. 5) and X_1' and X_2' correspond to the points 1 and 2 in the (k,ω) diagram. We now add a constant force in the x-direction. This will shift the states X_1' and X_2' from the points 1 and 2 to the points 1 and 2. Since the group velocity $\frac{dw}{dk}$ is negative at 1 and positive at 2, the two states will be pushed in opposite directions in space. Thus, we have a situation analogous to that of the Stern-Gerlach experiment in which the spin-up receives an "upward" kick and the spin-down receives a "downward" kick.

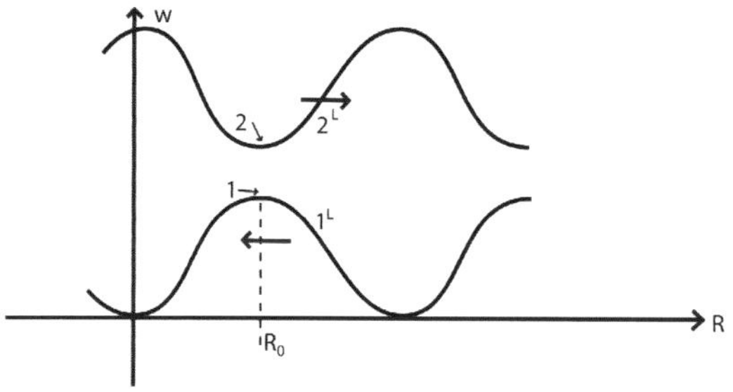

Figure 5.

5.2. Method 2: weak measurement at the time of the nonlocal interaction

What are the general issues involved in the second method to measure dynamical nonlocality?

First: if we start with the state $\psi_R + \psi_L$, i.e. a wavepacket around each slit, then the modular momentum is known but we cannot argue that the particle goes through one slit and is affected nonlocally by the other slit. Therefore we need to start with a state which is localized around one slit.

Second: but under these circumstances when the particle is localized around one slit, the modular variable is completely uncertain and therefore un-observable. How can we get around this fact in order to observe this nonlocality?

Third: if we are able to get around this fact, then how is causality not violated?

Weak measurements allows us to "have our cake and eat it" to a certain extent. To address the first issue, we use pre- and post-selection to arrange for a localized particle property (pre-selection). To address the second issue, we later post-select a definite state of modular momentum. We may perform a weak measurement in order to see the weak value of the modular momentum. This weak measurement has a negligible probability to kick a particle centered around the right slit to the left slit, so we still satisfy the first criteria. Finally, because we must rely on a post-selection and because of the nature of the weak measurement, it is impossible to violate causality with this method.

We proceed now to address each of these issues.

5.2.1. **Information gain without disturbance: safety in numbers** So far, Theorem I describes a completely general geometric property. How is it related to experiments? To actually make a measurement of an observable $\hat{A}$, we switch on an interaction [39] $H_{\text{int}} = \lambda g(t)\hat{q}\hat{A}$ with a normalized time profile $\int g(t)dt = 1$. The pointer, namely the momentum $\hat{p}_q$ conjugate to $\hat{q}$, shifts by $\lambda\langle\hat{A}\rangle$.

Now, the average of any operator $\langle\hat{A}\rangle \equiv \langle\Psi|\hat{A}|\Psi\rangle$ which appears in Theorem I, can be measured in three distinct ways [24, 21]:

1. Statistical method with disturbance: the traditional approach is to perform ideal-measurements of $\hat{A}$ on each particle, obtaining a variety of different eigenvalues, and then manually calculate the usual statistical average to obtain $\langle\hat{A}\rangle$.

2. Statistical method without disturbance: The interaction $H_{\text{int}} = -\lambda(t)\hat{q}\hat{A}$ is weakened by minimizing $\lambda\Delta q$. For simplicity, we consider $\lambda \ll 1$ (assuming without lack of generality that the state of the measuring device is a Gaussian with spreads $\Delta p_q = \Delta q = 1$). We may then set $e^{-i\lambda\hat{q}\hat{A}} \approx 1 - i\lambda\hat{q}\hat{A}$ and use Theorem I to show that the system state is:

$$\begin{aligned} e^{-i\lambda\hat{q}\hat{A}}|\Psi_{\text{in}}\rangle &= (1 - i\lambda\hat{q}\hat{A})|\Psi_{\text{in}}\rangle \\ &= (1 - i\lambda\hat{q}\langle\hat{A}\rangle)|\Psi_{\text{in}}\rangle - i\lambda\hat{q}\Delta\hat{A}|\Psi_{\text{in}\perp}\rangle \end{aligned} \tag{5.36}$$

Using the norm of this state $\| (1 - i\lambda\hat{q}\hat{A})|\Psi_{\rm in}\rangle \|^2 = 1 + \lambda^2\hat{q}^2\langle\hat{A}^2\rangle$, the probability to leave $|\Psi_{\rm in}\rangle$ un-changed after the measurement is:

$$\frac{1 + \lambda^2\hat{q}^2\langle\hat{A}\rangle^2}{1 + \lambda^2\hat{q}^2\langle\hat{A}^2\rangle} \longrightarrow 1 \quad (\lambda \to 0) \tag{5.37}$$

while the probability to disturb the state (i.e. to obtain $|\Psi_{in\perp}\rangle$) is:

$$\frac{\lambda^2\hat{q}^2\Delta\hat{A}^2}{1 + \lambda^2\hat{q}^2\langle\hat{A}^2\rangle} \longrightarrow 0 \quad (\lambda \to 0) \tag{5.38}$$

The final state of the measuring device is now a superposition of many substantially overlapping Gaussians with probability distribution given by $Pr(p_q) = \sum_i |\langle a_i|\Psi_{\rm in}\rangle|^2 \exp\left\{-\frac{(p_q-\lambda a_i)^2}{2\Delta p_q^2}\right\}$. This sum is a Gaussian mixture, so it can be approximated by a single Gaussian $\tilde{\Phi}^{\rm fin}_{\rm md}(p_q) \approx \langle p_q|e^{-i\lambda\hat{q}\langle\hat{A}\rangle}|\Phi^{\rm in}_{\rm md}\rangle \approx \exp\left\{-\frac{(p_q-\lambda\langle\hat{A}\rangle)^2}{\Delta p_q^2}\right\}$ centered on $\lambda\langle\hat{A}\rangle$.

It follows from eq. 5.38 that the probability for a collapse decreases as $O(\lambda^2)$, but the measuring device's shift grows linearly $O(\lambda)$, so $\delta p_q = \lambda a_i$ [21]. For a sufficiently weak interaction (e.g. $\lambda \ll 1$), the probability for a collapse can be made arbitrarily small, while the measurement still yields information. However, the measurement becomes less precise because the shift in the measuring device is much smaller than its uncertainty $\delta p_q \ll \Delta p_q$ (see fig 6).

3. Non-statistical method without disturbance is the case where $\langle\Psi|\hat{A}|\Psi\rangle$ is the "eigenvalue" of a single "collective operator," $\hat{A}^{(N)} \equiv \frac{1}{N}\sum_{i=1}^{N}\hat{A}_i$ (with $\hat{A}_i$ the same operator $\hat{A}$ acting on the i-th particle). Using this, we are able to obtain information about $\langle\Psi|\hat{A}|\Psi\rangle$ without causing disturbance (or a collapse) and without using a statistical approach because any product state $|\Psi^{(N)}\rangle$ becomes an eigenstate of the operator $\hat{A}^{(N)}$. To see this, we apply Theorem I to the N particle product state $|\Psi^{(N)}\rangle = |\psi\rangle_1|\psi\rangle_2....|\psi\rangle_N$ with all particles in the same state $|\psi\rangle$. We see that:

$$\hat{A}^{(N)}|\Psi^{(N)}\rangle = \frac{1}{N}\left[N\langle\hat{A}\rangle|\Psi^{(N)}\rangle + \Delta A\sum_i|\Psi_\perp^{(N)}(i)\rangle\right] \tag{5.39}$$

where $\langle\hat{A}\rangle$ is the average for any one particle and the N states $|\Psi_\perp^{(N)}(i)\rangle = |\psi\rangle_1|\psi\rangle_2...|\psi_\perp\rangle_i...|\psi\rangle_N$ are mutually orthogonal. With a normalized state, $|\Psi_\perp^{(N)}\rangle = \sum_i \frac{1}{\sqrt{N}}|\Psi_\perp^{(N)}(i)\rangle$, the last term of eq. (5.39) is $\frac{\Delta A}{\sqrt{N}}|\Psi_\perp^{(N)}\rangle$ and $|\frac{\Delta A}{\sqrt{N}}|\Psi_\perp^{(N)}\rangle|^2 \propto \frac{1}{N}$. The probability that measuring $\hat{A}_i/N$ changes the state of the i-th system is proportional to $1/N^2$ and therefore the probability that it changes the state of any system is proportional to $1/N$. Thus, as $N \to \infty$, $|\Psi^{(N)}\rangle$ becomes an eigenstate of $\hat{A}^{(N)}$ with value $\langle\hat{A}\rangle$ and not even a single particle has been disturbed (as $\hat{N} \to \infty$).

To perform an actual measurement in this case, we fix Δp_q (the width of the initial pointer momentum distribution) to be 1. We can then take $\lambda \gg 1$, allowing us to distinguish the result by having the shift, $\lambda\langle\hat{A}\rangle$, exceed the width $\Delta p_q = 1$ of the distribution of the pointer. In addition, fixing $\lambda \ll \sqrt{N}$ along with $|\hat{A}_i| < 1$ ensures that the measurement does not shift any particle into an orthogonal state. The coupling to any individual member of the ensemble is reduced by $\frac{1}{N}$. When N is very large, the coupling to individual systems is very weak, and in the limit $N \to \infty$, the coupling approaches zero. Although the probability that a measurement will disturb any member of the ensemble approaches zero as $\frac{1}{N}$, nevertheless, information about the average is obtained.

5.2.2. **Pre-selection, post-selection and weak measurements** Weak measurements have had a direct impact on the central "mystery" alluded to by Feynman concerning indeterminism, namely the fact that the past does not completely determine the future. This mystery was accentuated by an assumed "time-asymmetry" within quantum mechanics, namely the assumption that measurements only have consequences **after** they are performed, i.e. towards the future. Nevertheless, a positive spin was placed on quantum mechanic's non-trivial relationship between initial and final conditions by Aharonov, Bergmann and Lebowitz (ABL) [5] who showed that the new information obtained from future measurements was also relevant for the **past** of quantum systems and not just the future. This inspired ABL to re-formulate quantum mechanics in terms of *pre- and post-selected ensembles.* The traditional paradigm for ensembles is to simply prepare systems in a particular state and thereafter subject them to a variety of experiments. These are "pre-selected-only-ensembles." For pre-**and-post-selected-**ensembles, we add one more step, a subsequent measurement or post-selection. By collecting only a subset of the outcomes for this later measurement, we see that the "pre-selected-only-ensemble" can be divided into sub-ensembles according to the results of this subsequent "post-selection-measurement." Post-selection has also been a very useful technique (see fig. 6.a). For example, while it is standard lore that the wave and particle nature cannot manifest at the same time, weak measurements on pre- and post-selected ensembles *can* provide information about both the (pre-selected) interference pattern and about the (post-selected) direction of motion for each particle.

Because pre- and post-selected ensembles are the most refined quantum ensemble, they are of fundamental importance and have revealed novel aspects of quantum mechanics that were missed before, particularly the weak value which has been confirmed in numerous weak measurement experiments. Weak values have led to quantitative progress on many questions in the foundations of physics [10] [11] including interference, etc. [13] [28] field theory, in tunneling, in quantum information such as the quantum random walk, in foundational questions, in the discovery of new aspects of mathematics, such as Super-Fourier or super-oscillations[7].

What is the weak value? By adding a post-selection to the ordinary -yet weakened- von Neumann measurements mentioned in §5.2.1, the measuring device will register a weak value [7]:

$$\hat{A}_w = \frac{\langle \Psi_{\text{fin}} \,|\, \hat{A} \,|\, \Psi_{\text{in}} \rangle}{\langle \Psi_{\text{fin}} \,|\, \Psi_{\text{in}} \rangle} \tag{5.40}$$

with $|\Psi_{\text{in}}\rangle$ and $|\Psi_{\text{fin}}\rangle$ the initial and final (post-selected) states. The weak-value, A_{w}, is an unusual quantity and is not in general an eigenvalue of $\hat{A}$. We have used such limited disturbance measurements to explore many paradoxes (see, e.g. [10, 29]). A number of experiments have been performed to test the predictions made by weak measurements and results have proven to be in very good agreement with theoretical predictions.

Eq. 5.40 can also be motivated by inserting a complete set of states $\{|\Psi_{\text{fin}}\rangle_j\}$ into $\langle \hat{A} \rangle$

$$\langle \hat{A} \rangle = \langle \Psi_{\text{in}} | \hat{A} | \Psi_{\text{in}} \rangle = \sum_j |\langle \Psi_{\text{fin}} |_j \, \Psi_{\text{in}} \rangle|^2 \underbrace{\frac{\langle \Psi_{\text{fin}} |_j \, \hat{A} \,|\, \Psi_{\text{in}} \rangle}{\langle \Psi_{\text{fin}} |_j \, \Psi_{\text{in}} \rangle}}_{A_{\text{w}}^j \equiv weak\ value} \tag{5.41}$$

with $|\Psi_{\text{fin}}\rangle_j$ the states corresponding to the outcome of a final ideal measurement on the system (i.e. the post-selection). The average $\langle \hat{A} \rangle$ over all post-selections j is thus constructed out of

[7] Super-fourier phenomenon have also been useful in analyzing tachyons [3]. Sudarshan is a member of the lineage of Fourier: Marshak→ Bethe → Sommerfeld → Lindemann → Klein → Lipschitz → Dirichlet → (Fourier and Poisson)→ Lagrange → Euler → Bernoulli →Leibniz

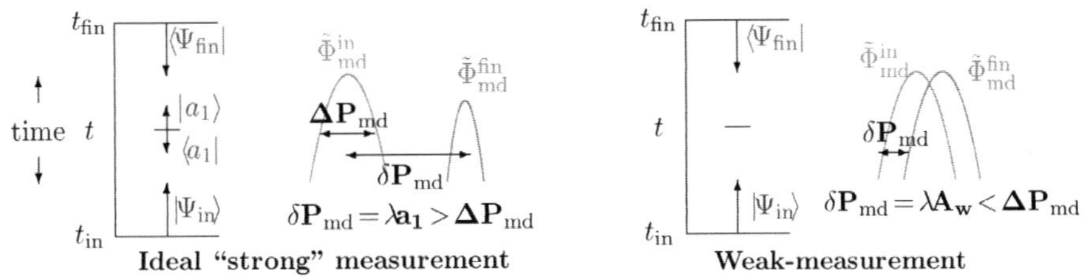

Figure 6. a) with an ideal or "strong" measurement at t (characterized e.g. by $\delta P_{\rm md}=\lambda a_1\gg\Delta P_{\rm md}$), then ABL gives the probability to obtain a collapse onto eigenstate a_1 by propagating $\langle\Psi_{\rm fin}|$ backwards in time from $t_{\rm fin}$ to t and $|\Psi_{\rm in}\rangle$ forwards in time from $t_{\rm in}$ to t; in addition, the collapse caused by ideal-measurement at t creates a new boundary condition $|a_1\rangle\langle a_1|$ at time $t\in[t_{\rm in},t_{\rm fin}]$; b) if a weak-measurement is performed at t (characterized e.g. by $\delta P_{\rm md}=\lambda A_w\ll\Delta P_{\rm md}$), then the outcome of the weak-measurement, the weak-value, can be calculated by propagating the state $\langle\Psi_{\rm fin}|$ backwards in time from $t_{\rm fin}$ to t and the state $|\Psi_{\rm in}\rangle$ forwards in time from $t_{\rm in}$ to t; the weak-measurement does not cause a collapse and thus no new boundary condition is created at time t.

pre- and post-selected sub-ensembles in which the weak value ($A^j_{\rm w}$) is multiplied by a probability to obtain the particular post-selection $|\Psi_{\rm fin}\rangle_j$.

To see more precisely how the weak value arises naturally from this weakened measurement with post-selection, we consider the final state of the measuring device after the above described procedure described in the third "non-statistical" method:

$$\begin{aligned}|\Phi^{\rm MD}_{\rm fin}\rangle &= \prod_{j=1}^{N}\langle\Psi_{\rm fin}|_j\exp(\frac{-i\lambda}{N}\hat{q}\sum_{\rm k=1}^{N}\hat{A}_k)\prod_{i=1}^{N}|\Psi_{\rm in}\rangle_i|\Phi^{\rm MD}_{\rm in}\rangle\\ &= \prod_{j=1}^{N}\langle\Psi_{\rm fin}|_j\exp(\frac{-i\lambda}{N}\hat{q}\hat{A}_j)|\Psi_{\rm in}\rangle_j|\Phi^{\rm MD}_{\rm in}\rangle \qquad (5.42)\end{aligned}$$

Since the particles do not interact with each other, we calculate one term and take the result to the Nth power. (In the following, we substitute the parity operator, $\hat{\mathsf{P}}$, for $\hat{A}$.) Using $\hat{\mathsf{P}}^2=1$, eq. 5.42 becomes:

$$\begin{aligned}|\Phi^{\rm MD}_{\rm fin}\rangle &= \left\{\langle\Psi_{\rm fin}|(\cos\frac{\lambda\hat{q}}{N}-i\hat{\mathsf{P}}\sin\frac{\lambda\hat{q}}{N})|\Psi_{\rm in}\rangle\right\}^N|\Phi^{\rm MD}_{\rm in}\rangle\\ &= (\langle\Psi_{\rm fin}|\Psi_{\rm in}\rangle)^N\left\{\cos\frac{\lambda\hat{q}}{N}-i\mathsf{P}_{\rm w}\sin\frac{\lambda\hat{q}}{N}\right\}^N|\Phi^{\rm MD}_{\rm in}\rangle \qquad (5.43)\\ &\approx \left\{1-i\mathsf{P}_{\rm w}\frac{\lambda\hat{q}}{N}+..\right\}^N|\Phi^{\rm MD}_{\rm in}\rangle\approx\exp(-i\lambda\hat{q}\mathsf{P}_{\rm w})|\Phi^{\rm MD}_{\rm in}\rangle \qquad (5.44)\end{aligned}$$

The first bracket of eq. 5.43 can be neglected since it does not depend on $\hat{q}$ and thus can only affect the normalization. Eq. 5.44 represents a shift in the pointer by the weak value, $\mathsf{P}_{\rm w}$, i.e. $\Phi^{\rm MD}_{\rm in}(p_q)\to\Phi^{\rm MD}_{\rm fin}(p_q-\lambda\mathsf{P}_{\rm w})$.

5.2.3. **Applying pre- and post-selection and weak measurements to interference phenomenon** In [28], these tools were used to perform measurements of dynamical nonlocality. To briefly summarize our procedure, we start with particles sent through the right slit. Before they encounter the double-slit, we perform a weak measurement of the modular momentum (which, again, **is** sensitive to the relative phase). We then choose whether to open the left slit or to close it. After the particles pass the double-slit setup, we perform an ideal measurement of the modular momentum and post-select only those particles in a particular eigenstate of this modular momentum. When we analyze the earlier weak measurement (assuming the post-selection is satisfied), we see two dramatically different results: one result if the left slit is closed and a very different result if the left slit is opened. The slit is open or closed only after the weak measurement has been completed and the results recorded.

All incoming particles are initially in the state $|\psi_{\rm R}\rangle$, so we would not expect that closing the left slit should have any effect on the result of any weak measurement (and in particular weak measurements performed prior to the opening or closing of the slits). Nevertheless, [28] predicted that the *earlier* weak measurement of the parity yields $\langle \hat{\mathsf{P}} \rangle_w = 0$ if the left slit is (*later*) closed and $\langle \hat{\mathsf{P}} \rangle_w = 1$ if the left slit is (*later*) opened. The quantum optics experiment suggested in [28] to measure the non-local exchange has been performed successfully [35].

How do we understand these two results? In principle, a weak measurement with a finite number of particles (N) shifts particles from the right slit to left slit so that the evolving wave-packet has $|\psi_{\rm L}\rangle$ components and therefore may "sense" whether the left slit is open or closed. However, in [28] it was shown that the weak measurement **can shift at most a few of the N particles from the right slit to the left slit.** But, how can the $\sim N$ particles which were not shifted, and did not go through the left slit still be influenced non-locally so that we will have the dramatic (and large) change from the left slit open case (in which *each* particle shifts the measuring device by $\langle \hat{\mathsf{P}} \rangle_w = 1$) to the slit-closed case (each particle shifts the measuring device by $\langle \hat{\mathsf{P}} \rangle_w = 0$)?

We do not see any reasonable way to use local interactions at the left-hand slit to account for the different subsequent behavior of the particles going through the right slit. We can, however, make sense of the results by considering the non-local behavior of modular variables. In particular, the results are calculated using the non-local exchange of modular momentum. The "open-slit" case is calculated by using *conservation* of modular momentum which follows from this nonlocal exchange. The use of this conservation principle is one of the crucial features that distinguishes this procedure from observations that are done with ordinary momentum.

6. Discussion

One of the main points of this article is to emphasize the utility of non-disturbing measurements in probing questions concerning quantum reality. In other publications [14], we have emphasized other classes of non-disturbing measurements, such as the weak measurement [24, 25, 21, 27]. Another purpose of this article is to highlight some new aspects of interference. Our use of the Deterministic Set of Operators (DSOs) in the Heisenberg picture lead us to a **physical** explanation for the different behaviors of a *single* particle when the distant slit is open or closed: instead of having a quantum wave that passes through all slits, we have a localized particle with *non-local* interactions with the other slit(s). Although particles localized around the right slit can exchange modular momentum (non-locally) with the "barrier" at the left slit, the uncertainty in quantum mechanics appears to be just right to protect causality.

While DSOs discussed here is a new conceptual point-of-view that has predicted novel, verified effects which **seem** counter-intuitive from the old perspective, the DSO framework is in fact a *re-formulation* of quantum mechanics, just as the Heisenberg and Schrödinger pictures are equivalent formulations of quantum mechanics. Therefore, experiments cannot prove the DSO framework over standard quantum mechanics (or vice-versa). The motivation to pursue such re-formulations, then, depends on their ability to generate new insights, new intuitions, new experiments, etc. In fact, the DSO framework fulfills several criterion which any re-formulation of quantum mechanics should satisfy in order to be useful and interesting:

- DSO is consistent with all the predictions made by standard quantum mechanics,
- DSO has revealed new features and effects of quantum mechanics that were missed before

(for example, DSOs suggests that a fundamental difference between classical mechanics and quantum mechanics is that the equations of motion for observables relevant to quantum mechanical interference phenomena can be dynamically **non-local**),

- DSO has stimulated discoveries in other fields,
- DSO suggested generalizations of quantum mechanics not easily articulated in the old language

6.1. Quantum Reality

We would like to conclude with a brief discussion concerning the relevance of these considerations to questions of the nature of quantum reality.

6.1.1. The Zeno effect of Sudarshan-Misra and its generalization by Aharonov-Vardi

The Zeno effect by Sudarshan and Misra [2] considers systems which naturally evolve away from a particular state. The fact that a rapid (dense) sequence of measurements prevents it from doing so has thus been nicknamed "watched pot doesn't boil." The generalization of the Zeno effect by Aharonov and Vardi (AV) [19] is related to DSOs. For example, the first example involving a spin-1/2 particle placed in a magnetic field $\vec{B}$ pointing in the x-direction and with $\sigma_z = +1$ initially. If at some time t later we rotate our apparatus and measure the spin in the yz-plane in a direction $\alpha = \omega t$, where ω is the Larmor frequency, then the result is predictable and we have a deterministic experiment. AV generated a time evolution determined by the *observer's* choice of a sequence of measurements, rather than the natural evolution of the system. We suggest that AV be called "watched pot with no fire can boil." By way of example, we consider again the spin-1/2 particle rotating clockwise with $\hat{\sigma}_x(\Delta t) = \hat{\sigma}_x(t=0)\cos\omega\Delta t + \hat{\sigma}_y(t=0)\sin\omega\Delta t$. However, we can force the particle to rotate in the *opposite* direction with certainty if measurements in a counter-clockwise rotating measuring device are made rapidly enough. [8]

What can Zeno and AV tell us about quantum reality? Bohr claimed that we cannot ignore the disturbance on the system (since $\hbar$ cannot be reduced to 0) and therefore we cannot assign a reality to a quantum system between classical measurements because there will always be counterfactual paradoxes similar to the Hardy paradox [10]. John Wheeler elegantly describes "observer-participants" by describing the game of 20 questions in its surprise version: [20] "While I was out of the room, they explain, they had agreed not to agree on a word. There was no word in the room when I entered." He concludes, "First, the word already existed in the room – we thought – independent of any question that we might or might not ask. But it didn't. Likewise the electron has a position and a momentum inside the atom – physics once thought – independent of any act of observation. But it doesn't." To what extent is AV, i.e. generation by an observer of an arbitrary time evolution, a manifestion of this idea? One could argue that the outcome of AV cannot be considered as a pre-existing property of the system, rather it was dictated by the choice of measurement of the observer.

6.1.2. **The relevance of non-disturbing measurements to the nature of quantum reality** Nevertheless, the basic observer-participancy mentioned above has fundamental underpinnings in quanum mechanics. For example, "contextuality" means that whenever we speak

[8] The probability to make the particles rotate in the opposite direction is $\{1-\frac{1}{N^2}\}^N \to e^{-\frac{1}{N}} \to 1$ as $N \to \infty$ where $\Delta t = \frac{T}{N}$. It can also be used to give the Feynman Path picture a physical meaning [19]. If AV is performed on any of the Feynman paths, then if the verification measurements are dense enough, then the Feynman path can be verified with probability one. If a measuring device is in a superposition such that one part of the superposition verifies one Feynman path and another superposition of the measuring device verifies a different Feynman path, then the relative phase between the 2 paths is the classical action utilized in the Feynman formalism. Therefore, AV can be used to give an operational meaning to Feynman's claim that each path has equal probability.

about an "element of reality" we must also include the conditions of observation. There cannot be any contextuality in a single DSO since (by eq. (2.1)) all the functions of any given member of DSO is also a member of the DSO. However, as shown in [30, 31], if we consider the DSO for the pre-selected state and another DSO for the post-selected state (one for the pre-selected vector and one for the post-selected vector), then by combining these DSOs, we can create a novel kind of disturbance which reveals contextuality. By analyzing the concept of contextuality (Bell-Kochen-Specker) in terms of pre-and-post-selection and DSOs, it is possible to assign definite values to observables in a new way [30, 31]. When measurements are performed which do not disturb the pre- and post-selection (i.e. weak measurements), then novel *experimental* aspects of contextuality can be demonstrated including a proof that every pre- and post-selection-paradox with definite predictions implies contextuality. In [30, 31], it is also shown how certain results of these measurements (eccentric weak values with e.g. negative values outside the spectrum), cannot be explained by a "classical-like" hidden variable theory. Additional connections between DSOs and pre- and post-selected systems are explored in [26]

We emphasize that the DSOs discussed here involve those variables which are relevant for a *single* particle without causing a disturbance and can therefore, in principle, be used to probe quantum reality without the usual reliance on ensembles. The third "non-statistical" method (using the collective observable discussed in §5.2.1) also leads to a broad new class of measureable Hermitian observables involving highly nonlocal properties (in both space and time). For example, consider again the spin-like observables discussed in §4.3 with $\hat{\sigma}^z$ telling us about which slit the particle is centered around, etc. The nonlocal relative phase can then be measured with:

$$\alpha \sim \frac{\frac{1}{N}\sum_{i=1}^{N}\hat{\sigma}_i^x}{\sqrt{1-(\frac{1}{N}\sum_{i=1}^{N}\hat{\sigma}_i^x)^2}} \tag{6.45}$$

We believe that this can only be measured using the technique of weak measurement involving the collective observable.

In the discussions above we referred to quantum states as being simply the mathematical tools for describing the system. However, states also have an ontological dimension. This is a highly debated issue, which even for the simple case of a standard state $|\Psi\rangle$ is very controversial. Does the state have a "reality" of its own, or is it just a mathematical tool for making predictions? Does the state actually collapse, or is the collapse simply our updating the mathematical description following the acquisition of new data (the results of new measurement). Is the state a physical entity (such as in Bohm's pilot wave model)? Discussing the ontological status of the second post-selected vector is bound to be even more controversial. It was not our intention to dwell too much on this issue here other than to emphasize that when applied to the Schroedinger picture, weak measurements dispense of the statistical aspect of the wavefunction (though still requiring an ensemble [21]).

Nevertheless, we conclude with a comment on a significant interpretative issue concerning weak values which was originally pointed out by Sudarshan [1]. Namely, there are two ways to view the "weak value" of an observeable in which the measuring device is shifted by strange values which can be any complex number or can be arbitrarily large: 1) the weak value is a result of a complex interference effect in the measuring device, 2) the measuring device is registering the weak value of the **system** as characterized by the 2-vectors. While this first interpretation of weak values is technically available, in a subsequent publication, we will show that it is often un-natural when compared to the second interpretation which regards the weak value as a property of the system. For example, when the system is post-selected, the measuring device measures a single value the weak value that depends only on the measured system and not on measuring device. With post-selection, there is no superposition of measuring device positions, i.e. there is no need to monitor the quantum phases in the measuring device. In fact the second is the only interpretation if the measuring device is classical.

References

[1] Duck I M, Stevenson P M, and Sudarshan E C G 1989 Phys. Rev.**D 40**, 2112

[2] Sudarshan E C G, Misra B 1977 J. Math Phys. 18(4) 756-763

[3] Sudarshan E C G, Bilaniuk O M P, and Deshpande V K 1962 Amer. J. Phys. 30, 718-723; Aharonov Y, Reznik, B, Stern B 1998 A. Phys. Rev. Lett. 81 2190-2193

[4] Aharonov Y, Bohm D 1959, Phys. Rev. **115**, 485

[5] Aharonov Y, Bergmann P G, and Lebowitz J L 1964 Phys. Rev. **134**, B1410, reprinted in *Quantum Theory and Measurement*, eds. Wheeler J A and Zurek W H 1983 (Princeton University Press) pp. 680-686

[6] Aharonov Y and Albert D Z 1980 Phys. Rev. **D12** 3316, ibid. 1981 **D24** 359

[7] Aharonov Y, Albert D Z, and Vaidman L 1988 Phys. Rev. Lett. **60**, 1351; Aharonov Y, Vaidman L 1990, Phys. Rev. A, **41**, 11; Aharonov Y, Vaidman L 1991, J. Phys. A **24**, 2315; Aharonov Y, Botero A 2005, Phys. Rev. A 72, 052111; Aharonov Y, Casher A, Albert D, Vaidman L 1987, Phys. Lett. A124, 199

[8] Aharonov Y, Vaidman L 1990 Phys. Rev. A, **41**, 11

[9] Aharonov Y and Vaidman L 1991 J. Phys. A **24**, 2315

[10] Aharonov Y, Massar S, Popescu S, Tollaksen J, and Vaidman L 1996 Phys. Rev. Lett., **77**, p. 983; Tollaksen, J 2007, Jrnl of Phys A: **40** 9033-9066; Aharonov Y, Botero A, Scully M 2001, Zeit Natur Sec A, **56**, Issue: 1-2, pg 5-15; Nussinov SN, Tollaksen J 2008, Phys. Rev. D, **78** 036007

[11] Aharonov Y, Botero A, Popescu S, Reznik B, Tollaksen J 2002, "Revisiting Hardy's paradox: counterfactual statements, real measurements, entanglement and weak values," Phys. Lett. A 301 (3-4): 130-138; This novel effect has been verified in several independent experiments such as Yokota, et al 2009, *New Journal of Physics*, vol 11, 033011 and Lundeen JS, Steinberg AM 2009, *Phys. Rev. Lett.*, 102:020404. This work has also been featured in popular science journals including the Asian edition of Scientific American (August, 2009), New Scientist "They said it couldnt be done but now we can see inside the quantum world," etc.

[12] Aharonov Y, Pendelton H and Petersen A 1969 Int. J. Theor. Phys., **2**, 213; Aharonov Y, Pendelton H and Petersen A 1970 Int. J. Theor. Phys., **3**, 443.

[13] Aharonov Y and Rohrlich D 2005 *Quantum paradoxes:* Weinheim Cambridge: Wiley-VCH

[14] Aharonov Y, Tollaksen J 2009, "New insights on Time-Symmetry in Quantum Mechanics," in VISIONS OF DISCOVERY: New Light on Physics, Cosmology And Consciousness, ed. R. Y. Chiao, M. L. Cohen, A. J. Leggett, W. D. Phillips, and C. L. Harper, Jr. Cambridge: Cambridge University Press (in press).

[15] Aharonov Y, in *Proc. Int. Symp. Foundations of QM*, Tokyo, pp. 10-19.

[16] Bohm D, Aharonov Y 1957, Phys. Rev. **108** 1070-1076

[17] Aharonov Y, Vaidman L 1990 Phys. Rev. A, **41**, 11

[18] Aharonov Y and Vaidman L 1991 J. Phys. A **24**, 2315

[19] Aharonov Y, Vardi Y 1980 Phys. Rev. D 21 (8): 2235-2240

[20] Wheeler J 1986 *How Come the Quantum?," New Techniques and Ideas in Quantum Measurement Theory*, ed. Daniel Greenberger, *Ann. NYAcad. Sci.***755**, 304

[21] Tollaksen, J 2007 *Non-statistical weak measurements, Quantum Information and Computation V*, edited by Eric Donkor, Andrew Pirich, Howard Brandt, Proceedings of SPIE Vol. 6573 (SPIE, Bellingham, WA), CID 6573-33

[22] Tollaksen J 2007 *Novel Relationships between Superoscillations and Modular Variables,* forthcoming in J. Phys. Conf. Series vol. 70, Editors: Brandt H, Kim Y S, and Man'ko M A

[23] Tollaksen J 2007 *Robust Weak Measurements on Finite Samples,* forthcoming in J. Phys. Conf. Series vol. 70,, Editors: Brandt H, Kim Y S, and Man'ko M A, quant-ph/0703038

[24] Tollaksen J 2007 "Robust Weak Measurements on Finite Samples," J. Phys. Conf. Series, vol. 70, 012015, Editors: Brandt H, Kim YS, and Man'ko MM, quant-ph/0703038. (featured in Hosten O, Kwiat P 2008, Science, "Observation of the Spin Hall Effect of Light via Weak Measurements," 319, 787 which was featured in Resch KJ 2008, Science, "Amplifying a Tiny Optical Effect," 319, 733, both of which were featured in Physics Today , April 2008, Day C, "Light exhibits a spin Hall effect," Volume 61, Issue 4, pp. 8-96, see also Popescu S 2009, "Weak measurements just got stronger," Physics 2, 32

[25] Tollaksen J, Ghoshal D 2007 *Weak Measurements, Weak Values and Entanglement, Quantum Information and Computation V*, Ed by Donkor E, Pirich A, Brandt H, Proc of SPIE Vol. 6573 (SPIE, Bellingham, WA), CID 6573-36

[26] Tollaksen J 2007, "Novel Relationships between Superoscillations and Modular Variables," Jrnl Phys: Conf, 70, Inst of Phys, 012016, Editors: H. Brandt, Y. S. Kim, and M. A. Man'ko; Aharonov T, Kaufherr T 2004, Phys Rev Lett 92, 7, 070404

[27] Tollaksen J 2007 *Quantum properties that are extended in time, Quantum Information and Computation V*, Ed by Donkor E, Pirich A, Brandt H, Proc of SPIE Vol. 6573 (SPIE, Bellingham, WA), CID 6573-35

[28] Tollaksen J, Aharonov Y, Casher A, Kaufherr T, Nussinov N 2004, "Quantum interference experiments, modular variables and weak measurements," arXiv:0910.4227.

[29] Tollaksen J 2001 PhD thesis, Boston University; thesis advisor: Yakir Aharonov.

[30] Tollaksen J 2007 *Probing Contextuality with pre-and-post-selection,* forthcoming in J. Phys. Conf. Series vol. 70,, Editors: Brandt H, Kim Y S, and Man'ko M A

[31] Tollaksen J 2007 *Pre-and-post-selection, weak-values, and contextuality,* to appear in Journal of Physics A: Mathematical and General; quant-ph/0602226

[32] Feynman R , Leighton R, Sands M 1965, The Feynman Lectures on Physics, Vol. III, AddisonWesley, p. 19.

[33] Popescu S, Rohrlich D 1992, "Generic quantum nonlocality," Phys. Lett. A 166, 293

[34] Scully MO, Englert BG, and Walther H 1991, Nature 351, 111; Storey EP, Tan SM, Collett MJ, and Walls DF 1994, Nature 367, 626; Wiseman H 2003, *Physics Letters A*, 311, 285; Mir R, Lundeen JS, Mitchell MW, Steinberg AM, Garretson, Wiseman H 2007, New Journal of Phys, **9**, 287

[35] Gray, Parks, Spence, Tollaksen, Troupe, forthcoming.

[36] Shimony A 1983, in *Foundations of Quantum Mechanics in Light of New Technology,* Proc of the Intnl Symp, Tokyo, eds. Kamefuchi S, Ezawa H, Murayama Y, Namiki M, Nomura S, Ohnuki Y, Yajima T 1984, (Tokyo: Phys Soc of Japan), pp. 22530; cf. p. 227; Shimony also states: "the first confirmation of entanglement ... antedated Bell's work, since Bohm and Aharonov [16]" see Shimony A "Bells Theorem," *Stanford Encyclopedia of Philosophy.*

[37] Feynman RP, Hibbs AR 1965, Quantum Mechanics and Path Integrals (New York: McGraw Hill) , pp. 29; Tonomura A, Endo J, Matsuda T, Kawasaki T, Ezawa H 1989, Am. J. Phys. 57 117.

[38] Batelaan N, Tonomura A 2009, "Testing Aharonov-Bohm effects," *Physics Today.*

[39] von Neumann J 1983 *Mathematical Foundations of Quantum Theory,* Princeton Univ Press, New Jersey

Journal of Physics: Conference Series **196** (2009) 012007 doi:10.1088/1742-6596/196/1/012007

The Symmetries of Nature

Sydney Meshkov

California Institute of Technology, Pasadena, CA 91125

Email: syd@ligo.caltech.edu

Abstract. The study of the symmetries of nature has fascinated scientists for eons. The application of the formal mathematical description of symmetries during the last century has produced many breakthroughs in our understanding of the substructure of matter. In this talk, a number of these advances are discussed, and the important role that George Sudarshan played in their development is emphasized

1. Personal remarks and history

I first met George in a hotel room in New York at the then Annual APS meeting, in either 1957 or 1958. We were introduced by Werner B. Teutsch, whom most of you, except Steve, Susumu and George, don't know or recall. Werner and I had been old friends from Penn- the University of Pennsylvania. Werner had done his thesis on Positronium with Vernon Hughes at Penn in 1954 and then went to the Institute for Advanced Study. After that, he went to Yale and Tufts. While at Tufts he met George, who was visiting at Harvard. They started to work together. Their first paper was in the very first issue of Physical Review Letters, with a distinguished set of co-authors. S. Weinberg, R.E. Marshak, S. Okubo [1]. The topic was, "Divergenceless currents and K-meson decay." I, too, had a paper in that first issue of Physical Review Letters. It was entitled, "Configuration mixing in the C^{14} ground state," co-authored with Elizabeth Baranger [2].

Through Werner, my friendship with George grew over the course of the years, and in 1964, George and Jack Leitner offered me a Professorship at Syracuse University. After much soul-searching, I declined and stayed at the National Bureau of Standards (now called NIST). Shortly thereafter, George came to UT-Austin, where we are today. In a strange twist, I, too, later followed George in becoming a Texan, with a 4 year stint at the SSC, starting in 1990. I had a wonderful time at the SSC, until its demise.

2. Symmetry

What is it?

a. Leon Lederman, in an article for "My Einstein," a recent book by John Brockman [3], discussing Dirac's prediction of antimatter, said, "Dirac's urge to elegance and beauty had uncovered a revolution in physics: the existence of antimatter. For every particle - electrons, protons, neutrons, quarks - there must be an antiparticle. What Dirac's epiphany illustrates is the deep influence of the concept of symmetry on the physics of the twentieth century because symmetry thrives in music and mathematics, its influence in physics not only sparked a revolution in theoretical science, but also acted as a unifying connection to the humanities."

b. In a slightly different vein, my friend Peter Kaus said, recently, "Symmetry is the magic word that distinguishes theory from coincidence."

c. Punching in "Symmetry" on Google yields 36,400,000 entries.
d. Wikipedia [4] lists the entries below.

1 Mathematical model for symmetry
1.1 Non-isometric symmetry
2 Directional symmetry
3 Reflection symmetry
4 Rotational symmetry
5 Translational symmetry
6 Glide reflection symmetry
7 Rotoreflection symmetry
8 Screw axis symmetry
9 Symmetry combinations
10 Color
11 Similarity vs. sameness
12 More on symmetry in geometry
13 Symmetry in mathematics
14 Symmetry in logic
15 Generalization of symmetry
16 Symmetry in physics
17 Symmetry in biology
18 Symmetry in chemistry
19 Symmetry in the arts and crafts
19.1 Architecture
19.2 Pottery
19.3 Quilts
19.4 Carpets, rugs
19.5 Music
19.5.1 Form
19.5.2 Pitch structures
19.5.3 Equivalency
19.6 Other arts and crafts
19.7 Aesthetics
20 Symmetry in games and puzzles
21 Symmetry in literature
22 Symmetry in telecommunications
23 Moral symmetry

e. Brief Review [4], [5]

If G is a symmetry group of a theory describing a physical system - i.e., the fundamental equations of the theory are invariant under the transformations of G - the states of the system transform into each other according to some representation of the group G. The group transformations are mathematically represented by operations relating the states to each other. These operations are the operators acting on the state space that correspond to the physical observables. The observables representing the action of the symmetries of the theory in the state space, and therefore commuting with the Hamiltonian of the system, play the role of the conserved quantities. The eigenvalues of the invariants of the symmetry group provide the labels for classifying the irreducible representations of the group.

An important role played by symmetry is that of classification - for example, the classification of crystals using their varied symmetry properties, or the classification of elementary particles by means of the irreducible representations of some symmetry groups.

The requirement of invariance with respect to a transformation group imposes severe restrictions on the form that a theory may take, limiting the types of quantities that may appear in the theory as well as the form of its fundamental equations. An example is Einstein's use of general covariance when searching for his gravitational equations.

The group theoretical treatment of physical symmetries, with the resulting possibility of unifying different types of symmetries by means of a unification of the corresponding transformation groups, has provided the technical resources for symmetry to play a powerful role in theoretical unification. We assume that symmetry means invariance under any kind of transformation. So an object is symmetric with respect to a given mathematical operation, if when applied to the object, this operation doesn't change the object.

f. One of the simplest symmetries that we know, an SU(2), is that which describes the functioning of a traffic light. We need one operator to tell us whether the light is red or green, another to take us from red to green and a third to take us from green to red.

3. Internal Symmetries

a. In this session there are 3 talks, in addition to mine.

George Sudarshan, No-Go Theorems and the Exclusion Principle

M.Y. Han, Duke University, Raleigh, NC

E.C.G. Sudarshan and Symmetry in Classical Dyamics, Optics and Quantum Mechanics
N. Mukunda, Indian Institute of Science, Bangalore, India

Why Quantum Dynamics is Linear, How One Qubit Almost Completely Reveals the Dynamics of Two, and Other Things Learned Following George's Lead
Tom Jordan, University of Minnesota, Duluth, MN

Inasmuch as they will give detailed descriptions of many topics dear to George's heart, I'll mainly concentrate on internal symmetries, with some excursions into the possibility of combining them with Lorentz Invariance.

b. George's contributions to the study of symmetries are legion. They started almost at the beginning of his career. Consider the 1957 paper by Marshak, Okubo, and Sudarshan, "Consequences of charge Independence for the Magnetic Moments and Masses of Sigma Hyperons. [6]"

In this paper they obtain a sum rule

$$\mu_+ + \mu_- = 2\mu_0 \qquad (1)$$

for the magnetic moments of the strange baryons, invoking charge independent interactions of strongly coupled isospin multiplets. This paper was written before SU(3) entered the scene. On their first page, there is a list of the multiplets considered. These were independent at the time, only later being grouped in SU(3) multiplets.

c. SU(3) versus G2

During the period 1960-1961 a big question was whether the proper classification group for mesons and baryons was G2 or SU(3). Behrends was a proponent of G2 [7], whereas Gell-Mann [8] and Ne'eman [9] wanted SU(3). The deciding factor was the prediction for the number of pseudoscalar mesons. At the time, there existed three pions and four kaons, all pseudoscalars. G2 predicted that there should be seven pseudoscalars, whereas SU(3) predicted that there should be eight, an additional $I = 0$, $Y = 0$ meson. The issue was settled with the discovery of the $\eta(548)$ meson. SU(3) was the correct choice.

d. Introduction to Internal Symmetries

In 1961, the game changed a bit. It looked like the pseudoscalar mesons could be accommodated in an octet of SU(3), but it wasn't totally clear whether the baryons should be in octets, or whether a competing model, the Sakata model [10], which had the physical proton, neutron and lambda particles as fundamental triplets, was the right way to go.

Fortunately, we (C. A. Levinson, H. J. Lipkin, S. Meshkov, A. Salam and R. Munir) [11], .were able to kill the Sakata model by looking at the prediction for proton anti-proton annihilation going into $K_L K_S$ compared to $K_L K_L$.

A good thing about the Sakata model was that groups like Ikeda, Ogawa, and Ohnuki [12], and Sawada and Yonezawa [13] produced tables where they combined BBB and $B\overline{B}$ from which I was able to abstract a complete set of SU(3) 8 x 8 Clebsch-Gordan coefficients.

This set of tables kept us in business for years.

e. SU(2) and SU(3) - broken

Once it was established that SU(3) was the way to go, there were some obvious problems. Just looking at the masses of mesons and baryons, in their respective multiplets, it was clear that there was a large symmetry breaking going on. This was explained by Gell-Mann [14] and Okubo [15], in 1962. They assumed that the symmetry breaker transformed like an $I = 0$, $Y = 0$ member of an octet. This neatly explained the observed splittings.

In that era we had something going for us – data - and lots of it. It was a fun time. I was working with Carl Levinson and Harry Lipkin, and then with Gaurang Yodh and George Snow.

We made copious use of Weyl reflections [16, 17] and applied them to decay widths and scattering amplitudes in hadronic processes. We invented the U-spin and V-spin subgroups of SU(3) [18] and observed that the photon is a U-spin scalar [19]. This was very useful in dealing with electromagnetic processes [19]. There were lots of interesting relationships and experimental tests among scattering amplitudes [20].

During this time, Sudarshan was firing on all cylinders, as well. By methods very different from ours - the common thread was the use of Weyl reflections - George and coworkers produced a host of relations, many of which he presented at the 1963 Athens, Ohio conference [21], where I gave our first public talk on U-spin and V-spin. Many of these relations were based on work done in collaboration with Allan Macfarlane and C. Dullemond [22], [23], [24] using the Shmushkevich method [25], [26].

The years 1963–1967 were years of prodigious physics output for George.

	Total	Symmetry
1963	7	3
1964	11	9
1965	18	12
1966	9	6
1967	11	6

f. Exciting times - SU(3), Quarks

In 1963 the Ξ^* (1520) was found at the mass predicted for a decuplet, where the spacing is linear, by the Gell-Mann Okubo mass formula. This led to the general belief that SU(3) was a good symmetry. In 1964 the Ω^- was found, again where it was predicted to be, and Gell-Mann [27] and Zweig [28] invented the quark model (Zweig called them aces). At that time there were only d, u, and s quarks.

g. SU(6) and Color

In 1964, quarks were given spin by Beg, Lee, Pais [29], Pais [30], Radicati and Gursey [31] and slightly later by Sakita and Wali [32].When quarks are given a spin, there is a spin statistics problem.

The SU(6) multiplets come from combining three quarks, 6x6x6 = 56 +70 + 70 + 20, where 56 = 10 x 4 + 8 x 2. SU(6) is broken into SU(3) flavor x SU(2) spin.

The problem is that 56 is a symmetric combination. Wally Greenberg, on leave from Maryland at IAS, invoked parastatistics, now called color, and explicitly wrote down the states in an SU(6) x O(3) model [33], though he didn't call it that. The symmetry problem for the 56 was solved by combining it with an antisymmetric color singlet giving a totally antisymmetric state.

Note that, also in 1964, George and Mahanthappa [34], in a paper entitled, "SU(6) x 0(3) Structure of Strongly Interacting Particles," also examined this problem, as did Richard Dalitz [35].

4. Combining Internal and space-time symmetries

It was clearly interesting to combine internal and space-time symmetries. This effort took place all over the world through 1964 and 1965. My memory - a bit hazy since it was 41 years ago, was going to the second Coral Gables Conference in January 1965 and hearing presentations by Salam, and several other groups claiming to have solved the problem. They didn't! George [36] discussed the problem but didn't claim to have solved it.

A bit later that year, I went to visit at Weizmann Institute and Harry Lipkin and I found that we could combine internal symmetries with a restricted version of the Lorentz transformation. We could do this for collinear processes such as decays, but not for scattering amplitudes.

We named the relevant SU(2) W-spin and called the combined symmetry $SU(6)_W$ [37,38]. W stood for Weizmann Institute. We did this for constituent quarks and learned that Dashen and Gell-Mann

[39] had done the same for current quarks. Barnes, Carruthers, and Von Hippel [40] also did analogous work. The W spin operators are invariant under Lorentz transformations in the z direction. The W-spin classification for a particle with arbitrary momentum in the z direction is the same as the classification at rest.

The generators of $SU(2)_W$ are:

$$W_z = \sigma_z/2$$

$$W_x = \beta\sigma_x/2$$

$$W_y = \beta\sigma_y/2$$

beta is the intrinsic parity of spin 1/2 particles in the rest frame. The virtue of this symmetry is that it correctly describes decays that are forbidden in the standard SU(6) approach.

5. Gauge symmetries, supersymmetry and neutrino symmetries

a. Gauge Symmetries

The 1970s was the era of Unified Gauge Theories. The works of Georgi, Glashow, Quinn and Weinberg [41], [42] stressed the role of the gauge group SU(5) and its breakdown into SU(3) x SU(2) x U(1). In addition, Georgi [43] and Fritzsch and Minkowski [44] as well as Ramond, Reiss, and Harvey [45] emphasized the role of SO(10).

b. Supersymmetry

A success! We already have half of the spectrum that results from symmetry breaking. Now we need the other half – the SUSY particles. We'll see what the LHC brings us.

c. Neutrino symmetries

Do the neutrinos follow the pattern of the quarks and charged leptons? Quark mass splittings have been described starting with a "Democratic" mass matrix, with symmetry S3 x S3 [46], [47], [48], [49] and adding successive breaking terms in S2 x S2 and S1 x S1. [50], [51]. This gives one large mass, split from two much smaller masses.

For neutrinos, we don't know what the pattern is yet. We do know that one state is roughly an equal mixture of ν_μ and ν_τ, and might like to think that this is the heaviest neutrino. Clearly this situation is different from the almost pure, top, bottom, and τ masses. Some attempts are made using S3 and S2 symmetries [52], [53].

6. Outlook

The use of symmetry principles and the group theory that describes them has been a monumental success in Physics. Invoking a new principle of group theory inertia , there is no reason for this to stop. Undoubtedly, new symmetry principles will arise and simplify our views of the universe. George Sudarshan has spent a lifetime doing just this, and we can gladly say, Thank You, George.

References

[1] Divergenceless currents and K-meson decay. Marshak R E, Okubo S, Sudarshan E C G, Teutsch W B, Weinberg S 1958 Phys. Rev. Lett. **1**, 25

[2] Configuration mixing in the C 14 ground state, Baranger E and Meshkov S 1958 Phys. Rev. Lett. **1**, 30

[3] Leon Lederman, in *My Einstein*, John Brockman 2006 published by Pantheon Books

[4] Wikipedia.org

[5] Symmetry and Symmetry Breaking, Stanford Encyclopedia of Philosophy, Dec. 2004

[6] Consequences of charge independence for magnetic moments and masses of Sigma hyperons, Marshak R E, Okubo S, Sudarshan E C G 1957 Phys. Rev. **106**, 599

[7] Weak-Coupling Currents and Symmetries of Strong Interactions, Behrends R E and Sirlin A 1961 Phys. Rev. **121**, 324
[8] The Eightfold Way: A Theory of Strong Interaction Symmetry, Gell-Mann M 1961 California Institute of Technology Report CTSL-20, unpublished; Gell-Mann M 1962 Phys. Rev. **125**, 1067
[9] Derivation of strong interactions from a gauge invariance, Ne'eman Y 1961 Nuclear Phys. **26**, 222
[10] On a composite model for new particles. Sakata S 1956 Prog. Theor. Phys. **16**, 686
[11] A reaction Forbidden by the Sakata Model of Unitary Symmetry, Levinson C A, Lipkin H J, Meshkov S, Salam A and Munir R 1962 Physics Lett. **1**, 125
[12] A possible symmetry in Sakata's model for bosons-baryons system, Ikeda M, Ogawa S, and Ohnuki Y 1959 Prog. Theor. Phys. **22**, 715 ; ibid. 1960 23, 1073
[13] Mass levels of baryons and mesons, Sawada S and Yonezawa M 1960, Prog. Theor. Phys. **23**, 662
[14] Strange Particle Physics. Strong Interactions, Gell-Mann M 1962 Proc. Intern. Conf. High Energy Phys (CERN), p. 805
[15] Unitary Symmetry in Strong Interactions, Okubo S 1962 Prog. Theor. Physics, **27**, 949
[16] Experimental tests of unitary symmetry, Levinson C A, Lipkin H J, and Meshkov S 1962 Nuovo Cim. **23**, 236
[17] Unitary symmetry of strong interactions, Levinson C A, Lipkin H J, and Meshkov S 1962 Phys. Lett. **1**, 44
[18] Verification of the Tenfold Assignment of the Baryon Resonances, Meshkov S, Levinson C A, and and Lipkin H J 1963 Phys. Rev. Lett. **10**, 361
[19] Unitary symmetry in Photoproduction and other electromagnetic Interactions, Levinson C A, Lipkin H J, and Meshkov S 1963 Phys. Lett. **7**, 81
[20] Comparison of a New SU(3) Prediction with Experiment, Meshkov S, Snow G A, and Yodh G B 1964 Phys. Rev. Lett. **12**, 87
[21] Consequences of SU3 Invariance, Sudarshan E C G 1963 Proc. of the Athens Conf. on Newly Discovered Resonant Particles, Athens
[22] Weyl Reflections in the Unitary Symmetry Theory of Strong Interactions, Dullemond C, MacFarlane A J, Sudarshan E C G 1963 Nuovo Cim. **30**, 845
[23] Relative Weights of the Decays of Certain Resonances in theories with Broken Symmetry, Dullemond C, MacFarlane A J, Sudarshan E C G 1963 Phys. Rev. Lett. **10**, 423
[24] Electromagnetic Properties of Stable Particles and Resonances According to the Unitary Symmetry Theory, Macfarlane A J and Sudarshan E C G 1964 Nuovo Cim. **31**, 1176
[25] A Discussion of Shmushkevich's method is given by Marshak R E and Sudarshan E C G in *Introduction to Elementary Particle Physics*, p. 185
[26] Generalized Shmushkevich Method: Proof of Basic Results, MacFarlane A J, Mukunda N, and Sudarshan E C G 1964 J. Math Phys. **5**, 576
[27] A Schematic Model of Baryons and Mesons, Gell-Mann M 1964 Phys. Lett. **8**, 214
[28] An SU(3) Model for Strong Interaction symmetry and its Breaking, George Zweig 1964, CERN-8419-TH-412
[29] SU(6) and Electromagnetic Interactions, Beg M A B, Lee B W, and Pais A1964 Phys. Rev. Lett. **13**, 514
[30] Implications of Spin-Unitary Spin Independence, Pais A 1964 Phys. Rev. Lett. **13**, 175
[31] Spin and Unitary Spin Independence of Strong Interactions, Gursey F and Radicati L A 1964 Phys. Rev. Lett. **13**, 173
[32] Phenomenological Approach to a relativistic SU(6) Theory, Sakita B and Wali K C 1965 Phys. Rev. Lett. **14**, 404
[33] Spin and Unitary-Spin Independence in a Paraquark Model of Baryons and Mesons, Greenberg O W 1964 Phys. Rev. Lett. **13**, 598

[34] SU(6) X 0(3) Structure of Strongly Interacting Particles, Mahanthappa K T and Sudarshan E C G 1964 Phys. Rev. Lett. **14**, 163

[35] Dalitz R H 1965 Proc. of the Oxford International Conference on Elementary Particles, Oxford, England

[36] Concerning space-time and symmetry groups, Mayer M E, Schnitzer H J, Acharya R, Han M Y, Sudarshan E C G 1964 Phys. Rev. **136**, 888

[37] W-spin and B-spin subgroups of SU(12), Lipkin H J and Meshkov S 1965 Phys. Rev. Lett. **14**, 670

[38] Spin Independence, W-spin, Parity, and SU(6) Symmetry, Lipkin H J and Meshkov S 1966 Phys. Rev. **143**, 1269

[39] Approximate symmetry and the algebra of current components, Dashen R and Gell-Mann M 1965 Phys. Lett. **17**, 142

[40] SU(6) and the Electromagnetic Form Factors, Barnes K J, Carruthers P and von Hippel F 1965 Phys. Rev. Lett. **14**, 82

[41] Unity of All Elementary-particle Forces, Georgi H and Glashow S L 1974 Phys. Rev. Lett. **32**, 438

[42] Hierarchy of Unteractions in Unified Gauge Theories, Georgi H, Quinn H R and Weinberg S 1974 Phys. Rev. Lett. **33**, 451

[43] Towards A Grand Unified Theory Of Flavor, Georgi H 1979 Nucl.Phys. **B156,** 126

[44] Unified Interactions Of Leptons And Hadrons, Fritzsch H and Minkowski P 1975 Annals Phys. **93** , 193

[45] CP Violation And Mass Relations In SO(10), Harvey J, Ramond P and Reiss D 1980 Phys. Lett. **92B**, 309

[46] Weak Interaction Mixing in the Six-Quark Theory, Fritzsch H 1978 Phys. Lett. **73B**, 317; 1979 Nucl. Phys. **B155**, 189

[47] Quark masses and Cabibbo angles, Harari H, Haut H, and Weyers J 1978 Phys. Lett. **78B**, 459

[48] A New View of quark and lepton mass hierarchy, Koide Y 1983 Phys. Rev. **D29**, 252

[49] A BCS Quark Mass Matrix, Kaus P and Meshkov S 1988 Mod. Phys. Lett. **A3**, 1251

[50] Generational Mass Generation, Kaus P and Meshkov S 1990 Phys. Rev. **D42**, 1863

[51] Flavor Democracy and the Lepton-Quark Hierarchy, Fritzsch H and Plankl J 1990 Phys. Lett. **B237**, 451

[52] Neutrino masses, mixing and hierarchy, Kaus P and Meshkov S 2004 Phys. Lett. **B580**, 236

[53] Neutrino masses and mixing from hierarchy and symmetry, Kaus P and Meshkov S 2005, Phys. Lett. **B611**, 147

Journal of Physics: Conference Series **196** (2009) 012008 doi:10.1088/1742-6596/196/1/012008

George Sudarshan, No-Go Theorems and the Exclusion Principle

M. Y. Han

Physics Department, Duke University, Durham, NC

E-mail: myhan@phy.duke.edu

Abstract. We review two areas of my work that were directly and indirectly initiated and inspired by George. One is the proofs of no-go theorems in combining the spacetime and internal symmetries in non-trivial ways and the other is how Georges firm conviction on the fundamentality of the spin-statistics theorem helped to expand the domain of applicability of the spin-statistics theorem into the arena of quarks and gluons, going far beyond the original application of the exclusion principle in atomic physics.

In order to provide deeper understanding of mass differences of particles belonging to spin-degenerate multiplets, attempts have been made to see if some non-trivial way of embedding the Lorentz group and internal symmetry groups such as SU(2) and SU(3) into a larger group. When a hint of no-go theorem (that such non-trivial embedding cannot be achieved) first appeared, George went to work and led many of us, including myself, into this area of research. A series of proofs of no-go theorems by George, myself and others eventually led to the definitive proof by Lochlainn O'Raifeartaigh that came to be known as the ORaifeartaigh theorem. Lochlainn also joined George at Syracuse at the same time I went there.

The second area in which George had significant influence on my work is his fervent belief in the fundamental importance of the spin-statistics relationship. First postulated by Worlfgang Pauli to explain the periodic table of elements, the relationship the exclusion principle has exceeded far beyond its original domain of validity. The relationship has been upheld across the scale molecular, atomic, and nuclear structures. What is less known is the fact the relationship continues to remain valid in scales smaller than nucleons. The spin-statistics relationship was one of the compelling reasons for Nambu and I to introduce a new set of then undiscovered degrees freedom for quarks inside nucleons. This new degrees of freedom came to be called the color charges of quarks and it led to the non-Abelian gauge field theory of the strong interactions known today as the quantum chromodynamics. Georges insight into the spin-statistics relationship was indispensable in the birth of the new theory of the strong force.

1. George's leadership toward the ORaifeartaigh Theorem

In the mid-1960s, about the same time frame in which the idea of quarks was beginning to get traction, there have been movements in attempts to combine relativistic invariance and internal symmetries in a non-trivial way. The idea was to embed the Lorentz group and internal symmetry groups such as SU(2) and SU(3) [flavor] into a larger group by which the mass differences within spin-degenerate multiplets such as shown by the Gell- Mann-Okubo mass formula could come about as a consequence of an exact symmetry under such a large group.

First inkling that this may not be achievable was put forward by W. D. McGlinn, PRL 12, 467 (1964). George immediately jumped into this line of research and led many of us, including

Journal of Physics: Conference Series **196** (2009) 012008 doi:10.1088/1742-6596/196/1/012008

myself, toward proving a series of such no-go theorems. By the way, the term no-go theorems was coined by Thomas Jordan, who will speak shortly in this session.

M.E. Mayer, H.J. Schnitzer, E.C.G. Sudarshan, R. Acharya and M. Y. Han, PR 136, B888 (1964). This work was done when George was spending a year at Brandeis University. We proved several theorems:

Theorem 1: Let X_ρ be the set of generators of the internal symmetry group S and P_μ, $M_{\mu\nu}$ the generators of the (proper orthochronous inhomogeneous) Lorentz group L. If the set of these generators is closed under commutation and if

$$[X_\rho, M_{\mu\nu}] = 0 \tag{1}$$

then

$$[[X_\rho, X_\sigma], P_\mu] = 0. \tag{2}$$

If S has no Abelian factor group we can further show that

$$[X_\rho, P_\mu] = 0. \tag{3}$$

We kept weakening the conditions but to no avail.

Theorem 2: Given a symmetry group S with the generators H_i, E_{alpha} in the Weyl basis, If H_i commutes with generators of L, then E_{alpha} must also commute with the generators of L

Theorem 3: If only E_{alpha} commutes with $M_{0\nu}$ (only the Lorentz transformation part), then the product of L and S reduces to a simple product.

After a year at Brandies, George moved from Rochester to Syracuse to form a new particle theory group there. As the first team to arrive at Syracuse with George were A. J. MacFarlane, Lochlainn ORaifeartaigh and a handful of research associates including myself. The work continued at Syracuse along the same line. In the fall of 1964, George and I independently wrote papers proving further theorems.

M. Y. Han, PR 138, B689 (1964) investigated the consequences of replacing the generators Mof the homogeneous Lorentz group by the Pauli-Lubansky operator

$$\omega_\mu = \frac{1}{2}\epsilon_{\mu\nu\lambda\sigma}M^{\nu\lambda}p^\sigma \tag{4}$$

Theorem 4: Let S_i be the set of generators of an internal symmetry group S. If S has no Abelian factor groups and if

$$[S_i, P_\mu] = 0, \quad \text{and} \quad [S_i, \omega_\sigma] = 0 \text{ for all } \sigma, \tag{5}$$

then

$$[S_i, M_{\mu\nu}] = 0, \tag{6}$$

and, again, $G = L \oplus S$.

George, in one of his defining papers on the subject, E.C.G. Sudarshan, J of Math Phys, 6, 1329 (1965) [This is one of Georges representative papers listed for this session of symmetry in the seven science quests.] proved a much more general theorem:

"Under the restriction to a symmetry group whose generators can be linearly expressed in terms of the generators of the Lorentz group and of an internal symmetry group, we show that charge conservation implies that the symmetry group is a direct product of the Lorentz group and the internal symmetry group."

George kept informing Lochlainn about the developments in this area of research, and to lesser extent, I also kept Lochlainn updated on every aspect of this line of research. I remember seeing Lochlainn holed up in his office for weeks with only one book on his desk, Lie Algebra by

Jacobson. After a while Lochlainn wrote up the results of his mathematical investigation and the result became known as the O'Raifeartaigh theorem. [Lochlainn O'Raifeartaigh, Phys. Rev. Letters 14, 575 (1965) and Phys. Rev. 139, B1052 (1965).]

O'Raifeartaigh's Theorem: Let R be the vector space of any irreducible representations of a group E (that may contain the Lorentz group L and an internal symmetry group S), and R_m any subspace of R which is an eigenspace of the mass operator $p_\mu p^\mu$ belonging to the discrete eigenvalues of m^2. Then the space of R_m is invariant with respect to E. Hence $R = R_m$, and there is no mass splitting, and the larger group is the direct product of the Lorentz group and an internal symmetry group.

George's and O'Raifeartaigh's theorems put an end to the search for a larger symmetry group that may contain the space-time and the internal symmetry groups in a non-trivial way so as to be able to explain the mass splitting as a consequence of a larger symmetry. Throughout this period, George led the effort with his insight and leadership.

2. George's influence toward the color charges of quarks

George's "religious" belief in the fundamentality of the spin-statistics relationship culminated in the book he co-authored with Ian Duck, Pauli and the Spin-Statistics Theorem (World Scientific, 1997).

The session that follows this one is devoted to Georges work in this area and here I would just like to mention how he influenced some of my work during the same period at Syracuse University. First some quotes from the book:

"Why is it that particles with half-integral spin are Fermi particles whose amplitudes add with the minus sign, whereas particles with integral spin are Bose particles whose amplitudes add with positive sign? We apologize for the fact that we cannot give you an elementary explanation. An explanation has been worked out by Pauli from complicated arguments of quantum field theory and relativity. He has shown that the two must necessarily go together, but we have not been able to find a way of reproducing his arguments on an elementary level. This probably means that we do not have a complete understanding of the fundamental principle involved." - Richard Feynman

"Everyone knows the Spin-Statistics Theorem but no one understands it." - George Sudarshan

"What has been proved is that the existing theories are consistent with the spin-statistics relation. What has not been demonstrated is a reason for the spin-statistics relation." - George Sudarshan

First postulated by Worlfgang Pauli to explain the periodic table of elements, the relationship the exclusion principle has exceeded far beyond its original domain of validity. The relationship has been upheld across the scale molecular, atomic, and nuclear structures. In fact, we are not aware of any instances in which the spin-statistics relationship has been violated in any part of the observed world. The fundamentality of the spin-statistics relationship rivals that of the Lorentz invariance and the CPT theorem.

What is not as widely known is that the powerful validity of the spin-statistics theorem extends into the part of the world that remains unobservable, namely, to the domain of quarks and gluons. In fact, one can go so far as to say that it is the spin-statistics relationship that gave rise to the present day theory of the strong nuclear force originating from the quark-gluon interaction. In other words, the very genesis of the quantum chromodynamics (QCD), the non-Abelian gauge field theory of the strong nuclear force, is none other than the spin-statistics relationship. Let me briefly sketch out this path from the spin-statistics to the present-day QCD.

When the three-quark states, it presented a serious conflict with the exclusion principle. All three states composed of three up quarks (uuu), three down quarks (ddd) and three strange quarks (sss) were classified as the completely symmetric states in the baryon decuplet, it was in a direct conflict with the principle. Two possible way of this dilemma were proposed:

1. CHANGE THE RULE: Relax the exclusion principle and allow quarks to obey less stringent rules by permitting them to obey the parafermi statistics of order 3. This is the paraquark approach proposed by Greenberg.

2. KEEP THE RULE: Insist on the exclusion principle for fermions and extend it to quarks. This required introduction of a new, three-fold degrees of freedom by which the three otherwise identical quarks in (qqq) states cane differentiate from each other, thereby making the states totally antisymmetric. This is the approach that Nambu and I decided to pursue as we introduced a new SU(3) degrees of freedom. This is what has come to be called the color SU(3).

Several years later, Murray Gell-Mann proposed a non-Abelian gauge field theory for the strong nuclear force among quarks; he named the new theory the quantum chromodynamics, QCD, the new three degrees of freedom was christened as the color charges and the octet of gauge bosons named as gluons. Gell-Mann made a conceptual jump in that the new SU(3) degrees of freedom should be regarded as the strong force analog of the electric charges simply, color charges of quarks named red, green and blue.

Soon thereafter Politzer, Gross and Wilczek discovered an important property of all non-Abelian gauge field theory (without any spontaneous symmetry breaking) called the asymptotic freedom (This had led to the trio to receive 2004 Nobel prize for physics).

The asymptotic freedom, which states that at short distances quarks behave like free particles, that is, the strength of the strong force decreases (as opposed to the Coulomb force which increases at short distances), has been supported by several experimental data in the deep-inelastic electron- and neutrino- proton scatterings as well as the hadron jets formed from the quark and gluon jets.

Although the asymptotic freedom allows, at short distances, perturbative calculations in QCD, the non-perturbative calculations in QCD are notoriously difficult and the theory has yet to yield useful information in the non-perturbative region. The defining formulas for QCD, in its compact form, are deceptively simple and can be expressed in just three lines that closely resemble those of QED, the quantum electrodynamics:

$$L = \frac{1}{4g^2} G^a_{\mu\nu} G^{\mu\nu}_a + \sum_j \hat{q}_j \left(i\gamma^\mu D_\mu + m_j \right), \tag{7}$$

where

$$G^a_{\mu\nu} \equiv \partial_\mu A^a_\nu - \partial_\nu A^a_\mu + i f^a_{bc} A^b_{mu} A^c_\nu, \tag{8}$$

and

$$D_\mu = \partial_\mu + i t^a A^a_\mu. \tag{9}$$

Here, $j =$ flavor and $a, b, c =$ color indices, f and t are the structure constants and generators of the color SU(3) , and the A's are the gluon fields.

The power of the spin-statistics relationship has been, once again, demonstrated to be valid in the domain of quarks and it provided the genesis for the current theory for the strong forces among quarks and gluons. George's faith in the spin-statistics relationship had a direct influence to the birth of the color charges of quarks, which in turn led to the only viable theory for the strong interaction to date.

Journal of Physics: Conference Series **196** (2009) 012009 doi:10.1088/1742-6596/196/1/012009

E. C. G. Sudarshan and Symmetry in Classical Dynamics, Optics and Quantum Mechanics

N. Mukunda

Indian Institute of Science and Indian Academy of Sciences, Bangalore, India

E-mail: nmukunda@gmail.com

Abstract. We review work initiated and inspired by Sudarshan in relativistic dynamics, beam optics, partial coherence theory, Wigner distribution methods, multimode quantum optical squeezing, and geometric phases. The 1963 No Interaction Theorem using Dirac's instant form and particle World Line Conditions is recalled. Later attempts to overcome this result exploiting constrained Hamiltonian theory, reformulation of the World Line Conditions and extending Dirac's formalism, are reviewed.

Dirac's front form leads to a formulation of Fourier Optics for the Maxwell field, determining the actions of First Order Systems (corresponding to matrices of Sp(2,R) and Sp(4,R)) on polarization in a consistent manner. These groups also help characterize properties and propagation of partially coherent Gaussian Schell Model beams, leading to invariant quality parameters and the new Twist phase.

The higher dimensional groups Sp(2n,R) appear in the theory of Wigner distributions and in quantum optics. Elegant criteria for a Gaussian phase space function to be a Wigner distribution, expressions for multimode uncertainty principles and squeezing are described.

In geometric phase theory we highlight the use of invariance properties that lead to a kinematical formulation and the important role of Bargmann invariants. Special features of these phases arising from unitary Lie group representations, and a new formulation based on the idea of Null Phase Curves, are presented.

1) The role of symmetry, the analysis of significant group structure and action in physical problems, are recurring themes in E.C.G. Sudarshan's work. Other speakers have surveyed this in the context of elementary particle physics. In consultation with Professor Thomas Jordan, I will cover mechanics, classical and quantum optics, and some aspects of quantum mechanics.

Some of the work to be described has been done by Sudarshan with collaborators. Others were initiated by him and some of us, carried further by us in later years. As time is limited, we can only afford a fleeting glance at each area.

2) In 1949 Dirac proposed several forms of relativistic dynamical descriptions of physical systems involving ten generators of Poincaré group action—the instant, point and front forms.[1] These are in correspondence with certain subgroups of the Poincaré group—E(3), SO(3,1) and $\mathcal{G}_{2,1}$ respectively, where the last is like the Galilei group in $2+1$ dimensions. In each case, the subgroup generators are simple or kinematic, the rest are dynamical or 'Hamiltonians'. Bakamjian and Thomas are among those who followed up Dirac's ideas quite early. In 1963, Currie, Jordan and Sudarshan examined the description of a system of relativistic point particles in the instant form, adding the important World Line Condition invented by them.[2] This ensures the existence of objective, observer-independent particle world lines in space time; for each

particle this is the requirement

$$\{K_j, q_k\} = q_j\{H, q_k\},$$

where K_j are the pure Lorentz transformation generators, H the Hamiltonian, and $\{,\}$ are classical Poisson brackets. They then showed that for a two-particle system the Poincaré algebra PB relations plus the World Line Conditions rule out the possibility of any interactions. This was generalised to any number of particles by Leutwyler in 1965.[3] This No Interaction Theorem may be found in many text books, and stood for many years.

In the late 1970's several authors—Komar, Rohrlich, Todorov, ...—independently suggested ways to overcome this theorem exploiting another beautiful idea of Dirac—constrained Hamiltonian mechanics.[4,5] However, none of them seemed to have paid proper attention to the existence and objectivity of particle world lines. In a comprehensive series of papers, Sudarshan and many of us gave a detailed analysis of this problem.[6] The World Line Conditions were developed in a manifestly covariant form, and combined with the Poincaré algebra, it was shown that interactions among relativistic point particles are permitted. But the major problem that remained was separability. An extension of Dirac's ten generator formalism to a set of eleven generators is involved here, the extra generator being an invariant Hamiltonian giving rise to manifestly covariant evolution equations. Several forms of these theories are available; in some, the translation subgroup of the Poincaré group acts trivially, while in a particular form the homogeneous Lorentz transformations act trivially.[7] In any case, one goes beyond the original Dirac forms of relativistic dynamics in that the evolution parameter is determined dynamically for each state of motion, and not once for all kinematically in each inertial frame. The problem with separability can be traced to this fact.

3) I turn next to another use of Dirac's generator formalism, this time the front form. Classical ray optics in the paraxial regime, also called Gaussian optics, is a very well developed subject. Here first order optical systems—FOS's—act on each incoming ray to yield a single outgoing ray. These systems, such as lenses, magnifiers, rotators and their combinations, are in correspondence with elements of the real symplectic groups Sp(2,R) and Sp(4,R) in the isotropic and anisotropic cases respectively. These groups appear because Fermat's principle is a variational principle. Extended to the scalar optical field, we have the subject of Fourier Optics. It was shown by Bacry and Cadilhac in 1981 that when the FOS's act on scalar waves, the metaplectic groups Mp(2) and Mp(4), double covers of Sp(2,R) and Sp(4,R), enter the picture.[8]

The question arises—is there a natural generalization of scalar Fourier Optics to the full classical Maxwell field, treating polarization consistently and retaining the roles of Mp(2) and Mp(4) in the description of FOS's? Dirac's front form showed the way—displayed suitably, the Poincaré algebra contains a 2+1 dimensional Galilei group structure, allowing us to extract two-dimensional canonical pairs of variables from the Poincaré generators.[9] Then by a simple rule paraxial scalar wave optics generalises consistently to paraxial Maxwell wave optics, i.e., to Fourier optics for the Maxwell field.[10] In this regime the actions of FOS's on polarization are completely determined. The two-dimensional canonical variables are identified starting from the combinations $K_1 - J_2, K_2 + J_1$, and P_1, P_2; once these are in hand, the groups Sp(2,R), Sp(4,R) and their metaplectic covers immediately follow. Thus we can say that Fourier Maxwell optics is a unique generalisation of the scalar theory.

Similar ideas have been used by Jagannathan and Khan in the case of the Dirac equation, leading to a satisfactory generalisation of traditional electron optics.[11]

4) Continuing with classical optics, partially coherent (scalar) beams are described using the so-called two-point correlation function. Here again, especially in the context of the widely studied Gaussian Schell Model—GSM—beams, one can exploit the groups Sp(2,R) and Sp(4,R) very effectively. Classification of such beams, their free propagation and passage through general FOS's, can all be handled in a systematic manner. The use of symmetry and group action yields results going well beyond what was known earlier in the literature. Here are some instances.

In the isotropic case, the IGSM family of beams, we have a three-parameter family, each beam represented by a time like vector in a fictitious $2+1$ dimensional Minkowski space $M_{2,1}$; and each FOS acts via a transformation of SO(2,1) in this picture.[12] The general anisotropic case, the AGSM family of beams, is considerably more elaborate.[13] The beams form a ten-parameter family, represented by antisymmetric tensors in a fictitious $3+2$ dimensional de Sitter space; and each FOS acts as an SO(3,2) de Sitter transformation on these tensors. Here we recall that Sp(2,R) and Sp(4,R) are two-fold covers of SO(2,1) and SO(3,2) respectively.

This analysis by Sudarshan, Simon and myself presented for the first time the two independent invariant quality parameters for the general AGSM beams, conserved under propagation and FOS action; their identification was possible only thanks to the underlying symmetry structure.

Simon and I carried this study further and showed that in the IGSM case a twist or handedness or helicity can be impressed on the two point function. This leads to the family of Twisted Gaussian Schell Model—TGSM—beams, which had not been noticed earlier in the literature.[14] They have subsequently been experimentally produced and theoretically analysed by others. The most interesting feature here is the upper bound on the twist parameter. The two-point functions in each transverse plane are given by

$$\Gamma_{TGSM}(\rho_1,\rho_2) = e^{-iu(x_1y_2-x_2y_1)/\lambdabar}\Gamma_{IGSM}(\rho_1,\rho_2),$$

where $\rho_1=(x_1,y_1)$, and $\rho_2=(x_2,y_2)$, are two points in the transverse plane, $\lambdabar=\lambda/2\pi$ and u is the twist strength. The bound in question is

$$|u| \leq \lambdabar/\sigma_g^2$$

where σ_g is the transverse coherence length. Thus such a phase cannot be impressed on a fully coherent beam for which $\sigma_g\rightarrow\infty$!

Simon and I have also completely analysed this twist phase in the general AGSM context.[15]

5) A related very fruitful idea of Sudarshan from 1979 is the adaptation of Wigner distribution methods to paraxial partially coherent optics.[16] In this way the complete range of wave optical phenomena can be cast into the ray language, provided that one allows for what he called 'bright' and 'dark rays'. The actions of FOS's via the integral Huyghens kernel on the field amplitude reduce to much simpler point transformations by matrices on the generalised rays, the matrices belonging to Sp(2,R) or Sp(4,R) as the case may be. This approach has been exploited in the beam problems described earlier; it also leads to the famous ABCD Law, a matrix version of the fractional Mobius transformation, accompanying FOS action.[17]

6) The higher dimensional real symplectic groups Sp(2n,R) play an important role in the non-relativistic quantum mechanics of multimode systems, especially in connection with Wigner distributions. For example, the Wigner quality of a real phase space function $W(q,p)$—whether it is the Wigner distribution associated with some bonafide quantum state—is very subtle and difficult to capture in finite terms in general. What is true is that under Sp(2n,R) action, or more precisely action by its metaplectic double cover Mp(2n), the transformation rule is very simple as mentioned earlier:

$$\begin{aligned}
&S\in Sp(2n,R)\rightarrow \bar{U}(S) \text{ on Hilbert space, determined upto sign;}\\
&\text{Density matrix } \hat{\rho}\rightarrow\hat{\rho}'=\bar{U}(S)\hat{\rho}\bar{U}(S)^{-1}\Leftrightarrow\\
&\text{Wigner distribution } W'(q,p)=W(S^{-1}(q,p)),
\end{aligned}$$

where we view (q,p) as a 2n-component real column vector. In the particular case of Gaussian phase space distributions we showed that the Wigner quality is easily expressible using Williamson's Theorem.[18]

$$W(q,p)=\pi^{-n}(\det G)^{1/2}\exp\left\{-(q\ p)G\begin{pmatrix} q\\ p\end{pmatrix}\right\},$$

G real symmetric 2n dimensional positive definite;

Williamson: $S^T G S = \mathrm{diag}(\kappa_1, \ldots, \kappa_n, \kappa_1, \ldots, \kappa_n)$, suitable $S \in Sp(2n, R)$,

$$\kappa_j > 0, j = 1, 2, \ldots, n;$$

$W(q,p)$ is a Wigner distribution $\Leftrightarrow 0 < \kappa_j \leq 1, j = 1, \ldots, n.$

These conditions can also be expressed explicitly in terms of the polynomial Sp(2n,R) invariants formed from the matrix G.

These groups are also relevant in multimode quantum optics problems, since they preserve the fundamental Heisenberg commutation relations whether written in real or complex forms. Here I may overlap a little bit with tomorrow's session on quantum optics. For a general state $\hat{\rho}$ of an n-mode quantum optical system, the variance matrix or second order moments matrix is defined in terms of hermitian quadrature components as

$$V = \begin{pmatrix} V_1 & V_2 \\ V_2^T & V_3 \end{pmatrix},$$

$$(V_1)_{jk} = Tr(\hat{\rho}\ \hat{q}_j\ \hat{q}_k) \qquad (V_2)_{jk} = \frac{1}{2} Tr(\hat{\rho}\{\hat{q}_j, \hat{p}_k\}),$$
$$(V_3)_{jk} = Tr(\hat{\rho}\ \hat{p}_j\ \hat{p}_k) \qquad j, k = 1, 2, \ldots, n.$$

(For simplicity we have assumed that the mean values of $\hat{q}_j$ and $\hat{p}_j$ vanish). The uncertainty principles can be stated in an explicitly Sp(2n,R) invariant manner[19]:

$$V + \frac{i}{2}\beta \geq 0,$$
$$\beta = \begin{pmatrix} 0 & 1_{n\times n} \\ -1_{n\times n} & 0 \end{pmatrix} = \text{Symplectic metric matrix}$$

(The content here is the same as in the conditions for a Gaussian phase space function to be a Wigner distribution.) Thus V can arise from some quantum state if and only if this matrix condition is satisfied. Now we can ask: when does V describe a squeezed state? The delicate point here is that squeezing is a property invariant under $U(n)$ alone and not under the full group Sp(2n,R), where $U(n)$ is the total photon number conserving maximal compact subgroup of Sp(2n,R). A squeezed state is of course non classical in the sense specified by Sudarshan's diagonal coherent state representation of 1963.[20] This signature of non-classicality can be expressed in terms of the variance matrix V as:[19]

$$\hat{\rho} \text{ is a squeezed state} \Leftrightarrow$$
$$\text{Some diagonal element of } S(U)VS(U)^T \text{ for some } U \in U(n) < \frac{1}{2}$$
$$\Leftrightarrow \ell(V) = \text{least eigen value of } V < \frac{1}{2}.$$

Here $S(U)$ is the 2n-dimensional real form of the unitary matrix U of n dimensions, as it occurs in the defining representation of Sp(2n,R). It is worth emphasizing that while the uncertainty principles are Sp(2n,R) invariant, squeezing is only $U(n)$ invariant; and even though in general V cannot be brought to diagonal form using only elements of $U(n)$, the squeezing condition is expressible in terms of the least eigen value of V.

All these results show the close connections between group structures in canonical mechanics, classical optics, and certain quantum optical signatures of non-classicality.

7) My last item in this retrospective concerns contributions to the theory of the geometric phase in quantum mechanics. I think both Professor Simon and I were introduced to this subject

by a talk given by Sudarshan in Kottayam, a town in his home state of Kerala, in 1986. We three did some initial work together, then Simon and Professor Marmo and some others including me took it somewhat further. I will only present a brief sketch of our ideas.

Let me first mention briefly what Sudarshan, Simon and I did in 1986. We found a way of exploiting Hamilton's representation of elements of SU(2) by turns—great circle arcs on S^2—for polarization optics problems and then connected up with geometric phases for two-level systems.[21] We also generalized SU(2) turns to screws for SU(1,1)[22]; later some of us treated SL(2,C) in the same spirit.[23] Within polarization optics per se Simon and I were able to give a minimal 3-component scheme to synthesize the most general SU(2) polarization gadget.[24] At this point I turn to geometric phases, and come back to turns later.

Michael Berry's original 1983 discovery of the geometric phase was in the context of cyclic evolution in the framework of the quantum adiabatic theorem.[25] Later this was generalised to non-adiabatic and even non-cyclic evolution by Aharonov and Anandan, and by Samuel and Bhandari.[26] Then Simon and I were able to give a purely kinematical treatment, not involving quantum dynamics at all.[27] It turned out that the geometric phase is the simplest kinematic expression associated with any Hilbert space curve $\mathscr{C}$ which possesses invariance under two independent transformations:

(i) Local phase changes of $\mathscr{C}$,

(ii) Reparametrisations of $\mathscr{C}$.

This approach led to the geometric phase for the non-cyclic case in an immediate and extremely elementary manner:

$$\begin{aligned}
&\mathscr{C} = \{\psi(s) \in \mathscr{H} \| \|\psi(s)\| = 1, s_1 \leq s \leq s_2\} \subset \mathscr{H} \rightarrow \\
&\varphi[\mathscr{C}] = \varphi_{\text{total}}[\mathscr{C}] - \varphi_{\text{dyn}}[\mathscr{C}], \\
&\varphi_{\text{total}}[\mathscr{C}] = \arg(\psi(s_1), \psi(s_2)), \\
&\varphi_{\text{dyn}}[\mathscr{C}] = -i \int_{s_1}^{s_2} ds \left(\psi(s), \frac{d\psi(s)}{ds}\right)
\end{aligned}$$

Thanks to its invariances, $\varphi[\mathscr{C}]$ is both geometric and a ray-space quantity, so its argument is actually the ray space image of the Hilbert space curve $\mathscr{C}$. A direct connection to the Bargmann invariants then became apparent. These invariants are expressions of the form

$$\Delta^{(3)}(\psi_1, \psi_2, \psi_3) = (\psi_1, \psi_2)(\psi_2, \psi_3)(\psi_3, \psi_1), \psi'\text{s} \in \mathscr{H},$$

with generalisations to four or more vectors. They are cyclically symmetric, and also invariant under independent phase changes in each vector, so they are actually ray space quantities. The key result is a connection between the phase of a Bargmann invariant defined by its vertices, and the (kinematic) geometric phase associated with a geodesic triangle with the same vertices:

$$\arg \Delta^{(3)}(\psi_1, \psi_2, \psi_3) = -\varphi[\text{geodesic triangle with vertices } \psi_1, \psi_2, \psi_3].$$

The geodesics here are determined by the Fubini–Study ray space metric.

In later work two significant results emerged and are worth mentioning:

a) Unitary irreducible representations of Lie groups lead naturally to associated geometric phases which can be systematically analysed. Reminiscent of the Wigner–Eckart theorem, the 'dynamical' parts of these phases split into (sums of) two factors: an algebraic piece determined by the initial fiducial vector and the generators of the representation; and a geometric piece involving the integral of a Maurer–Cartan form along a curve in the group manifold.[28] This result has been exploited to survey all possible geometric phases that can arise from representations of SU(3), phases for three-level systems, etc.[29].

b) The role of ray space geodesics in relating Bargmann invariants and geometric phases was mentioned earlier. It turns out that the most natural Hilbert space curves relevant for geometric phases are not geodesics but something we have called Null Phase Curves.[30] They are much more general and numerous than geodesics which form only a very special example. In general, while there is a unique geodesic connecting any two points, there are infinitely many null phase curves connecting them. These curves are characterised entirely in terms of Bargmann invariants, and they help us express the connection between these invariants and geometric phases in the most general form:

$$\begin{aligned}&\mathscr{C} = \{\psi(s) | s_1 \leq s \leq s_2\} \text{ is a Null Phase Curve} \Leftrightarrow \\ &\arg \Delta^{(3)}(\psi(s), \psi(s'), \psi(s'')) = 0 \text{ for all } s, s', s''; \\ &\arg \Delta^{(3)}(\psi_1, \psi_2, \psi_3) = -\varphi[\mathscr{C}_{12} \cup \mathscr{C}_{23} \cup \mathscr{C}_{31}], \\ &\mathscr{C}_{12} = \text{any Null Phase Curve from } \psi_1 \text{ to } \psi_2, \ldots\end{aligned}$$

Unlike geodesics, these curves are not defined as solutions to any (finite order) ordinary differential equations, but have in a sense only a non-local characterisation. But it is they that are really relevant for geometric phases, and so deserve further study. An approach to geometric phases based entirely on Bargmann invariants and null phase curves has been developed[31].

I conclude this account of our work on the geometric phase with two points. Simon and I gave a complete analysis of the connection between turns and two-level geometric phases[32]; and we also showed that the classical optical Guoy phase, expressed as the phase of a Bargmann invariant, is probably the earliest instance of the geometric phase![33].

Conclusion

It is a privilege to have been invited to be here and pay tribute to Professor E. C. G. Sudarshan. He taught me many things—but sadly no Sanskrit—and we worked closely together for close to thirty years. He strengthened an already existing profound and deep admiration and respect I had for everything Dirac said and did. I might mention some other work we did together with Jacob Kuriyan in the area of unitary representations of non-compact simple Lie groups. We called it the Master Analytic Representation—a combination of Weyl's unitary trick and Dirac's analytic continuation of the dimension of a group representation.

For all this and so much else—thanks, George, and wish you good health and happiness.

References

[1] Dirac P A M 1949 *Rev. Mod. Phys.*, **21**, 392
[2] Currie D G, Jordan T F, and Sudarshan E C G 1963 *Rev. Mod. Phys.*, **35**, 350
[3] Leutwyler H 1965 *Nuovo Cimento*, **37**, 556
[4] Dirac P A M 1950 *Can. J. Math.*, **2**, 147 (1950)
[5] Komar A 1978 *Phys. Rev.*, **D18**, 1881, 1887, 3617; Rohrlich F 1979 *Ann. Phys. (NY)*, **117**, 292; Todorov I T 1979 *Istanbul Lectures*
[6] Mukunda N and Sudarshan E C G 1981 *Phys. Rev.*, **D23**, 2210; Sudarshan E C G, Mukunda N, and Goldberg J N 1981 *Phys. Rev.*, **D23**, 2218; Goldberg J N, Sudarshan E C G, and Mukunda N 1981 *Phys. Rev.*, **D23**, 2231; Balachandran A P, Marmo G, Mukunda N, Nilsson J S, Simoni A, Sudarshan E C G, and Zaccaria F 1984 *J. Math. Phys.*, **25**, 167
[7] Balachandran A P, Marmo G, Mukunda N, Nilsson J S, Simoni A, Sudarshan E C G, and Zaccaria F 1982 *Nuovo Cimento*, **67A**, 121
[8] Bacry H and Cadilhac M 1981 *Phys. Rev.*, **A23**, 2533
[9] Sudarshan E C G, Simon R, and Mukunda N 1983 *Phys. Rev.*, **A28**, 2921; Mukunda N, Simon R, and Sudarshan E C G 1983 *Phys. Rev.*, **A28**, 2942
[10] Mukunda N, Simon R, and Sudarshan E C G 1985 *J. Opt. Soc. Am.*, **A2**, 416
[11] Jagannathan R, Simon R, Sudarshan E C G and Mukunda N 1989 *Phys. Lett.*, **A134**, 457; Jagannathan R and Khan S A 1996 *Quantum theory of the optics of charged particles*, in Advances in Imaging and Electron Physics vol. **97**, ed. Hawkes P W, Academic Press, San Diego, pp. 257–358

[12] Simon R, Sudarshan E C G, and Mukunda N 1984 *Phys. Rev.*, **A29**, 3273
[13] Simon R, Sudarshan E C G, and Mukunda N 1985 *Phys. Rev.*, **A31**, 2419
[14] Simon R and Mukunda N 1993 *J. Opt. Soc. Am.*, **A10**, 95
[15] Simon R and Mukunda N 1998 *J. Opt. Soc. Am.*, **A15**, 2373
[16] Sudarshan E C G 1979 *Phys. Lett.*, **73A**, 269; *Physica*,**96A**, 315
[17] Simon R, Mukunda N, and Sudarshan E C G 1988 *Optics Commun.*,**65**, 322
[18] Simon R, Sudarshan E C G, and Mukunda N 1988 *Phys. Lett.*, **A124**, 223; *Phys. Rev.*, **A37**, 3028
[19] Simon R, Mukunda N, and Biswadeb Dutta 1994 *Phys. Rev.*, **A49**, 1567
[20] Sudarshan E C G 1963 *Phys. Rev. Lett.*, **10**, 277
[21] Simon R, Mukunda N, and Sudarshan E C G 1989 *Pramana*, **32**, 769
[22] Simon R, Mukunda N, and Sudarshan E C G 1989 *Phys. Rev. Lett.*,**62**, 1331; *J. Math. Phys.*, **30**, 1000
[23] Simon R, Chaturvedi S, Srinivasan V, and Mukunda N 2006 *Int. J. Theoret. Phys.*, **45**, 2075
[24] Simon R and Mukunda N 1990 *Phys. Lett.*, **138**, 474; **143A**,165
[25] Berry M V 1984 *Proc. R. Soc. London*, **A392**, 45
[26] Aharonov Y and Anandan J 1987 *Phys. Rev. Lett.*, **58**, 1593; Samuel J and Bhandari R 1988 *Phys. Rev. Lett.*, **60**, 2339
[27] Mukunda N and Simon R 1993 *Ann. Phys. (NY)*, **228**, 205
[28] Mukunda N and Simon R 1993 *Ann. Phys. (NY)*, **228**, 269
[29] Khanna G, Mukhopadhyay S, Simon R, and Mukunda N 1997 *Ann. Phys. (NY)*, **253**, 55; Arvind, Mallesh K S and Mukunda N 1997 *J. Phys.*, **A30**, 2417
[30] Rabei E M, Arvind, Mukunda N, and Simon R 1999 *Phys. Rev.*, **A60**, 3397
[31] Mukunda N, Arvind, Ercolessi E, Marmo G, Morandi G, and Simon R 2003 *Phys. Rev.*, **A67**, 042114
[32] Simon R and Mukunda N 1992 *J. Phys.*, **A25**, 6135
[33] Simon R and Mukunda N 1993 *Phys. Rev. Lett.*, **70**, 880

Journal of Physics: Conference Series **196** (2009) 012010 doi:10.1088/1742-6596/196/1/012010

Why quantum dynamics is linear

Thomas F. Jordan

Physics Department, University of Minnesota, Duluth, Minnesota 55812

E-mail: tjordan@d.umn.edu

Abstract. A seed George planted 45 years ago is still producing fruit now. In 1961, George set out the fundamental proposition that quantum dynamics is described most generally by linear maps of density matrices. Since the first sprout from George's seed appeared in 1962, we have known that George's fundamental proposition can be used to derive the linear Schrodinger equation in cases where it can be expected to apply. Now we have a proof of George's proposition that density matrices are mapped linearly to density matrices, that there can be no nonlinear generalization of this. That completes the derivation of the linear Schrodinger equation. The proof of George's proposition replaces Wigner's theorem that a symmetry transformation is represented by a linear or antilinear operator. The assumption needed to prove George's proposition is just that the dynamics does not depend on anything outside the system but must allow the system to be described as part of a larger system. This replaces the physically less compelling assumption of Wigner's theorem that absolute values of inner products are preserved.

The history of this question is reviewed. Nonlinear generalizations of quantum mechanics have been proposed. They predict small but clear nonlinear effects, which very accurate experiments have not seen. This begs the question. Is there a reason in principle why nonlinearity is not found? Is it impossible? Does quantum dynamics have to be linear? Attempts to prove this have not been decisive, because either their assumptions are not compelling or their arguments are not conclusive. The question has been left unsettled. The simple answer, based on a simple assumption, was found in two steps separated by 44 years.

1. Introduction

Can we prove that quantum dynamics has to be linear? That the Schrodinger equation has to be linear? What assumptions are needed? Is there a reason in principle that quantum dynamics is linear? If not, quantum dynamics could be a linear approximation of a nonlinear theory; experiments might reveal small nonlinear effects. This possibility is too interesting to give up easily. We want to see if it can work. We will accept that it is impossible only if we see an undeniably convincing reason.

Wigner and Bargmann proved[1, 2] that quantum dynamics must be linear if it does not change absolute values of inner products of state vectors. Bialynicki-Birula and Mycielski[3] proposed a nonlinear Schrodinger equation that inspired precise experimental tests[4, 5, 6]. Weinberg[7, 8] proposed a general nonlinear form of quantum mechanics, which led to more experimental tests[9, 10, 11, 12]. The seriousness of these proposals and their experimental tests shows that the Wigner-Bargmann proof of linearity was not decisive; its assumption was not accepted. A more recent idea that relativity requires quantum dynamics to be linear [13, 14, 15] has not been conclusive. The question has been left unsettled. This history is described in Sections II-IV.

There is a simple answer, based on a simple assumption. It was found in two steps separated by 44 years. They are steps back to simpler and more compelling assumptions. A proof of the assumptions of the Wigner-Bargmann proof has been known[16, 17, 18, 19] since 1962. This earlier step is reviewed in Section V. It assumes that the maps of density matrices in time are linear. For this step, it is also assumed that density matrices are mapped one-to-one onto density matrices. An alternative is to assume that pure states are mapped one-to-one onto pure states and that entropy does not decrease.

In a step taken in 2006, it is proved that the maps of density matrices in time are linear [20]. It is assumed, as in the earlier step, that at each time the physical quantities and states are described by the usual linear structures of quantum mechanics, so the question is only about how things change in time. Beyond that, the proof assumes only that the dynamics does not depend on anything outside the system, but must allow the system to be described as part of a larger system. The proof is described in Section VII. The assumption is discussed in Section VIII.

2. Proposals and tests for nonlinear quantum dynamics

Is quantum mechanics a linear approximation of a more fundamental nonlinear theory? With small nonlinear effects not yet seen? How might they be seen?

One possibility, proposed by Bialynicki-Birula and Mycielski[3], is a nonlinear Schrodinger equation

$$i\frac{\partial\psi}{\partial t} = H\psi = [\frac{-1}{2m}\nabla^2 + V - b\,ln(|\psi|^2)]\psi. \tag{1}$$

The size of the nonlinearity is set by a small positive number b. The ln function makes it simple to normalize wave functions and to describe a system composed of noninteracting subsystems. It gives the familiar relation between energy and frequency. And there are soliton solutions, wave packets that do not spread. Altogether, this is a very attractive candidate for a nonlinear quantum mechanics. A proposal [4] to test it with neutron interferometry was realized first with a two-slit experiment [5] that gave

$$b < 3.4 \times 10^{-13} eV \tag{2}$$

and then with Fresnel diffraction [6] that lowered the limit to

$$b < 3.3 \times 10^{-15} eV. \tag{3}$$

This is wave mechanics pure and simple. It is about how a wave function behaves if it satisfies this nonlinear wave equation.

What makes the dynamics nonlinear is that the Hamiltonian depends on the wave function; the Hamiltonian depends on the state. For a wave function of position, this gives a nonlinear Schrodinger equation like (1). For a spin described by Pauli matrices $\Sigma_1, \Sigma_2, \Sigma_3$, it means the Hamiltonian can be a matrix like

$$H = \epsilon\langle\Sigma_3\rangle\Sigma_3. \tag{4}$$

The Hamiltonian is a matrix that depends on the state, because Σ_3 is a matrix and the mean value $\langle\Sigma_3\rangle$ depends on the state. The equations of motion for the mean values $\langle\Sigma_1\rangle$, $\langle\Sigma_2\rangle$, $\langle\Sigma_3\rangle$ for this Hamiltonian are the same as for any Hamiltonian that is a number times a spin matrix,

$$\frac{d}{dt}\langle\Sigma_1\rangle = -2\epsilon\langle\Sigma_3\rangle\langle\Sigma_2\rangle$$

$$\begin{aligned} \frac{d}{dt}\langle\Sigma_2\rangle &= 2\epsilon\langle\Sigma_3\rangle\langle\Sigma_1\rangle \\ \frac{d}{dt}\langle\Sigma_3\rangle &= 0 \end{aligned} \tag{5}$$

These equations of motion are nonlinear because the Hamiltonian contains a mean value. They describe precession with a frequency that depends on the state.

This is a representative case of Weinberg's nonlinear quantum mechanics [7, 8], which brings in nonlinear dynamics for spin. Weinberg's framework is very broad. It lets ordinary linear quantum mechanics be changed in various ways. For example, the representation of physical quantities by linear operators can be changed. I looked at that possibility and decided it is not a fruitful way to go; for a spin it gives nothing new [21]. I concluded, as I think everyone has, that it is more reasonable to assume that at each time everything is the same as in ordinary linear quantum mechanics, so the question is just how things change in time. Making that assumption, and following a suggestion of Polchinski[22] to make Weinberg's theory applicable to a system composed of noninteracting subsystems, I showed that Weinberg's theory is just the statement that the Hamiltonian depends on the state, that the Hamiltonian, like our example (4), involves mean values as well as operators [23, 24]. Actually, this gives a bit more than Weinberg's theory; it includes some simple examples that Weinberg's theory can not describe [24].

Weinberg's proposal inspired four precise spectroscopic tests [9, 10, 11, 12]. Two looked for the effect of a term like (4) in the Hamiltonian. The other two looked for the effect of a similar term for spin 3/2. None found a nonlinear effect. They put limits down to

$$\begin{aligned} |\epsilon| &\lesssim 1.6 \times 10^{-20} eV \\ &< 2 \times 10^{-27} \text{ of binding energy per nucleon} \end{aligned} \tag{6}$$

on the size of the nonlinear interaction.

Are these searches for something that doesn't exist? Is nonlinear quantum dynamics impossible? Does quantum dynamics have to be linear?

3. The Wigner-Bargmann proof of linear dynamics

Can we prove that quantum dynamics has to be linear?
Yes, if we assume that

$$\begin{aligned} \textit{pure states} &\leftrightarrow \textit{pure states} \\ \psi &\leftrightarrow \psi' \ \textit{is one to one} \\ \textit{and}\ \ |\langle\psi|\phi\rangle|^2 &= |\langle\psi'|\phi'\rangle|^2. \end{aligned}$$

That is what Wigner and Bargmann[1, 2] showed. They proved that the change from ψ to ψ' can be made with an operator that is either linear or antilinear. The product of two antilinear operators is linear, so if the change can be made in two similar steps, like the change over a time interval that can be split into halves, the operator must be linear. If we assume a group property, that the change over a time interval is the change over part of the interval followed by the change over the rest of the interval done the same way, and we assume that changes of probabilities in time are continuous, we get the linear Schrodinger equation [25].

These proofs were made long before the proposals and tests for nonlinear quantum mechanics. Why were theories developed and experiments done to look for something that had been proved impossible? I think the reason is that the assumption of

$$|\langle\psi|\phi\rangle|^2 = |\langle\psi'|\phi'\rangle|^2 \tag{7}$$

is not compelling. What does it say? States represented by ψ and ϕ at one time go to states represented by ψ' and ϕ' at a later time. Given the state represented by ψ at the earlier time, the probability that if you make a measurement at the earlier time you will find the state represented by ϕ is the same as the probability that if you wait and make a measurement at the later time you will find the state represented by ϕ'. That's reasonable. But does it have to be true? Can you imagine how it might not be true? Bialynicki-Birula and Mycielski describe one way. Weinberg describes another. The Hamiltonian depends on the state is how I would say it. Why not?

4. Relativity?

In the discussion following Weinberg's proposal and its experimental tests there was a new idea.

Can we prove that quantum dynamics has to be linear? Does relativity require it? Does nonlinearity imply signals faster than light?

I have seen three papers that argue it does. Noting[20] that two of them[13, 15] do not apply to Weinberg's theory, I am interested in only one[14].

It works with the specific example of Weinberg's theory described by the Hamiltonian (4) and equations of motion (5) that we have already seen. The solutions of these equations of motion are that $\langle\Sigma_3\rangle$ is a constant and

$$\begin{aligned}\langle\Sigma_1\rangle(t) &= \langle\Sigma_1\rangle(0)cos(2\epsilon\langle\Sigma_3\rangle t) - \langle\Sigma_2\rangle(0)sin(2\epsilon\langle\Sigma_3\rangle t)\\ \langle\Sigma_2\rangle(t) &= \langle\Sigma_2\rangle(0)cos(2\epsilon\langle\Sigma_3\rangle t) + \langle\Sigma_1\rangle(0)sin(2\epsilon\langle\Sigma_3\rangle t).\end{aligned} \tag{8}$$

Consider two different sets of initial conditions: In set (a)

$$\langle\Sigma_3\rangle(0) = \pm 1. \tag{9}$$

Then

$$\langle\Sigma_1\rangle(0) = 0 = \langle\Sigma_2\rangle(0), \tag{10}$$

because the sum of the squares can't be larger than 1, so

$$\langle\Sigma_1\rangle(t) = 0 = \langle\Sigma_2\rangle(t). \tag{11}$$

The initial conditions (b) are that

$$\langle\Sigma_3\rangle(0) = \pm\frac{1}{\sqrt{2}} = \langle\Sigma_1\rangle(0). \tag{12}$$

Then $\langle\Sigma_2\rangle(0)$ is 0 and

$$\langle\Sigma_2\rangle(t) = \pm\frac{1}{\sqrt{2}}sin(\pm\sqrt{2}\epsilon t) = \frac{1}{\sqrt{2}}sin(\sqrt{2}\epsilon t). \tag{13}$$

Even for $50-50$ mixtures of the $+$ and $-$ cases, the result is different for the initial conditions (a) and (b). That, it is argued, allows signals faster than light.

The argument[14] uses two particles with spin 1/2 in the state with total spin 0. The particles move away from their common source in different directions to separated locations where their spins are measured, one by Alice and one by Bob. This is the much-used set-up that comes from Bohm's 1951 version of the 1935 argument of Einstein, Podolsky and Rosen [26, 27].

If Alice and Bob measure the components of their spins in the same direction, then when Alice gets +, Bob gets −, and when Alice gets −, Bob gets +. This is experimentally verifiable; you can watch it happen over and over.

Suppose now that when a spin comes to Bob, he runs it through the nonlinear dynamics described by the Hamiltonian (4) for a time t and then measures the component of the spin in the y direction. Suppose he does this for each of a long string of spins while Alice measures the z component of each spin that comes to her. Alice gets + half the time and − half the time, on the average. The argument[14] is that then $\langle\Sigma_3\rangle$ is 1 for half the spins that Bob receives and $\langle\Sigma_3\rangle$ is -1 for the other half. The spins that come to Bob are described by the initial conditions (a). After he runs a spin through the nonlinear dynamics and is ready to measure the y component, Bob has a spin in a state where $\langle\Sigma_2\rangle$ is 0. He gets + half the time and − half the time for the y component.

Suppose that instead Alice measures the component of each spin in the direction 45° between z and x. She gets + half the time and − half the time. The argument[14] is that then the spins that come to Bob are + in the 45° direction half the time and − in the 45° direction half the time. They are described by the initial conditions (b). After Bob runs a spin through the nonlinear dynamics and is ready to measure its y component, he has a spin in a state where $\langle\Sigma_2\rangle$ is $(1/\sqrt{2})sin(\sqrt{2}\epsilon t)$. He can get + more than half the time or − more than half the time, depending on his choice of t. The argument[14] concludes that Bob can tell whether Alice is measuring spin components in the z direction or in the 45° direction. Alice can send a signal to Bob. Since there is no restriction on the distances and times, the signal can be faster than light.

It's not necessarily so. If there is not time for a signal at the speed of light between the measurements made by Alice and Bob, then moving observers will disagree about which measurement happens first. The $50-50$ probabilities for Alice's results are calculated from the state of total spin 0 for the two spins, without considering what Bob does. The probabilities for Bob's results can be calculated the same way, without considering what Alice does. Then there is no signal [23, 28, 29].

If there is time for a signal at the speed of light between the measurements made by Alice and Bob, so Alice can call and tell Bob what is coming, then the states of Bob's spins are prepared as the argument describes, and the probabilities for Bob's results are what the argument says. Then there is a signal, but it is not faster than light.

What can be said [28, 29], and this is where the argument gets some traction, is that in the presence of the nonlinear dynamics and the absence of signals faster that light, there must be an abrupt change in Bob's results as gradual changes of distances or times change the separation between the measurements made by Alice and Bob from timelike to spacelike. Is this weird? Yes. Does it prove that this nonlinear dynamics is impossible? No. That it does not is shown particularly clearly in a delightful paper written as a fantasy about a fictional character who finds a way to detect the changes in Bob's results [30].

5. Proof of the Wigner-Bargmann assumptions

There is a simple solution. It was found in two steps separated by 44 years. They are steps back to simpler assumptions. A proof of the assumptions of the Wigner-Bargmann proof has been known[16, 17, 18, 19] since 1962. A proof of the main assumption used in it was found recently[20]. We review the earlier step in this section. The recent step is described in Section VII and VIII.

Can we prove that quantum dynamics has to be linear?

Yes, if we assume, or can prove, that the change of density matrices is linear.

If $\rho \rightarrow \rho'$ is linear and one to one,
then pure states $\leftrightarrow$ pure states
and $|\langle\psi|\phi\rangle|^2 = |\langle\psi'|\phi'\rangle|^2$.

The proof is simple. We assume the dynamics applies to all states. In an interval of time, it maps every density matrix ρ to a density matrix ρ'. The assumption that the map is linear means that if ρ_1 and ρ_2 are mapped to ρ_1' and ρ_2', then

$$\rho = p\rho_1 + (1-p)\rho_2, \tag{14}$$

for a number p between 0 and 1, is mapped to

$$\rho' = p\rho_1' + (1-p)\rho_2'. \tag{15}$$

We assume the map is one to one, so it has an inverse. The inverse map is linear; the proof that the inverse of a linear operator is linear[25, Theorem 7.1] applies with attention restricted to density matrices.

Pure states are mapped to pure states. A pure state can't go to a mixed state: if ρ_1' and ρ_2' in Eq.(15) are distinct, so are ρ_1 and ρ_2 in Eq.(14); thus if ρ' is for a mixed state, so is ρ. The inverse map also takes pure states to pure states, so the set of all pure states is mapped one-to-one onto itself.

For each vector $|\psi\rangle$ of length 1, let $|\psi'\rangle$ be a vector of length 1 such that $(|\psi\rangle\langle\psi|)'$ is $|\psi'\rangle\langle\psi'|$. For each density matrix ρ there are orthonormal vectors $|\psi_j\rangle$ and positive numbers p_j whose sum $\sum_j p_j$ is 1 such that

$$\rho = \sum_j p_j|\psi_j\rangle\langle\psi_j|. \tag{16}$$

The linearity implies that

$$\rho' = \sum_j p_j|\psi_j'\rangle\langle\psi_j'|. \tag{17}$$

Since $\langle\psi_j'|\psi_j'\rangle$ is 1,

$$\mathrm{Tr}[(\rho')^2] = \sum_{jk} p_j p_k |\langle\psi_j'|\psi_k'\rangle|^2 \geq \sum_j (p_j)^2 = \mathrm{Tr}[\rho^2]. \tag{18}$$

The same result for the inverse implies that

$$\mathrm{Tr}[(\rho')^2] = \mathrm{Tr}[\rho^2]. \tag{19}$$

Let $|\psi\rangle$ and $|\phi\rangle$ be vectors of length 1 and let

$$\rho = \frac{1}{2}|\psi\rangle\langle\psi| + \frac{1}{2}|\phi\rangle\langle\phi|. \tag{20}$$

Then

$$Tr[\rho^2] = \frac{1}{2} + \frac{1}{2}|\langle\psi|\phi\rangle|^2 \tag{21}$$

and

$$\rho' = \frac{1}{2}|\psi'\rangle\langle\psi'| + \frac{1}{2}|\phi'\rangle\langle\phi'|, \tag{22}$$

so Eq.(19) implies that

$$|\langle\psi'|\phi'\rangle|^2 = |\langle\psi|\phi\rangle|^2. \tag{23}$$

Wigner and Bargmann will take it from there.

Linear maps of density matrices can also be used to describe processes where pure states are mapped to mixed states, different states are mapped to the same state, the map is not onto all states or, generally, the map has no inverse that applies to all states. An assumption is needed to separate these processes from dynamics described by the Schrodinger equation, which has an inverse for all states. For this proof we assumed that the map of density matrices is one to one.

An alternative is to assume that it is the set of all pure states that is mapped one-to-one onto itself, instead of the set of all density matrices. Then the inequality (18) is obtained as before. If it is assumed that the entropy does not decrease[31], then

$$\mathrm{Tr}[(\rho')^2] \leq \mathrm{Tr}[\rho^2]. \tag{24}$$

This implies Eqs.(19) and (23).

6. What is linear

The result of quantum dynamics is the time dependence of mean values for Hermitian operators representing physical quantities. This includes the time dependence of probabilities, which are mean values for projection operators. The result is the same whether it is obtained from the Schrodinger picture or the Heisenberg picture. The mean value $\langle P\rangle$ for a Hermitian operator P is $\mathrm{Tr}\,[P\rho]$ where ρ is the density matrix that represents the state. In the Schrodinger picture, the time derivative of $\langle P\rangle$ is

$$\frac{d}{dt}\langle P\rangle = Tr[P\frac{d\rho}{dt}]. \tag{25}$$

The linearity of the map of density matrices means that for a density matrix ρ that is a mixture described by Eq.(14), the time derivative is

$$\frac{d\rho}{dt} = p\frac{d\rho_1}{dt} + (1-p)\frac{d\rho_2}{dt}, \tag{26}$$

so for a mean value

$$\begin{aligned}\frac{d\langle P\rangle}{dt} &= \frac{dTr[P\rho]}{dt} = Tr[P\frac{d\rho}{dt}] \\ &= pTr[P\frac{d\rho_1}{dt}] + (1-p)Tr[P\frac{d\rho_2}{dt}].\end{aligned} \tag{27}$$

This means that the time derivative of a mean value depends on the state in a linear way. It can not be a product of mean values as it is in the equations of motion (5) for our example of Weinberg's theory. The time derivative of an operator can be a product of operators; the equations of motion in the Heisenberg picture can be nonlinear. But the time derivative of a mean value has to be a linear function of mean values [20].

7. Proof of linear maps for density matrices

Can we prove that the change of density matrices is linear? Yes, if we assume that the system can coexist with another without interaction.

In 1961, George Sudarshan set out the fundamental proposition that quantum dynamics is described most generally by linear maps of density matrices [32, 33]. Now we have a proof of it. This is the recent step [20]. It is the conclusion of the chain of steps, the proof of the assumption of the proof of the assumption of the Wigner-Bargmann proof. The assumption we use to prove it is discussed in the next section. Here is the proof.

Suppose the system S that we are considering is one of two separate systems S and R. We consider the larger system composed of the two subsystems S and R. We assume there can be something else, the other system R, and we assume that it makes no difference. We assume the dynamics of S does not depend on anything outside S. Not on what happens in R, or on the state of R, or on any correlations of the state of S with the state of R. We assume that S can be described as part of the larger system of S and R combined, but we assume that the dynamics of S does not depend on the situation of S in the larger system.

Suppose the state of the larger system of S and R combined is represented by the density matrix

$$\overline{\Pi} = p\rho_1|\alpha\rangle\langle\alpha| + (1-p)\rho_2|\beta\rangle\langle\beta| \tag{28}$$

where $|\alpha\rangle$ and $|\beta\rangle$ are orthonormal vectors for R that do not depend on the time and, as before, ρ_1 and ρ_2 are density matrices for S and p is a number between 0 and 1. The reduced density matrix $\mathrm{Tr}_R\,\overline{\Pi}$, which is the density matrix ρ for S, is described by Eq.(14). The probability $\langle P\rangle$ for a proposition represented by a projection operator P for S is the sum of joint probabilities

$$\begin{aligned}\langle P\rangle &= \mathrm{Tr}_{SR}[P\overline{\Pi}] = \mathrm{Tr}_{SR}[P|\alpha\rangle\langle\alpha|\overline{\Pi}] + \mathrm{Tr}_{SR}[P|\beta\rangle\langle\beta|\overline{\Pi}] \\ &= \langle P|\alpha\rangle\langle\alpha|\rangle + \langle P|\beta\rangle\langle\beta|\rangle = p\,\mathrm{Tr}_S[P\rho_1] + (1-p)\,\mathrm{Tr}_S[P\rho_2].\end{aligned} \tag{29}$$

Suppose a measurement is made on R that distinguishes the states represented by $|\alpha\rangle$ and $|\beta\rangle$. The probability is p that the result is $|\alpha\rangle$ and $1-p$ that the result is $|\beta\rangle$. If the result is $|\alpha\rangle$, the probability for the proposition represented by P is $\mathrm{Tr}_S[P\rho_1]$, and if the result is $|\beta\rangle$, the probability for P is $\mathrm{Tr}_S[P\rho_2]$. This can be verified experimentally by repeating the process of preparing the state represented by $\overline{\Pi}$, measuring to distinguish the states of R, and testing various propositions for S. The density matrices ρ_1 and ρ_2 describe physically distinct possibilities. The times of the events can be changed as long as the measurement on R is early enough. The time dependence of ρ_1 and ρ_2 will account for changes in the results. The time derivative of the probability for the proposition represented by P is $\mathrm{Tr}_S[Pd\rho_1/dt]$ or $\mathrm{Tr}_S[Pd\rho_2/dt]$ depending on the result of the measurement on R. Altogether, with the probabilities for both results being considered, the time derivative of the probability for P is

$$pTr_S[P\frac{d\rho_1}{dt}] + (1-p)Tr_S[P\frac{d\rho_2}{dt}] = Tr_S[P(p\frac{d\rho_1}{dt} + (1-p)\frac{d\rho_2}{dt})]. \tag{30}$$

The probability $\langle P\rangle$ for the proposition represented by P is also

$$\langle P\rangle = \mathrm{Tr}_{SR}[P\overline{\Pi}] = \mathrm{Tr}_S[P\,\mathrm{Tr}_R\,\overline{\Pi}] = \mathrm{Tr}_S[P\rho]. \tag{31}$$

Its time derivative is $\mathrm{Tr}_S[Pd\rho/dt]$. This is always correct. It may be the only possibility at hand. For example, suppose the state of S and R combined is represented by the density matrix

$$\Pi = p\rho|\alpha\rangle\langle\alpha| + (1-p)\rho|\beta\rangle\langle\beta| = \rho[p|\alpha\rangle\langle\alpha| + (1-p)|\beta\rangle\langle\beta|]. \tag{32}$$

The dynamics for S must be the same for Π as for $\overline{\Pi}$.

The time derivative of the probability for the proposition represented by P is always $\mathrm{Tr}_S[Pd\rho/dt]$. There are situations where it must also be described by Eq.(30) to fit observations of events in a larger system. The dynamics in S can not depend on the situation of S in a larger system. Therefore

$$Tr_S[P\frac{d\rho}{dt}] = Tr_S[P(p\frac{d\rho_1}{dt} + (1-p)\frac{d\rho_2}{dt})]. \tag{33}$$

From this equality for various projection operators P, we conclude that

$$\frac{d\rho}{dt} = p\frac{d\rho_1}{dt} + (1-p)\frac{d\rho_2}{dt}. \tag{34}$$

8. The assumption at the bottom

What we need to assume is just that the system we are considering can coexist with another system without interaction, that the dynamics we are considering can be independent of something else in the universe, that the system we are considering can be described as part of a larger system without interaction with the rest of the larger system.

This is like the implicit assumption we make every day when we consider a part of the universe and assume the rest has no effect. We take that assumption for granted. Without it there would be no physics at all.

Actually that implicit assumption is stronger than the assumption we make here. We implicitly assume that *everything* outside the considered system has no effect. Here we assume only that there is *something* out there that has no effect. The limit to the validity of the implicit assumption comes from the part of the universe outside the considered system that has the largest effect. The limit on the validity of the assumption we make here comes from the part that has the smallest effect.

The assumption we make here is close in spirit to the idea of envariance[34], but the proof of Section VII does not use quantum entanglement. The proof does not even require quantum mechanics; it could be done in classical mechanics as well. The correlations involved are classical. The substance would be the same in classical mechanics; only the language would be different.

Can we prove that quantum dynamics has to be linear?
Yes, if we assume that the system can coexist with another without interaction.

Acknowledgment

I am grateful to Wojciech Zurek for a suggestion that brought me back to this subject with my eyes open.

References

[1] Wigner E P 1959 Group Theory (Academic, New York), appendix to Ch. 20 and first section of Ch. 26
[2] Bargmann V 1964 J. Math. Phys. 5, 862
[3] Bialynicki-Birula I and Mycielski J 1976 Ann. Phys. (N. Y.) 100, 62
[4] Shimony A 1979 Phys. Rev. A 20, 394
[5] Shull C, Atwood D K, Arthur J, and Horne M A 1980 Phys. Rev. Lett. 44, 765
[6] Gahler R, Klein A G, and Zeilinger A 1981 Phys. Rev. A 23, 1611
[7] Weinberg S 1989 Ann. Phys. (N.Y.) 194, 336
[8] Weinberg S 1989 Phys. Rev. Lett. 62, 485
[9] Bollinger J J, Heinzen D J, Itano W M, Gilbert S L, and Wineland D J 1989 Phys. Rev. Lett. 63, 1031
[10] Chupp T E and Hoare R J 1990 Phys. Rev. Lett. 64, 2261
[11] Walsworth R L, Silvera I F, Mattison E M, and Vessot R F C 1990 Phys. Rev. Lett. 64, 2599
[12] Ma jumder P K, Venema B J, Lamoreaux S K, Heckel B R, and Fortson E N 1990 Phys. Rev. Lett. 65, 2931
[13] Gisin N 1989 Helv. Phys, Acta 62, 363
[14] Gisin N 1990 Phys. Lett A 143, 1

[15] Simon B, Buzek V, and Gisin N 2001 Phys. Rev. Lett. 87, 170405
[16] Jordan T F, Pinsky M A, and Sudarshan E C G 1962 J. Math. Phys. 3, 848
[17] Kadison R V 1965 Topology 3 Suppl. 2, 177
[18] Hunziker W 1972 Helv. Phys. Acta 45, 233
[19] Simon B 1976 in Studies in Mathematical Physics: Essays in Honor of Valentine Bargmann, edited by Leib E H, Simon B, and Wightman A S (Princeton U.P., Princeton, NJ), pp. 327349
[20] Jordan T F 2006 Phys. Rev. A 73, 022101
[21] Jordan T F 2006 Phys. Lett. A 151, 215.
[22] Polchinski J 1991 Phys. Rev. Lett. 66, 397
[23] Jordan T F 1993 Ann. Phys. (N.Y.) 225, 83
[24] Jordan T F 1994 Phys. Rev. A 49, 5086
[25] Jordan T F 2007 Linear Operators for Quantum Mechanics (Wiley, Dover)
[26] Bohm D 1951 Quantum Theory (Prentice-Hall, Englewood)
[27] Einsten A, Podolsky B, and Rosen N 1935 Phys. Rev. 47, 777 (1935)
[28] Jordan T F and Sariyianni Z E 1999 Phys. Lett. A 263, 263
[29] Czachor M and Doebner H D 2002 Phys. Lett. A 301, 139
[30] Kent A 2005 Phys. Rev. A 72, 012108
[31] Peres A 1989 Phys. Rev. Lett. 63, 1114
[32] Sudarshan E C G, Mathews P M, and Rau J 1961 Phys. Rev. 121, 920
[33] Jordan T F and Sudarshan E C G 1961 J. Math. Phys. 2, 772
[34] Zurek W H 2005 Phys. Rev. A. 71, 052105

Journal of Physics: Conference Series **196** (2009) 011008 doi:10.1088/1742-6596/196/1/011008

Introduction to "Spin and Statistics" : the third quest

Samir Bose

Physics Department, University of Notre Dame, Notre Dame, Indiana

E-mail: Bose.1@nd.edu

George Sudarshan was convinced that the spin-statistics relation must be related to the symmetry relation of their scalar products. This has been one of Sudarshan's seven quests. Sudarshan's proof based on this idea was first published in 1968. Further refinements of the proof followed, and the results culminated in the publication of "Wolfgang Pauli and the Spin-Statistics Relation" with Ian Duck in 1997.

$$\phi_1 \star \phi_2 = -\phi_2 \star \phi_1 ; A \star B = B \star A$$

G Bhamathi and E C G Sudarshan, birthday party at the Seven Science Quest Symposium, 2006

Journal of Physics: Conference Series **196** (2009) 011008 doi:10.1088/1742-6596/196/1/011008

Speakers at Spin and Statistics: Sudarshan's third quest

Samir Bose

Ian Duck

Luis Boya

Anil Shaji

$$\phi_1 \star \phi_2 = -\phi_2 \star \phi_1 ; A \star B = B \star A$$

Journal of Physics: Conference Series **196** (2009) 012011 doi:10.1088/1742-6596/196/1/012011

The connection between spin and statistics

Luis J. Boya

Departamento de Física Teórica, Universidad de Zaragoza E - 50 009, Zaragoza, Spain

E-Mail: luisjo@unizar.es

Abstract. After some brief historical remarks we recall the first demonstration of the Spin-Statistics theorem by Pauli, and some further developments, like Streater-Wightman's, in the axiomatic considerations. Sudarshan's short proof then follows by considering only the time part of the kinetic term in the lagrangian: the identity SU(2)=Sp(1,C) is crucial for the argument. Possible generalization to arbitrary dimensions are explored, and seen to depend crucially on the type of spinors, obeying Bott´s periodicity 8. In particular, it turns out that for space dimentions (7, 8, or 9) modulo 8, the conventional result does not automatically hold: there can be no Fermions in these dimensions.

1. Introduction

The idea of this work came from discussions with Sudarshan in Zaragoza some years ago. George was surprised to learn that the properties of spinors crucial for the ordinary connection between spin and statistics would not hold in arbitrary dimensions, because of periodicity 8 of real Clifford algebras. After many discussions and turnarounds, I am happy today to present this collaboration, and I also apologize for the delay, for which I am mainly responsible. In any case, the priviledge to work and discuss physics with Sudarshan is a unique experience, for which I have been most fortunate.

2. Historical remarks

In retrospect, the first hint of the Bose statistics for "photons" comes from Plack´s radiation law [1] (1900): the quantum hypothesis corpuscularizes the radiation, but by itself it leads only to Wien´s law (Eintein, 1905). Without the strange correlations (first noticed by Ehrenfest in 1905) between "photons", the correct law does not obtain.

Independent light quanta lead to Wien´s law (1896) for the radiation density ρ of a warmed black body with frequency ν at temperature T

$$\rho(\nu, T) = A\, \nu^3 \cdot \exp(-\beta\, \nu / T) \qquad (1\text{-}1)$$

which is already unsatisfactory: for high T the density $\rho(\nu, T)$ tends to a constant, whereas from thermodynamics one should expect a dependence linear with T [2]. This is indeed the case for the correct Planck´s formula

$$\rho(\nu, T)\, d\nu = 2 \cdot 4\pi\, (\nu^2/c^2)\, (h\, \nu)\, [\exp(h\nu/kT) - 1]^{-1}\, d(\nu/c) \qquad (1\text{-}2)$$

which fits experiments wonderfully (then and now). The correlation aspects of (1-2) can be interpreted also as wave-particle duality (the fluctuation formula of Einstein, 1909): it is intriguing that the Bose correlations manifest themselves in different ways.

There is no hint of the exclusion principle (statistics for Fermions) until Bohr´s unsuccesful attempts (ca. 1922) to understand the periodic system of elements with the help of the correspondence principle; that the task was impossible was pointed out by Pauli, who later enunciated the correct principle. In January 1925, W. Pauli stated the *Aequivalenz Verbot* (exclusion principle). From some terms missing in atomic spectra, in particular, from the absence of the singlet state of orthohelium 1^3S_1, Pauli concluded that the electrons in each atom are characterized for four quantum numbers each, and no two of them can have the four numbers equal. The periodic system, the concept of valence, chemical compunds, etc., was inmediately understood. Indeed, today we realize that the exclusion principle is the true *differentiating principle* in Nature.

Still within the Old Quantum Theory, Fermi applied the exclusion principle [3] to the statistical properties of metals, obtaining satisfactory results (January, 1926). It is remarkable that the density formulas of Planck, Fermi and the classical one by Boltzmann, are so similar:

$$\rho \propto 1/[\,(\exp E/kT) + \varepsilon] \qquad (1\text{-}3)$$

with $\varepsilon = 0$, -1 or +1 for Boltzmann, Bose, or Fermi statistics respectively.

In June 1925 Matrix Mechanics was established by Heisenberg, and the mystery of statistics was cleared up very soon alter. The next year, both Heisenberg and Dirac independently pointed out that the unobservability of the phase of the wavefunction allows naturally for an extra sign under the permutation of two identical particles:

$$\Psi(1, 2) = \pm\, \Psi\,(2, 1) \qquad (1\text{-}4)$$

For the moment, there is only an empirical rule: integer spin objects go with the + sign, half-integer ones with the – sign. For many particles, there is a neat group-theoretical interpretation:

For n "Fermions", the permutation group S_n is represented modulo the alternating group, so $S_n/A_n = Z_2$, whereas for "bosons", there is only the Identity representation. Other representations (parastatistics) seem to be naturally excluded.

If we accept that two anomalies *should* go together, there is even a hint for the spin-statistics connection: For spin 1/2 particles, the wavefunction picks up a – sign under a 2π rotation, AND the odd transposition picks up also a – sign; the two oddities go together: Fermi statistics for spinors. For spin integer, neither of this happens. This is just a clue, not a proof, of the theorem we seek on spin and statistics.

Early uses of statistics were P. Jordan's work with second quantization, where the Fermi case forced *anti*conmutation relations for quantum fields (Jordan, 1927; Jordan and Klein, 1927; Jordan and Wigner, 1928). Ehrenfest and Oppenheimer proved (1931) that the rule was compatible with compound systems, and in this way quantum statistics was succesfully applied to nuclear physics. Indeed, one of the first arguments for the proton-neutron model of nuclei was the experimental statistics for Nitrogen-14.

Howver, nobody noticed, at the time, than the Fermi/Bose rules allowed a neat presentation of *supersymmetry*: Define Bose/Fermi operators a, b and hamiltonian H by

$$2H = \{a^\dagger, a\} + [b^\dagger, b] \qquad \text{with } [a, a^\dagger] = \{b, b^\dagger\} = 1 \qquad (1\text{-}5)$$

Then diagonalizing

$$H = (N_a + 1/2) + (N_b - 1/2) \quad \text{with } N_{a,b} = a^\dagger a, \quad b^\dagger b$$
(1-6)

and $N_a = 0, 1, 2, \ldots$ but $N_b = 0$ or 1. *Supersymmetry* is shown in that there is no zero-point energy, there is Bose-Fermi symmetry in the spectrum (each level is double degenerate), and also in that the SUSY operator Q is a square root of the energy:

$$Q := b^\dagger a + (b^\dagger a)^\dagger \qquad \text{implies} \qquad Q^2 = H$$
(1-7)

In the 30s there was some important work (Heisenberg, Dirac, Pauli, and Weisskopf) in which the two ways of quantization were used consistently, and even the symmetry/antisymmetry alternative helped in taming divergences (perturbation-theory infinities).

3. The conventional proof

Pauli prepared his first proof of the spin-statistics connection for relativistic quantum fields (an outgrowth of his study, with Fierz, of higher-spin wave equations), for the 1939 Solvay conference. The work [4] appeared in Phys Rev in 1940.

The proof relies heavily on the properties of SL(2,C), the covering group of the Lorentz group; the irreducible linear representations (*irreps*) are labeled as D_{jk}, with 2j and 2k integers. For j+k integer, we have tensor irreps, type (+1) for both integers, type (-1) if both are half-integers. Spinors have j + k half-integer, again type $(+\varepsilon)$ for j integer, type $(-\varepsilon)$ for k integer. The four clsasses multiply as Klein´s *Vierergruppe*.

The essence of Pauli´s argument is this: describe particles with fields $\varphi(\mathbf{x},t)$; for tensors (class +1 and –1), the *particle* density j_0 is undefined ($\propto \varphi^* \partial_0 \varphi$), but the *energy* density T_{00} is defined, $\propto |\partial_0 \varphi|^2$. So the field can describe two types of charges, but with positive energy each.

For spinors, the opposite is true, as it is plain already from the Dirac equation:

$$\text{Particle density definite, } \rho = \psi^* \psi \qquad (2\text{–}1)$$

$$\text{Energy undefined, } E = \pm \sqrt{(k^2 + m^2)} \qquad (2\text{-}2)$$

The deep and general reason for the proof is this: there is an *allowed* symmetry (close to the later CPT symmetry) in the equations of motion which changes the sign of the complex particle density for tensors, and changes the sign of the energy density for Fermions. All this holds in the *first quantized* theory.

Then Pauli argues that second quantization comes to the rescue and it is necessary in both cases (pair creation was well established by 1939; in particular, the Pauli-Weisskopf 1934 "anti-Dirac" paper established systematically pair creation for bosons), but in the tensor case conmutators will respect indefinite charge and definite energy, whereas in the spinor case anticonmutators will restore indefinite charge (so electrons and positrons are described by components of the same field), and recover definite energy (as excitations from the vacuum of either charge sign, hence positive energy: second quantization oversteps hole theory).

Implementation of the rule implies also causality: physical magnitudes (= observables) conmute outside each other´s light cones.

Pauli´s argument deals only with *free* particles, without interactions, but does not restrict only to first order equations. The final remark of Pauli´s Phys. Rev. paper [4] is worth repeating:

" ... in our opinion, the connection between spin and statistics is one of the most important application of special relativity theory "

Is this true? Keep tuned!

Pauli should indeed feel very proud of his proof, because it is a satisfactory solution of a problem which he himself triggered with the discovery of the exclusion principle.

For other contributions to the issue in the early 40s see the book by Sudarshan and Duck (1997) [3].

4. Some further developments

It was to be expected that the "renormalization" revolution in Quantum Field Theory in 1947/49 would shed some light on the statistics issue; indeed that was the case.

J. Schwinger in his 1951 papers [5,6], seems to derive the theorem by imposing a sort of Θ = CPT invariance. Time reversal is antiunitary and it is interpreted also as rotation in a fictitious 4-5 biplane; after some analysis JS concluded that this implies *anti*comutation rules for spinors; the fifth dimension is hidden already in the γ_5 matrix (the whole Dirac algebra generates in fact the conformal group, O(2, 4)). An important advance of this work is that the conmutation/anticonmutation alternatives are *derived*, not postulated. Still, in Schwinger's approach no interactions are contemplated.

In the later work of Lüders and Zumino (from 1954 on) a reversal of arguments occured: the spin-statistics connection was used to *prove* the CPT Theorem. In his contribution to a *Festschrift* for Niels Bohr, Pauli (1955) incorporated interactions [7]. He proved the CPT theorem based in three main asumptions: 1) The usual Spin-Statistics connection, 2) Invariance under the connected Lorentz group, and 3) Locality: Fields are *finite*-dimensional tensors and spinors, and interaction terms occur with at most *finite* order derivatives.

We are approaching now 1956, when Parity symmetry was seen to be violated. Sudarshan took a leading role in establishing the V – A form of the weak interacions, incorporating automatically parity violation. It is curious that the early arguments emphasize the masslesness of the neutrinos, whereas the true argument is that only one helicity (of either neutrinos or other Fermions) enters in the interaction. So the argument holds even today, with massive neutrinos.

Among the further work up to the Streater-Wightman book [8] (1964) we note the Lüders-Zumino work of 1958. They disentangled the spin-statistics theorem from the CPT theorem. They started from the axioms of field theory, including uniqueness of the vacuum and positive metric in state space. Burgoyne´s proof (1958) is similar, but includes arbitrary spins. The march of the proof is: first prove the Spin-Statistics connection, then the CPT invariance follows fairly easily.

Axiomatization takes a leading role in the full treatment exposed in the Streater-Wightman book. Indeed, the authors claim that

"In this book, we have eliminated all theorems whose proof are non-existent."

They establish four group of axioms, including asymptotic completeness, a postulate necessary for the collision theory of Ruelle and Haag. They proceed to prove first the CPT theorem, shown to be equivalent to some identities among the vacuum expectation values, identities that necessarily hold in every field theory of *local* fields. Excluding explicitely parastatistics, they then show that the usual connection holds, in the sense of conmutators/anticonmutators of field operators at spacelike distances. The proof, irreproducible, relies heavily in analysis with several complex variables.

R. Jackiw has a pessimistic attitude [9] towards those "axiomatic" approaches, and we simpathize:

"All the axiomatic progress in Quantum Field Theories in 1+1 dimensions has been unable to discover the most important new phenomenon: these theories might describe: *solitons*, as special excitations of the field, with capital physical implications " (1976).

In Spain we say: " Matando moscas a cañonazos"...

5. SUDARSHAN´S SHORT PROOF

The essence of Sudarhan´s proof [10] of the spin-statistics theorem (since 1968) is that the whole affair is really a nonrelativistic problem, or rather that it is the rotation part SO(3) of the Poincarè group which is crucial for the proof. In fact, spinors as irreps of the covering groups

SU(2) of the 3-dimensional rotation group already exist! For $n \geq 2$, SO(n) is connected but for $n > 2$ its universal cover, Spin(n), is a double covering. The essential nontrivial homotopy group of symmetries of Nature is already revealed in the pure space rotation part. This topological feature, allowing genuine projective representations, seems to be the culprit for the existence of Fermions and the attendant exclusion principle. Although there are cases in which the covering group SU(2) shows up already in classical physics [11], the crucial projective nature of implementation of symmetry in Quantum Mechanics makes up for at least two types of realizations of the rotation group.

We shall see that although the argument can go in a nonrelativistic setting, one needs really a full quantum field theory for the proof to proceed. For the actual proof, Sudarshan and Duck [3] describe the theory through a Lagrangian, and look at the kinematic part only, as Schwinger already did in 1951, see above [5]. The crucial observation is a Cartan identity among Lie groups:

$$\mathrm{Spin}(3)=\mathrm{SU}(2)=\mathrm{Sp}(1,C)\,(\text{unitary symplectic}) \qquad \text{(IV-1)}$$

Hence, the group action leaves invariant an antisymmetric bilinear form. But *even* products of this form leave a symmetric bilinear form invariant; indeed for the square, we get the natural quadratic form for the s=1 irrep of the SO(3) group. The complete argument follows now:

2j odd: D_j irreps of SU(2) are pseudoreal, *symplectic*, faithtful, even (2j+1) dimensional
(IV-2)

2j even: D_j irreps of SU(2) are real, faithful of SO(3), *orthogonal*, odd (2j+1) dimensional (IV-3)

Let us try to contruct a Lagrangian, which should be a scalar: from above, the spinorial part should pair the fields antisymmetrically to obtain scalars, whereas the tensor part should pair them symmetrically, for the same reason. Write now all fields as a "long" field $\xi(x)$, somewhat in the Schwinger spirit, hermitean ($\xi = \xi^\dagger$). As Lagrangian density we take, with $A' = \partial A/\partial t$

$$\mathcal{L} = (1/2)\, \Sigma_{rs}\, K^0{}_{rs}\, (\xi_r\, \xi_s' - \xi_r'\, \xi_s) + \mathcal{L}_2 + \mathcal{L}_3 + \mathcal{L}_4 \qquad \text{(IV-4)}$$

where $\mathcal{L}_2$ includes the space derivatives in the kinetic term, $\mathcal{L}_3$ is the mass term, and $\mathcal{L}_4$ includes the interaction terms.

For our analysis the time derivative term of the kinetic part $\mathcal{L}_1$ will suffice - another advantage of Sudarshan´s treatment. Antisymmetrization avoids a posible total derivative, and the labels r,s include, of course, all fields in the game, Bose and/or Fermi.

Now for *tensor* fields, scalar (invariant) implies symmetrical products, and the antisymmetry in the ξ's forces the numerical kernel K^0 to be antisymmetric as well. For spinor fields, of course, the opposite conclusion obtains:

$K^0{}_{rs}$ is (r, s) antisymmetric for *both* r,s tensor labels (IV-5)

$K^0{}_{rs}$ is (r, s) symmetric if both r,s are spinor labels (IV-6)

Of course, there is no spinor-tensor mixing: either superselection rules, or better, $D_{tens} \otimes D_{spin}$ will *never* contain the scalar D_0 .

To clinch the argument one has to show, in this lagrangian formalism, that for *conmuting* fields the numerical kernel K^0_{rs} must be antisymmetric (in r,s), so conmutators should go with bosons, whereas for *anticonmuting* fields the kernel should be antisymmetric (in r,s), so Fermi fields should anticonmute.

To show this, Sudarshan relies on Schwinger´s quantum action principle (1951) [5], which unifies equations of motion and Noether's theorem for conserved quantities in the most beautiful way. Let us consider variations on the action brought about by variations on the fields ξ only, so we need to consider only the space integral:

$$\xi \longrightarrow \xi + \delta\xi_k \Rightarrow$$

$$2\,\delta I(\delta\xi) = \int d^3x\, \Sigma\, K_{rs}{}^0\, (\xi_r\, \delta\xi_s - (\delta\xi_r)\, \xi_r) \quad \text{(IV-7)}$$

Since the infinitesimal variation of the action transforms the field variables infinitesimally, we require

$$[\xi_a\, ,\, \delta I(\delta\xi)\,] = i\, h\, \xi_a \quad \text{(IV-8)}$$

or

$$2ih\, \delta\xi_a = \int d^3x\, [\xi_a\, ,\, \Sigma\, K_{rs}{}^0\, (\xi_r\, \delta\xi_s - \delta\xi_r\, \xi_s\,)] \quad \text{(IV-9)}$$

Let us discuss now the two alternatives. If the fields ξ are *bosonic* fields, $\delta\xi$ is supposed to commute with everything. It follows straightforwardly, expanding the conmutator, keeping the variations dx at the end, and adding space labels $\xi(\mathbf{x})$ etc., that (see Sudarshan & Shaji [12])

$$[\xi_a\, ,\, \delta I(\delta\xi)\,] \neq 0 \Rightarrow K_{rs}{}^0 = -\, K_{sr}{}^0\ (\text{Bose}) \quad \text{(IV-10)}$$

which imply the usual canonical conmutation relations

$$[\xi_a(y),\, \Pi_j(x)] = i\, h\, \delta^3\, (x-y)\, \delta_{aj} \quad \text{(IV-11)}$$

introducing the conjugate momenta

$$\Pi_k \int \partial\mathcal{L}/\partial\xi_k{}' = K_{kr}{}^0\, \xi_r \quad \text{(IV-12)}$$

For anticonmuting fields, i.e. supposing the variations $d\xi$ anticonmute with Fermi fields, it is fairly clear we shall obtain the opposite result [12] after reordering viz:

$$\{\xi_a, \delta I(\delta\xi)\} \neq 0 \Rightarrow K_{rs}^0 = + K_{sr}^0 \text{ (Fermi)} \quad \text{(IV-13)}$$

which, in turn, leads to antoconmutation, i.e. the symmetric rule between fields and conjugate momenta.

"We have seen that the rotational invariance of the kinematical part of the Lagrangian density requires that the matrix K_{rs}^0 be antisymmetric for integral spin fields while it will be symmetric for half-integral spin fields. The action principle of Schwinger requires K_{rs}^0 to be antisymmetric for conmuting (Bose) fields, and symmetric for anticonmuting (Fermi) fields:

The Spin-Statistics connection obtains!"

Negative-norm states might appear, and have to be declared "outlaws". Also, even in our nonrelativistic setting the coupling of creation and annihilation operators appear with the same strength (Kirchoff´s principle).

The proof is remarkable; among other things, it nearly fulfills the expectation that "the two pathologies go together", as stated above: Fermions are strange animals in two compensating ways!

Incidentally, all the modern work of Sudarshan and collaborators on the spin-statistics issue seems to be prompted by the remarks of Feynman, collected by Neuenschwander as a question [13], about the lack of a simple and neat proof of this fundamental relation. An answer close to our "two pathologies cancel" is in [14].

6. Higher dimensions

Does the spin-statistics connection hold in arbitrary dimensions? In one and two (= 1, 1), the situation is special and we shall not touch on it; in particular in (1, 1), Fermi-Bose conversion is posible. The quantized solitons in that case are automatically Fermions. What about higher dimensions, which appear so conspicuously in modern physics, related to supersymmetry, string theory, etc.? Very few papers address directly that question. We do [15], which we follow here.

We recall first the behaviour of the rotation or spin groups in higher space dimensions. We have, modulo 8:

Spin(3)	Spin(4)	Spin(5)	Spin(6)	Spin(7,8,9)	Spin(10)
SU(2)=Sp(1)	$SU(2)^2$	Sp(2)	SU(4)	SU(4,5,5)	SU(6)
H	H	H	C	R,R,R	C

An irrep is type R if equivalent to a real one; type C if complex, and type H if equivalent to conjugate but not real.

Now the H (or quaternionic) type yields the Id irrep in the *anti*symmetric (symplectic) square of the spin irrep, like the simple SU(2) case. Hence the situation in dim (3, 4, 5) mod 8 is identical to the conventional 4-dim (1, 3) with space group SO(3).

In the real R type case, one obtains the Id irrep in the symmetric part of the square of the spin irrep. The complex case, type C, is somewhat intermediate (see below).

We conclude that the Id irreps for the space rotation group in dimensions 7, 8 and 9 mod 8 are always symmetric, so we cannot have spinor field theories. It does go through in dimensions 3, 4, and 5. In SO(4), there is another complication, as the center of the spin group is $Z_2 \times Z_2$ (Klein´s Vierergruppe), and there are no faithful irreps: hence, it is not automatic that the spinor wavefunction acquires a "-" sign under a 2π space rotation.

As for the complex case, dim 2, 6 mod 8, one can always arrange the Id irrep to be in the antisymmetric part, but it is not mandatory, as the product is not the square, but the *modulus* square (D times conjugate). If written in real form, we get the Id irrep twice, one in the symmetric, another in the antisymmetric part. The "right" statistics is optional. As for the reasoning from the Action principle, it seems not to depend on dimension, so we do not reconsider it any further.

We see that in Superstring theory, which works in 10 = (1, 9) dimensions, the rotation group is in the "bad" class, and so is the little group for massless particles (O(8)). We have not developed consequences of this in the actual presentations in string theory. As for 11-dim supergravity and also in M-Theory, one works in 11 = (1, 10) dimensions: good connection is posible, if optional. The massless case (group O(9)) is bad. Finally, in F-Theory (Vafa, 1996) we are in 12 = (2, 10) dimensions, so good connection is optional for O(10); and O(11), if pertinent, is good (H type).

7. Concluding Remarks

We see that the proof, when it holds, even in higher dimensions, is really very simple. It needs only the temporal part of the kinetic term of the Lagrangian, and it is very much in the spirit and query of Neuenschwander [13], namely, to obtain a simple and natural proof. And from Bott´s 8-periodicity, we find the special cases of space dimensions 8n-1, 8n, and 8n+1, where the usual proof does not go through.

References

[1] Duck I M and Sudarshan E C G 2000 *100 years of Planck´s Quantum*, World Scientific
[2] Boya L J 2003 *The Thermal Radiation Formula of Planck*, Rev. Acad. Cien. Zgza **58** 91-114, arXiv: physics/0402064
[3] Duck I M and Sudarshan E C G 1940 *Pauli and the Spin-Statistics Theorem*, World Scien. 1997
[4] Pauli W 1940 *The connection between Spin and Statistics*, Phys. Rev. **58** (1940) 716-378. Reprinted in [5]
[5] Schwinger J 1956 *Quantum Electrodynamics*, Selected papers. Dover
[6] Schwinger J 1951 *Theory of Quantized fields* I, Phys. Rev. **82** (1951), 914-927. Reprinted in [5]
[7] Pauli W 1955 *Exclusion Principle, Lorentz group and reflexion of space-time and charge*. Niels Bohr *Festschrift*, Pergamon 1955
[8] Streater R and Wightman A 1964 *PCT, Spin-Statistics and all That*, Benjamin
[9] Jackiw R 1995 *Theoretical and Mathematical Physics* Selected Papers. World Scientific, p. 449
[10] Sudarshan E C G 1968 *On the Spin-Statistics Connection*: Proc. Ind. Acad. Sci **67**, 284 . *See also* id. 1975 Jour. Ind. Ins. Scien., Statis. Phys. Suppl. 123
[11] Boya L J and Sudarshan E C G 2005 *Relation between rigid body rotations, isotropic cones and spinors*. Found. Phys. Lett. **18** 53-63.
[12] Shaji A and Sudarshan E C G *Nonrelativistic proof of the Spin-Statistics Theorem*, arXiv: quant-phys. 0306033.
[13] Neunschwander D E 1994 *The spin-statistics theorem*, Am. J. Phys **62**, 972.
[14] Gould R 1995 *Answer to Question # 7* Am. J. Phys **63**, 109
[15] Boya L J and Sudarshan E C G 2007 *The Spin–Statistics Theorem in Arbitrary Dimensions,* to appear in Int. J. Theor. Phys

Journal of Physics: Conference Series **196** (2009) 012012 doi:10.1088/1742-6596/196/1/012012

Reflections on the spin-statistics theorem

Ian Duck

Physics Department, Rice University, Houston, Texas

E-mail: iduck@mail.rice.edu

Abstract. This poem was inspired by George Sudarshan's work.

Spin-statistics is a monumental mystery
That determines microscopic possibility.
    No one could explain it,
    No one understood it,
Not, of course, til Sudarshan and me.

This law holds true just perfectly
Unspoiled by atomic instability.
    Our only explanation?
    Just pure postulation!
Except, of course, for Sudarshan and me.

Oh, all the best scored brilliantly
From Pauli to Wightman to infinity.
    But their proofs are so abstruse
    They are of very little us.
Not use, of course, by Sudarshan and me.

Proofs appear quite frequently
But all are flawed essentially.
    Take two electrons … we are told
But then such proofs unfold.
Not that, of course, of Sudarshan and me.

The common error we see
Takes exchange physically:
    Two-particle translation
    Followed by rotation.
Not done, of course, by Sudarshan and me.

The problem here, quite basically,
Is equating states ide ntically
    Which are only equivalent
    With phase indeterminant.
Not us, of course, nor Sudarshan and me.

The fatal error is the plea:
Take two electrons … not three?
    Canonical quantum mechanics
    Contains no such handy tricks.
Not use, of course, by Sudarshan and me.

Operators like *L* and *P*
Transport electrons symmetrically.
    So to rotate just two
    You really can't do.
Nor done, of course, by Sudarshan and me.

Our proof is simplicity
But does require facility
    With second quantization-
    Exit the whole population!
Not you, of course, nor Sudarshan and me.

We need a bit of relativity,
A Hamiltonian and field hermiticity,
    Symmetry of the mass matrix,
    And a couple of other tricks.
Made clear, of course, by Sudarshan and me.

At the end, it's rewarding to see
Just how very blessed are we
    Who study this quantum jewel.
    We understand Pauli's rule!
With thanks, of course, to Sudarshan and me.

But many still do not agree
That our proof is error free
    Or even sensible,
    Let alone comprehensible.
Ever thus, bless us Lord, Sudarshan and me.

References

[1] Sudarshan E C G 1968 *The Fundamental Theorem on the Connection Between Spin and Statistics*, in proceedings of the Nobel Symposium 8, N. Swarthholm, ed. (Almquist and Wiksell, Stockholm)
[2] Ian Duck and Sudarshan E C G 1998 *Toward an Understanding of the Spin-Statistics Theorem*, Am. J. Phys. 66, 4, pp. 284-302
[3] Ian Duck and Sudarshan E C G 1997 *Pauli and the Spin-Statistics Theorem* (World Scientific)

Journal of Physics: Conference Series **196** (2009) 012013 doi:10.1088/1742-6596/196/1/012013

Sudarshan's non-relativistic approach to the spin-statistics connection

Anil Shaji

Information Physics Group, Department of Physics and Astronomy, University of New Mexico, Albuquerque NM

E-mail: shaji@unm.edu

Abstract. George Sudarshan was quite a bit ahead of the times when he sketched out a non-relativistic proof of the spin-statistics connection forty years back. A recent revival of interest in the role of relativistic arguments in proving the spin-statistics connection prompted us to revisit Sudarshan's original arguments and present them in simpler fashion. The proof is designed to work in three space dimensions. Sudarshan's observation was that if we impose $SU(2)$ invariance, along with a few other restrictions, on the kinematic part of the Lagrangian of a field theory which is derivable from a Weiss-Schwinger type of the Principle of Least Action, then the spin-statistics connection automatically follows. The assumptions and simplifications that go into constructing the proof will be carefully laid out so as to identify the limitations of this approach.

1. Introduction

Is it possible to understand the spin-statistics connection without using relativistic arguments? A few years back this might have been considered merely a pedagogical question; perhaps even a misguided one. After all, Pauli, in his seminal paper on this topic [1], considers his proof of the spin-statistics theorem a triumph of the special theory of relativity and says:

> In conclusion we wish to state, that according to our opinion the connection between spin and statistics is one of the most important applications of the special relativity theory.

Despite Pauli's assertions and inspired by Prof. Sudarshan's ideas [2, 3, 4, 5], we venture to ask what really is the role of special relativity (if any) in understanding the connection between spin and statistics? Is it possible to deduce the theorem starting from just the postulates of quantum theory of fields without involving special relativity?

In Pauli's original proof and subsequent elaborations and completions of it [6, 7, 8, 9], relativity enters in the following three ways.

(i) Quantum fields are considered to be finite dimensional representations of the Lorentz group.

(ii) Observables associated with these fields vanish at space like separations.

(iii) The quantized fields have positive energy and charge density.

Unfortunately none of these three assertions have a direct counterpart in non relativistic quantum field theories. Even energies need not be positive, they just have to be bounded from below. What we recognize here is that somehow adapting Pauli's approach for showing the spin-statistics connection and then taking the non-relativistic limit of such a formulation is very unlikely to give us anything meaningful. This also means that if one were to follow such an approach, it will be easy to conclude rather prematurely that non-relativistic proofs of the spin-statistics connection cannot exist [10]. The question we have to ask then is whether there is an alternate starting point from which we can deduce the spin statistics connection. A starting point which has very little, and if possible, nothing to do with relativistic considerations.

What is our motivation for trying to prove the spin-statistics connection without using special relativity? It comes from experimentally accessible, non-relativistic, few-body, systems that require the spin-statistics relation to explain their behavior (see for instance [11] and references therein). We note here that the spin-statistics connection (equivalently the Pauli exclusion principle) was originally postulated in the context of atomic and nuclear physics. The original proof by Pauli and others were also developed with this domain of applicability in mind. It turned out that the domain of applicability of the theorem is much larger than that, extending in both directions to particle physics as well as to semi-classical quantum systems.

The spin-statistics connection seems crucial to understanding the behavior of several physical systems for which relativistic considerations seem quite insignificant. In fact, relativistic considerations are not just insignificant, but sometimes including them makes the understanding quite complicated. Examples of such systems include electrons in metals, phonons in solids and Bose condensates, atoms in optical lattices, Cooper pairs, entangled states and so on. Simply put, it seems that understanding the spin-statistics connection without relativity can give us better insights into the nature of such systems than going the other way and constructing a fully relativistic theory of these systems to include the spin-statistics connection in a natural way.

Spectacular advances in experimental techniques over the past two decades or so, driven partly by the advances in quantum information theory [12], means that quantum systems like the ones above are not just things that we observe in nature but rather systems that can be constructed, controlled and manipulated at will. Non-relativistic theories seem to adequately describe most of these systems and spin-statistics connection has to be inserted "by hand" when formulating these theories. With increasing experimental control over these systems, it is now reasonable to ask if we can at all tailor make a quantum system in which quasi-particles exhibit the wrong spin-statistics connection. This is not as unreasonable as it sounds because, after all, anyonic quasi-particles is quite actively talked about in the context of topological quantum computation [13, 14]. It is worth mentioning that in quantum information theory, the spin-statistics connection is often treated as an aid in doing certain information processing jobs [15].

Several approaches have been tried to obtain the spin-statistics connection without using relativity [16, 17, 18, 19]. We do not mention them here; instead, refer the reader to [4, 5] for detailed discussions on many of them.

This paper is structured as follows: In Sec. 2 we outline the steps needed to prove the spin-statistics connection using Sudarshan's approach. Sections 3, 4, 5, and 6 contain the details of the proof. We discuss briefly the implications of the assumptions that led to the proof in Sec. 7 and concluding remarks are in Sec. 8.

2. Outline of the proof

Our proof of the spin-statistics connection uses second quantization and field theory for describing quantum many body systems. We have to be careful though to make sure that the fields we consider could very well be non-relativistic ones. By using field operators in our discussion we can describe particle exchange that is central to the spin-statistics relation in very general terms. We carefully and deliberately avoid the elaborate devices based on rotation operators and adiabatic transport used by some authors to obtain the spin-statistics connection. This is because we believe that the spin-statistics connection must arise from the properties of the physical system independent of the *processes* that it goes through.

The objective is to establish a relationship between two distinct properties associated with quantum fields denoted by ξ:

- **Spin:** The spin is specified by how the field transforms under space rotations.
- **Statistics:** The statistics depends on whether the fields obey commutation or anti-commutation relations.

Starting from the spin we relate it to the statistics that the quantum field should obey through the steps that are outlined below.

(i) Continuing the theme of behavior under rotations, we construct a Lagrangian density $\mathfrak{L}$ for the fields that is a scalar with a few other restrictions on its kinematic part.

(ii) Once we have the Lagrangian, we make an action integral I from it and then using a Weiss-Schwinger [20, 21] type of least action principle we construct a complete theory of the non-relativistic fields we are interested in.

(iii) In this generalized principle of least action the surface variation of the action is the generator of transformations in the field quantities themselves [20]. This lets us write down the following equation involving products of fields, their variations, and other quantities that appear in $\mathfrak{L}$

$$[\xi^{(n)}, \delta I_{\delta\xi}] = \delta\xi^{(n)}. \tag{1}$$

(iv) We simplify the above equation assuming either commutation or anti-commutation relations for the fields ξ. We then find that if the initial assertion that the Lagrangian be a scalar is to hold then the commutation and anti-commutation relations have to go along with one type of spin or the other.

In other words, consistency checks on this sequence of steps gives us the spin-statistics relation. In the next three sections we flush out the details of these steps and make sure that the we have not used any relativistic considerations in each step.

3. The Lagrangian

In what follows it is assumed that we are dealing with a standard version of quantum field theory which is derivable from a Principle of Least Action based on a Lagrangian. Our arguments leading to the spin-statistics connection are confined at the outset to *three space dimensions.* The three dimensional space admits multicomponent wave functions that have specific transformation properties under rotations. The rotations that we consider are not restricted to the group of proper rotations in three dimensional space, $SO(3)$, but rather to its covering group $SU(2)$. The extension to $SU(2)$ is needed for our discussion of the spin-statistics connection because we have to include both integral and half-integral spin representations of the rotation group and be able to treat them as proper representations. Since $SU(2)$ is a subgroup of the covering group of the Galilei group, we are able to keep our arguments entirely non-relativistic.

Along the lines of Schwinger's proof [20] of the spin-statistics connection the following four conditions are imposed on the *kinematic part* of the Lagrangian density.

(i) The Lagrangian $\mathfrak{L}$ is *invariant under $SU(2)$ (and $SO(3)$) transformations* and corresponds to a field theory for fields, ξ, which are finite dimensional representations of the $SU(2)$ covering of the three dimensional rotation group.

(ii) The Lagrangian is expressed in the *hermitian field basis*; $\xi = \xi^\dagger$.

(iii) It is at most linear in the *first time derivatives* of the field and the time derivatives occur only in the kinematic term.

(iv) The *kinematic term is bilinear* in the field ξ

The last two requirements ensure that the Euler-Lagrange equations of motion derived from the Lagrangian are first order linear differential equations of the Hamiltonian form. The Dirac Lagrangian is already in the required form while for massive spin-0 and spin-1 fields satisfying the Klein-Gordon equation, the Duffin-Kemmer [22, 23] form of the equation is such a suitable one. A short discussion on how to convert a generic bilinear kinematic Lagrangian with higher order time derivatives in it into a form that is suitable for the present discussion is given in [24].

We now focus on the kinematic part of the Lagrangian written in a form that satisfies the four requirements laid out above. This is acceptable because we do not expect the spin-statistics connection to come out of interactions and other dynamical effects. The generic (Schwinger) Lagrangian satisfying the four requirements has the form

$$\begin{aligned}\mathfrak{L} &= \frac{1}{2}K^{0}_{rs}(\xi^{(r)}\dot{\xi}^{(s)} - \dot{\xi}^{(r)}\xi^{(s)}) \\ &\quad + \text{ Terms containing space derivatives } + \text{ Mass terms} \\ &\quad + \text{ Other non-derivative terms including interactions.}\end{aligned} \tag{2}$$

The first term that is explicitly written down in the *kinematic* term. In the Lagrangian, r and s are spin indices that are summed over. The kinematic term is explicitly antisymmetrized with respect to the time derivative so that we avoid the possibility of the two terms together forming a total derivative. The terms containing the space derivatives need not be of this particular from and can contain higher derivatives and higher powers of the derivatives. No restrictions are placed on the mass term or on any interaction terms that may be present except to the extend that condition (iii) rules out velocity dependent potentials in the theories we are considering.

The number of Hermitian fields that appear in the Lagrangian depend on the spin and possible charge of the physical field that we are considering. For a spin-0 field with no internal flavor index (charge) we need only one Hermitian field component, i.e. $r = s = 1$. If we have a charged scalar field, then we need a pair of Hermitian fields with $r,\ s = 1$ and an extra flavor index $\alpha = 1, 2$. The Lagrangian will then contain two pieces corresponding to each value of α. For spin-1/2 and no charge, $r,\ s = 1, \dots 4$ as in the Majorana theory. The number of hermitian fields have to be doubled to eight, $r, s = 1 \dots 4$, $\alpha = 1, 2$ when we are considering the Dirac theory for a spinor carrying a charge. The Hermitian fields are used here (following Schwinger) expressly for the purpose of demonstrating the spin-statistics connection. For most other purposes this choice is unnecessarily cumbersome.

3.1. Singular Lagrangians

If we require that the kinematic term of the Lagrangian be written in the form in (2) then we have to consider the possibility that K^0_{rs} is a singular matrix. This would happen when some of the equations that we write down for the fields $\xi^{(r)}$ in the theory are not equations of motion (no time derivatives) but rather are equations of constraint. In the discussion that follows we will talk about the symmetry properties of K^0_{rs} with respect to the indices r and s. We will also talk about the commutation or anticommutation relation of the fields ξ in terms of K^0_{rs} and its inverse. Since we will be considering theories with primary constraints K^0_{rs} will almost always be singular and because of this, *when we talk about the symmetry properties of K^0_{rs} or about the inverse of K^0_{rs} it must be understood that we are talking only about the non-singular part of the matrix.*

4. Sudarshan's observation that leads to the spin-statistics connection

The matrix K^0_{rs} appearing in the Lagrangian in (2) is a numerical matrix that has no direct dependence on space and time coordinates. It depends on the spin labels only. It follows that if the Lagrangian has to be rotationally invariant (scalar) then the products of fields and their

Journal of Physics: Conference Series **196** (2009) 012013 doi:10.1088/1742-6596/196/1/012013

derivatives that appear in the Lagrangian are themselves scalars. It will turn out that K^0_{rs} must have specific symmetry properties with respect to the spin indices (r, s) for the Lagrangian to be nontrivial.

Obtaining the correct spin-statistics connection hinges on Sudarshan's observation [2, 3] that for the $SO(3)$ group of proper rotations (or rather its covering group $SU(2)$) in three dimensions, the representations belonging to *integral spin* have a bilinear scalar product that is *symmetric* in the indices of the factors. For example the scalar product of two real vectors is

$$(V^{(1)}, V^{(2)}) = \sum_{j,k=1,2,3} V_j^{(1)} V_k^{(2)} \delta_{jk} = \sum_{j,k=1,2,3} V_j^{(2)} V_k^{(1)} \delta_{jk} = (V^{(2)}, V^{(1)}). \tag{3}$$

On the other hand, *half-integral spin* representations have *antisymmetric* bilinear scalar products. For instance, for the spin-1/2 case the scalar product of two Hermitian spinors is

$$\begin{aligned}(\psi^{(1)}, \psi^{(2)}) &= \sum_{r,s=1}^{4} \psi_r^{(1)} \psi_s^{(2)} (i\beta_{rs}) = \sum_{r,s=1}^{4} \psi_s^{(1)} \psi_r^{(2)} (i\beta_{sr}) \\ &= -\sum_{r,s=1}^{4} \psi_r^{(2)} \psi_s^{(1)} (i\beta_{rs}) = -(\psi^{(2)}, \psi^{(1)}),\end{aligned} \tag{4}$$

where

$$\beta_{rs} = \begin{pmatrix} 0 & \hat{\sigma}_2 \\ \hat{\sigma}_2 & 0 \end{pmatrix}$$

is an imaginary antisymmetric matrix corresponding the Majorana representation of the spinors.

The requirement that the Lagrangian be $SU(2)$ invariant (in the Hermitian field basis) means that integral spin fields appear in the Lagrangian in symmetrized scalar combinations while half integral fields appear in antisymmetrized scalar combinations. These requirements put restrictions on the symmetry properties of the invariant matrices K^0_{rs} appearing in the Lagrangian.

To investigate the symmetry requirements on K^0_{rs} we start from the Kinematic term of the Lagrangian in (2) i.e.

$$\mathfrak{L}_{kin} = \sum_{r,s} \frac{1}{2} K^0_{rs} (\xi^{(r)} \dot{\xi}^{(s)} - \dot{\xi}^{(r)} \xi^{(s)}). \tag{5}$$

$\mathfrak{L}_{kin}$ may be rewritten in the following form

$$\mathfrak{L}_{kin} = \frac{1}{2} \sum_{r,s} \xi^{(r)} \Lambda_{rs} \xi^{(s)} \quad ; \quad \Lambda_{rs} \equiv K^0_{rs} (\partial_t^{(r)} - \partial_t^{(s)}).$$

For tensor fields the scalar product constructed out of $\xi^{(r)}$ and $\xi^{(s)}$ is symmetric in the indices r and s. It follows that if the Lagrangian is to be non-trivial, Λ_{rs} should also be symmetric in r and s.

$$\Lambda_{rs} = \Lambda_{sr} \quad \Rightarrow \quad K^0_{rs} (\partial_t^{(r)} - \partial_t^{(s)}) = -K^0_{sr} (\partial_t^{(r)} - \partial_t^{(s)}).$$

We see that because of the antisymmetry of the time derivative term, K^0_{rs} has the opposite symmetry of Λ_{rs} and so for tensor fields, K^0_{rs} must be an antisymmetric matrix.

For spinor fields, since the scalar product is antisymmetric, Λ_{rs} also has to be antisymmetric. Consequently for spinor fields, K^0_{rs} must be a symmetric matrix.

Starting from the observation that for *integral spin* fields the numerical matrix K^0_{rs} must be antisymmetric in the spin indices r and s while it must be symmetric for *half integral spin* fields due to rotational invariance of the Lagrangian; if we can now show by independent means that for *commuting fields* K^0_{rs} must be antisymmetric and for *anticommuting fields* it must be symmetric we will obtain the proper spin-statistics connection. This is achieved using the Principle of Least Action for field quantities as formulated by Weiss and Schwinger [21, 20].

5. The principle of least action

The generator of infinitesimal transformations on the eigenstates of a complete set of commuting operators of a quantized system/field is obtained by the extended variation of the quantities contained in the action integral [20]

$$I = \int_{\sigma_2}^{\sigma_1} (dx)\, \mathfrak{L}[x]. \tag{6}$$

The integral is over an n-dimensional (space-time) domain with coordinates labeled by x_k ; $k = 1, \ldots, n$ and bounded by the $n-1$ dimensional surfaces σ_1 and σ_2. In the general case σ_i are $(n-1)$ dimensional equal time slices in the n-dimensional space-time. As we already stated, in this discussion we consider *only* theories in three space and one time dimensions. Accordingly σ_i are *three dimensional, equal time slices.* In equation (6), for $\mathfrak{L}$ we use the specific form of the kinematic part of Lagrangian density that we wrote down in equation (2).

The Principle of Stationary Action is then formulated for operator dynamical variables as the statement that the action integral operator is unaltered by infinitesimal variations of the field quantities in the interior of the region bounded by σ_1 and σ_2; it being dependent only on the variation of the complete commuting set of operators attached to the bounding surfaces.

The variation of the action integral can then be shown to be decomposable into two terms;

$$\delta I = \int_{\sigma_1}^{\sigma_2} dt d^3x\, \delta\mathcal{L}[x] + \int d^3x\, (\mathfrak{L}[x, t_2] - \mathfrak{L}[x, t_1])\delta x. \tag{7}$$

The first term is an integral over the space-time domain $\mathcal{D}$ bounded by σ_1 and σ_2. The second term is a "surface" term integrated only over the three spaces at times t_1 and t_2. Setting the variation of the action in the interior of the space-time domain $\mathcal{D}$ to zero leads to the Euler-Lagrange equations of motion for the fields while the surface variation term is treated as the generator of infinitesimal transformations on the system.

We need not go into the details of this development here but as Schwinger, pointed out, $\delta\mathfrak{L}$ should be treated carefully in the quantum case since the order of the operators that appear in $\mathfrak{L}$ must not be altered in the course of effecting the variation. Accordingly, we recognize that the commutation properties of $\delta\xi^{(r)}$ that appear in the variation of the action based on the Lagrangian density in equation (2) are involved in the consequences of the extended variation of the action integral. In other words we must, at this point, make the explicit assumption that the commutation properties of $\delta\xi$ with respect to $\xi^{(r)}$ and the structure of the Lagrange function must be connected in a consistent fashion.

We now choose to simplify matters as much as possible and as before focus on $\mathfrak{L}_{kin}$. We are interested in the variation of the action that is brought about by the variation in the field quantities ξ themselves. We will not therefore consider the variation of the action integral operator that is brought out by a change in the coordinates x or time t. The variations that we consider here have the form

$$\xi_k \to \xi_k + \delta\xi_k\,, \quad \dot{\xi}_k \to \dot{\xi}_k + \frac{d(\delta\xi_k)}{dt}. \tag{8}$$

The ensuing variation in the action is then given by

$$\delta I_{\delta\xi} = \frac{1}{2}\int dt\, d^3x \left(\frac{\partial\mathfrak{L}}{\partial\xi^{(r)}}\delta\xi^{(r)} + \frac{\partial\mathfrak{L}}{\partial\dot{\xi}^{(r)}}\delta\dot{\xi}^{(r)}\right). \tag{9}$$

The integral is over a three dimensional space and one dimensional time domain. Also, the above equation must again be understood in a symbolic sense with the exact position of the variation $\delta\xi$ in each term to be fixed later in an appropriate fashion. Using the identity,

$$\frac{\partial \mathfrak{L}}{\partial \dot{\xi}^{(r)}}\delta\dot{\xi}^{(r)} = \frac{\partial}{\partial t}\left(\frac{\partial \mathfrak{L}}{\partial \dot{\xi}^{(r)}}\delta\xi^{(r)}\right) - \frac{\partial}{\partial t}\left(\frac{\partial \mathfrak{L}}{\partial \dot{\xi}^{(r)}}\right)\delta\xi^{(r)},$$

and using Gauss' theorem we rewrite the variation in the action integral as

$$\delta I_{\delta\xi} = \frac{1}{2}\int dt\, d^3x \left\{\frac{\partial \mathfrak{L}}{\partial \xi^{(r)}} - \frac{\partial}{\partial t}\left(\frac{\partial \mathfrak{L}}{\partial \dot{\xi}^{(r)}}\right)\right\}\delta\xi^{(r)} + \frac{1}{2}\int_\sigma d^3x \frac{\partial \mathfrak{L}}{\partial \dot{\xi}^{(r)}}\delta\xi^{(r)}. \tag{10}$$

Setting the first term (variation in the interior of the domain $\mathcal{D}$ bounded by σ) to zero we obtain the Euler-Lagrange equations for the hermitian fields $\xi^{(r)}$. It is the surface variation terms that are of interest to us as the generators of infinitesimal transformations on the field quantities themselves. For the specific form for $\mathfrak{L}_{kin}$ that we are considering in equation (5), the (surface) variation term is given by

$$\delta I_{\delta\xi} = \frac{1}{2}\int_\sigma d^3x \sum_{r,s} K^0_{rs}(\xi^{(r)}\delta\xi^{(s)} - \delta\xi^{(r)}\xi^{(s)}). \tag{11}$$

Since $\delta I_{\delta\xi}$ is the generator of infinitesimal transformations of the field quantities ξ attached to the surface σ, using Eq. (1) we obtain

$$\frac{1}{2}\int_\sigma d^3x \left[\xi^{(n)}, \sum_{r,s} K^0_{rs}(\xi^{(r)}\delta\xi^{(s)} - \delta\xi^{(r)}\xi^{(s)})\right] = i\hbar\delta\xi^{(n)}. \tag{12}$$

Expanding out the commutator we obtain

$$\frac{1}{2}\int_\sigma d^3x \sum_{r,s} K^0_{rs}(\xi^{(n)}\xi^{(r)}\delta\xi^{(s)} - \xi^{(n)}\delta\xi^{(r)}\xi^{(s)} - \xi^{(r)}\delta\xi^{(s)}\xi^{(n)} + \delta\xi^{(r)}\xi^{(s)}\xi^{(n)}) = i\hbar\delta\xi^{(n)}. \tag{13}$$

Now we are in a position to assume that the fields appearing in equation (13) are either commuting fields or anticommuting fields and investigate what restrictions (if any) each assumption places on the matrix K^0_{rs}.

5.1. *Commuting fields*

Let us assume now that the fields ξ are bosonic fields and we consider variations in them. The significant step is that we assume $\delta\xi$ *commutes* with everything. Using this assumption we can rewrite the left hand side of equation (13) as

$$\frac{1}{2}\int_\sigma d^3x \sum_{r,s} K^0_{rs}\{(\xi^{(n)}\xi^{(r)}\delta\xi^{(s)} - \xi^{(r)}\xi^{(n)}\delta\xi^{(s)}) - (\xi^{(n)}\xi^{(s)}\delta\xi^{(r)} - \xi^{(s)}\xi^{(n)}\delta\xi^{(r)})\} \tag{14}$$

Putting in the space coordinates $\mathbf{x}$ and $\mathbf{y}$ of the fields explicitly we can rewrite the previous expression in terms of commutation relations of the fields as

$$\frac{1}{2}\int_\sigma d^3x \sum_{r,s} K^0_{rs}\{[\xi^{(n)}(\mathbf{y}), \xi^{(r)}(\mathbf{x})]\delta\xi^{(s)}(\mathbf{x}) - [\xi^{(n)}(\mathbf{y}), \xi^{(s)}(\mathbf{x})]\delta\xi^{(r)}(\mathbf{x})\}. \tag{15}$$

Exchanging the indices i and j in the second term we can rewrite the expression in equation (15) as

$$[\xi^{(n)}, \delta I_{\delta\xi}] = \int_\sigma d^3x \sum_s \delta\xi^{(s)}(\mathbf{x})\left[\xi^{(n)}(\mathbf{y}), \frac{1}{2}\sum_r (K^0_{rs} - K^0_{sr})\xi^{(r)}(\mathbf{x})\right] \tag{16}$$

From equation (16) we see explicitly that $[\xi^{(n)}, \delta I_{\delta\xi}] = 0$ if K^0_{rs} is symmetric. So to satisfy the condition that the surface variation of the action is the generator of infinitesimal transformations we need K^0_{rs} to be an *antisymmetric* matrix if the fields $\xi^{(r)}$ (and $\delta\xi^{(r)}$) are bosonic commuting fields. The possibility that K^0_{rs} be neither symmetric nor antisymmetric ($K^0_{rs} \neq \pm K^0_{sr}$) in the spin indices is already excluded by the rotational invariance of the Lagrangian density that we require.

5.2. *Anticommuting fields*

If we assume fermionic anticommutation relations instead of commutation relations for the fields ξ and assume that $\delta\xi$ *anticommutes* with everything, the analogue of equation (14) that we obtain from (13) is

$$\frac{1}{2}\int_\sigma d^3x \sum_{r,s} K^0_{rs}\{(\xi^{(n)}\xi^{(r)}\delta\xi^{(s)} + \xi^{(r)}\xi^{(n)}\delta\xi^{(s)}) + (\xi^{(n)}\xi^{(s)}\delta\xi^{(r)} + \xi^{(s)}\xi^{(n)}\delta\xi^{(r)})\} \tag{17}$$

This can be simplified as before to

$$[\xi^{(n)}, \delta I_{\delta\xi}] = \int_\sigma d^3x \sum_s \delta\xi^{(s)}(\mathbf{x}) \left\{\xi^{(n)}(\mathbf{y}), \frac{1}{2}\sum_r (K^0_{rs} + K^0_{sr})\xi^{(r)}(\mathbf{x})\right\} \tag{18}$$

In this case we see that $[\xi^{(n)}, \delta I_{\delta\xi}] = 0$ if K^0_{rs} is antisymmetric. Thus for anticommuting fermionic fields we require K^0_{rs} to be a *symmetric* matrix to be consistent with the dynamical principle.

6. The spin-statistics connection

We have seen that the rotational invariance of the kinematic part of the Lagrangian density requires that the matrix K^0_{rs} be antisymmetric for integral spin fields while it be symmetric for half-integral fields. The Schwinger Principle of Least Action on the other hand requires that K^0_{rs} be antisymmetric for commuting (bosonic) fields and symmetric for anticommuting (fermionic) fields. This leads us to the conclusion that *fields with integral spin must be bosonic while fields with half-integral spin must be fermionic.* This is the spin-statistics connection.

7. Discussion

We use hermitian fields in $\mathfrak{L}_{kin}$ in our proof of the spin-statistics connection. In hermitian fields, the creation and annihilation parts always appear together. For relativistic fields, the requirement of locality assures us that the two parts always appear together. In the non-relativistic case, imposing hermiticity on the fields, automatically realizes the *Krichoff's principle.* When we study the thermodynamics of radiation in equilibrium, we assume that there is a continual emission and absorption of the radiation by matter. Furthermore we know that the emissivity and absorptivity of matter are proportional to each other: this is Kirchoff's Principle. In terms of ordinary quantum mechanics what this means is that when we couple harmonic oscillator degrees of freedom, the coupling is in terms of the coordinate $q \sim (a + a^\dagger)/2$ rather than coupling the creation and annihilation operators separately and independently. In an analogous manner we realize Kirchoff principle for the non-relativistic fields by. The requirement that the probabilities of creating and destroying the quanta of the fields that we are considering are proportional to each other ensures that the creation and annihilation fields do not appear independently in the Lagrangians that we are considering.

In our discussion of the Principle of Least Action for quantum mechanical operators and fields in section 5, we considered the variation of the action when the fields ξ_r are varied by $\delta\xi_r$. To get commutation or anticommutation relations for the fields and to obtain the correct spin-statistics connection we made the choice that $\delta\xi_r$ commutes with every quantity that appears in the variation of the action when ξ_r denote commuting fields and that $\delta\xi_r$ anticommutes with everything if ξ_r are anticommuting quantities. This is required if we want to keep the varied fields $\xi_r' = \xi_r + \delta\xi_r$ such that they have *exactly the same commutation or anticommutation properties as* ξ_r.

Choosing the arbitrary variation $\delta\xi_r$ to be a quantity that commutes with everything is easy enough because all one has to do is to make sure that $\delta\xi$ is a c-number. On the other hand, choosing $\delta\xi_r$ such that it anticommutes with all the field quantities that appear in $\delta I_{\delta\xi}$ is not so straightforward. For an even number of anticommuting fields, ξ_r, one possible choice to make $\delta\xi_r$ proportional to the product of all the fields, i.e

$$\delta\xi_r \sim \epsilon_r \, \Pi_{i=1}^{2n} \xi_i \quad ; \quad r = 1, 2, \ldots 2n.$$

$\delta\xi_r$ will then anticommute with all the quantities that appear in the variation of the action [25].

Restricting $\delta\xi_r$ to be either commuting or anticommuting with everything is essential for the canonical quantization of commuting or anticommuting fields [26, 27]. But it is possible to make the choice of $\delta\xi_r$ more general. If we consider the fields ξ_r defined on the equal time slice σ as forming a complete set of operators on σ, a natural generalization of $\delta\xi_r$ is to choose it as a linear combination of ξ_r,

$$\delta\xi_r(\mathbf{x}) = \sum_s \varepsilon_{rs}\xi_s(\mathbf{x}) \tag{19}$$

where ε_{rs} are infinitesimal c-number coefficients. The basic commutation relation in (12) can be rewritten as

$$\frac{1}{2}\int_\sigma d^3x \left[\xi_n, \sum_{k,r,s} K^0_{rs}(\xi_r\varepsilon_{sk}\xi_k - \varepsilon_{rk}\xi_k\xi_s)\right] = i\hbar\sum_k \varepsilon_{nk}\xi_k. \tag{20}$$

Without working out the details we can immediately see that eq. (20) will lead to *trilinear* commutation or anticommutation relations among ξ_r. These generalized commutation and anticommutation relations, due to Wigner and Green [28, 29], lead to reducible para-Bose and para-Fermi systems.

The arguments leading to the spin-statistics connection rely on the symmetry, or antisymmetry, that is required independently by $SU(2)$ invariance of the Lagrangian and by the Principle of Least Action, of the numerical matrix K^0_{rs}. It might seem possible that the introduction of internal symmetry indices (flavors) on the fields ξ_r can change the symmetry requirements on K^0_{rs}. It turns out if we try to do this we will end up with negative norm states in the theory [24].

8. conclusion

We have shown that for the general class of fields that can be canonically quantized, Sudarshan's ideas lead to a proof of the spin-statistics connection that does not depend on relativistic arguments. All that is required is the $SU(2)$ invariance of the kinematic part of the Lagrangian. The four requirements on $\mathfrak{L}_{kin}$ listed in Sec. 3 for the proof to work are quite reasonable and justified especially when we are dealing with non-relativistic systems. It is important to note that the requirements are placed just on the kinematic part. This means that we obtain the spin-statistics connection using just the properties of the system and not using the behavior of the system under any dynamical process.

Acknowledgments

Prof. E. C. G. Sudarshan, to whom this paper is dedicated, has been the inspiration and source for most of the ideas presented in this paper. I am greatly indebted to him for patiently clarifying the steps that lead to the proof of the spin-statistics connection presented here. I also thank Dr. Tom Jordan for pointing out the need for a clear elucidation of a non-relativistic derivation of the spin-statistics connection and initiating the work presented here. This work was supported in part by Office of Naval Research grant No. N00014-03-1-0426.

References

[1] W Pauli. The connection between spin and statistics. *Phys. Rev.*, 58:716, October 1940.

[2] E C G Sudarshan. The fundamental theorem on the relation between spin and statistics. *Proc. Indian Acad. Sci.*, A67:284, 1968.

[3] E C G Sudarshan. The fundamental theorem on the connection between spin and statistics. In Nils Svartholm, editor, *Proc. Nobel Symposium 8: Elementary Particle Theory, Relativistic Groups and Anlyticity*, page 379, Stockholm, 1968. Almqvist and Wiksell.

[4] I M Duck and E C G Sudarshan. *Pauli and the spin-statistics theorem.* World Scientific, Singapore, 1998.

[5] I M Duck and E C G Sudarshan. Toward an understanding of the spin-statistics theorem. *Am. J. Phys.*, 66(4):284, 1998.

[6] G Luders and B Zumino. Connection between spin and statistics. *Phys. Rev.*, 110(6):1450, June 1958.

[7] N Burgoyne. On the connection of spin and statistics. *Nuovo Cimento*, 8(4):607, 1958.

[8] A S Wightman. Quantum field theory in terms of vacuum expectation values. *Phys. Rev.*, 101:860, 1956.

[9] D Hall and A S Wightman. A theorem on invariant analytic functions with application to relativistic quantum field theory. *Mat. Fys. Medd. K. Dan. Vidensk. Selsk*, 31(5):1, 1957.

[10] C R Hagen. Spin and statistics in galilean covariant field theory. *e-print, arXiv:quant-ph/0403039*, 2004.

[11] Y Omar. Particle statistics in quantum information processing. *e-print arXiv:quant-ph/0412213*, 2004.

[12] M A. Nielsen and I L. Chuang. *Quantum Computation and Quantum Information.* Cambridge University Press, Cambridge, 2000.

[13] A Kitaev. Fault-tolerant quantum computation by anyons. *e-print arXiv:quant-ph/9707021*, 1997.

[14] M H Freedman, A Kitaev, M J Larsen, and Z Wang. Topological quantum computation. *Bull. Am. Math. Soc.*, 40(1):31, 2002.

[15] S Bose, A Ekert, Y Omar, N Paunković, and V Vedral. Optimal state discrimination using particle statistics. *Phys. Rev. A*, 68:52309, 2003.

[16] A P Balachandran, A Daughton, Z-C Gu, R D Sorkin, G Marmo, and A M Srivastava. Spin-statistics theorems without relativity or field theory. *International Journal of Modern Physics A*, 8:2993–3044, 1993.

[17] M V Berry and J M Robbins. Indistinguishability for quantum particles: spin, statistics and the geometric phase. *Proc. Roy. Soc. Lond. A*, 453:1771, 1997.

[18] M Peshkin. Spin and statistics in nonrelativistic quantum mechanics: The spin-zero case. *Phys. Rev. A*, 67:42102, 2003.

[19] M Peshkin. Reply to "comment on 'spin and statistics in nonrelativistic quantum mechanics: The spin-zero case' ". *Phys. Rev. A*, 68:46102, 2003.

[20] J Schwinger. The theory of quantized fields. i. *Phys. Rev.*, 82(6):914, June 1951.

[21] P Weiss. On the hamilton-jacobi theory and quantization of a dynamical continuum. *Proc. Roy. Soc. Lond. A*, 169(936):102, December 1938.

[22] R Y Duffin. On the characteristic matrices of covariant systems. *Phys. Rev.*, 54(12):1114, December 1938.

[23] N Kemmer. The particle aspect of meson theory. *Proc. Roy. Soc. Lond. A*, 173(952):91, November 1939.

[24] A Shaji and E C G Sudarshan. Non-relativistic proofs of the spin-statistics connection. *e-print, arXiv:quant-ph/0306033*, 2003.

[25] E Cartan. *The theory of spinors.* M. I. T. Press, Cambridge, Mass., 1967.

[26] S S Schweber. *An introduction to relativistic quantum field theory.* Harper & Roe, New York, 1962.

[27] S Weinberg. *The Quantum Theory of Fields*, volume 1. Cambridge University Press, Cambridge, 1995.

[28] E P Wigner. Do the equations of motion determine the quantum mechanical commutation relations? *Phys. Rev.*, 77(5):711, March 1950.

[29] H S Green. A generalized method of field quantization. *Phys. Rev.*, 90(2):270, April 1953.

Journal of Physics: Conference Series **196** (2009) 011009 doi:10.1088/1742-6596/196/1/011009

Introduction to "Quantum Coherence" : the fourth quest

Harry Swinney

Department of Physics, The University of Texas at Austin

Email: swinney@physics.utexas.edu

Abstract: Opening remarks at the Quantum Coherence session.

Quantum theory of optical coherence was formulated by Sudarshan in his classic paper of 1963 in Phys Rev Letters, in which he presented the Quantum Optical Equivalence Theorem. The first talk will be Dr. Mehta.

$$\rho = \int \varphi(z)|z\rangle\langle z|d^2z$$

E C G Sudarshan with FOGS (Friends of George Sudarshan), 2006

Journal of Physics: Conference Series **196** (2009) 011009 doi:10.1088/1742-6596/196/1/011009

Speakers at Quantum Coherence: Sudarshan's Fourth quest

Harry Swinney

C L Mehta
(G Bhamathi, G Sudarshan, H Swinney, C L Mehta)

Jeff Kimble

N Mukunda presenting talk by R Simon

$$\rho = \int \varphi(z)|z\rangle\langle z|d^2z$$

Journal of Physics: Conference Series **196** (2009) 012014 doi:10.1088/1742-6596/196/1/012014

Sudarshan Diagonal Coherent State Representation: Developments and Applications

C.L. Mehta

Indian Institute of Technology, New Delhi, India

Email: mehtacl@hotmail.com

Abstract: A review is presented of the early developments in quantum optical coherence. Some properties of the coherent states particularly their over completeness which led to the discovery of diagonal coherent state representation by Sudarshan will be discussed. We then consider some of the important applications of this diagonal coherent state representation.

Classical theory of optical coherence was developed by E. Wolf[1] by considering optical field as part of a stochastic process. Coherence functions were essentially correlations between field components at different space time points. The second order coherence function may thus be written as

$$\Gamma(\underline{r}_1,\underline{r}_2,\tau) = \left\langle V^*(\underline{r}_1,t)V(\underline{r}_2,t+\tau)\right\rangle_c , \tag{1a}$$

$$\langle(\ldots)\rangle_c = \int(\ldots)P(\{V\})d^2\{V\}, \tag{1b}$$

where V(**r**,t) is the analytic signal associated with the electromagnetic field component at the point **r** and at time t. The averages $\langle(\ldots)\rangle_c$, are the stochastic averages over the given ensemble. Near about the same time L. Mandel[2] studied photo-electron counting statistics and derived the counting formula as a Poisson transform of the integrated light intensity distribution

$$P(n,T) = \int P(W)\frac{W^n}{n!}e^{-W}dW , \tag{2a}$$

$$W = \int_0^T I(t)\,dt . \tag{2b}$$

Most features of the photo-counting distribution were well understood by Mandel's formula. Several effects known at that time such as propagation of coherence, Brown-Twiss experiments on bunching of photons and others had been adequately explained by classical approach to light fluctuations. With the advent of lasers, a need arose for the quantum description of electromagnetic fields associated with arbitrary light beams. For an optical field described by a density operator ρ one then considers the analogous quantum coherence functions:

$$\Gamma(\underline{r}_1,\underline{r}_2,\tau) = \left\langle E^{(-)}(\underline{r}_1,t)\,E^{(+)}(\underline{r}_2,t+\tau)\right\rangle_q \tag{3}$$

with $\langle(...)\rangle_q = Tr[\rho(...)]$ and $E^{(-)}$ and $E^{(+)}$ being the creation and annihilation parts of the field operator E. If we take up the usual quantization of electromagnetic field and, for simplicity, restrict ourselves to one mode case only, we can use the harmonic oscillator number states to describe such fields. These states are the eigenstates of the number operator a^+a :

$$a^+a|n\rangle = n|n\rangle. \tag{4}$$

The number operator being hermitian, the states $|n\rangle$ are orthogonal

$$\langle m\,|\,n\rangle = \delta_{mn} \tag{5}$$

and form a complete set

$$\sum_n |n\rangle\langle n| = 1, \tag{6}$$

so that an arbitrary state $|\psi\rangle$ can be expressed in the form

$$|\psi\rangle = \sum \psi_n |n\rangle \tag{7}$$

with uniquely determined ψ_n given by

$$\psi_n = \langle n|\psi\rangle. \tag{8}$$

Glauber[3] realized the importance of using eigenstates of the annihilation operator

$$a|z\rangle = z|z\rangle, \tag{9}$$

$$|z\rangle = e^{-\frac{1}{2}|z|^2} \sum \frac{z^n}{\sqrt{n!}}|n\rangle \tag{10}$$

as the basis states in place of the number states. These so called coherent states are in fact the minimum uncertainty product states first introduced by Schroedinger[4]. Since a is not a hermitian operator, its eigenvalues are in general complex numbers $z = x + iy = r\,e^{i\theta}$ and the eigenstates are not orthogonal

$$\langle z'|z\rangle \neq 0 \;\; for\; z \neq z'. \tag{11}$$

However they form a complete set,

$$\frac{1}{\pi}\int |z\rangle\langle z|\, d^2 z = 1\,, \tag{12}$$

where $d^2z = dx\, dy = r\, dr\, d\theta$ denotes integration over the complex z plain. An arbitrary state $|\psi\rangle$ may therefore be expressed linearly in terms of $|z\rangle$:

$$|\psi\rangle = \int \psi(z)\,|z\rangle\, d^2 z \tag{13}$$

with $\psi(z) = (1/\pi)\langle z|\psi\rangle$. Further since the states $|z\rangle$ are not linearly independent, the function $\psi(z)$ is not unique. This assertion is readily verifiable from the obvious relation such as

$$\int z\,|z\rangle\, d^2 z = 0\,. \tag{14}$$

Coherent states may also be used to express operators in the form [cf. Eq.(3) of ref. 3]

$$\rho = \int F(z,z')\,|z\rangle\langle z'|\, d^2 z\, d^2 z' \tag{15}$$

and the function F(z,z') is doubly "non-unique". It was noted by Sudarshan[(5)] that these coherent states are not only complete but are in fact over complete. Sudarshan for the first time made use of this over completeness property and realized the possibility of a diagonal coherent state representation

$$\rho = \int \phi(z)\,|z\rangle\langle z|\; d^2 z\,. \tag{16}$$

The generalized function $\phi(z)$ in this expression is then unique. Thus for example the density operator ρ in the number representation

$$\rho = \sum \rho_{nn'}\,|n\rangle\langle n'| \tag{17}$$

will have the diagonal coherent state representation $\phi(z)$, with $z = r\,e^{i\theta}$, given by [cf. Eq. (6) of ref. 5],

$$\phi(z) = \sum_{n,n'} \frac{\sqrt{n!\,n'!}\,\rho_{nn'}}{(n+n')!\; 2\pi r}\, e^{r^2 + i(n'-n)\theta} \left(-\frac{\partial}{\partial r}\right)^{n+n'} \delta(r)\,. \tag{18}$$

This diagonal coherent state representation of the density matrix had far reaching applications. The expectation value of any normally ordered operator $a^{+\mu} a^{\nu}$ in the statistical state represented by this density operator takes on the form

$$\left\langle a^{+\mu}\, a^{\nu}\right\rangle = \int z^{*\mu}\, z^{\nu}\, \varphi(z)\, d^2 z\,. \tag{19}$$

This is the same as the expectation value of the complex classical function $z^{*\mu} z^{\nu}$ in a statistical state described by the probability distribution function $\phi(z)$. This property therefore enables the various coherence functions to appear in a similar way as in the Wolf- Mandel formulation. Eq. (16) may also be used to express $\rho_{n,n'}$ in terms of $\phi(z)$ and in particular we obtain for $\rho_{n,n}$ the expression

$$\rho_{n,n} = \langle n \mid \rho \mid n \rangle = \int \phi(z) e^{-|z|^2} \frac{|z|^{2n}}{n!} d^2 z . \tag{20}$$

Interpreting $|z|^2$ as the intensity W of the field, Eq. (20) reproduces Mandel's Poisson transform relation Eq. (2a).

Sudarshan also noted that the hermiticity of ρ implies that the distribution $\phi(z)$ is real. However the positivity of ρ does not imply positivity of ϕ and therefore *brings out a clear distinction between classical and quantum ensembles*. It may be emphasized that for all fields, the diagonal coherent state representation is formally as if it was a classical ensemble, but with the understanding that the ensemble probability distributions could be non-positive. With this non-positivity of the diagonal coherent state representation in general, one can deduce specific quantum effects such as anti-bunching, squeezing etc. Glauber's off-diagonal representation of ρ in terms of $|z\rangle\langle z'|$ as in Eq. (15) can in no way lead to this similarity or difference between the classical and quantum descriptions.

It is to be noted that $\phi(z)$ is unique, though there may be different equivalent forms for $\phi(z)$. Sudarshan's original expression for $\phi(z)$ as in Eq.(18) contains derivatives of delta function and as such appears to be singular. In a subsequent publication Mehta and Sudarshan[6] discussed the characteristic function approach to the diagonal coherent state representation, which gives a well defined and rigorous meaning to $\phi(z)$ as a generalized distribution. Consider the three characteristic functions defined as:

$$X_N(\alpha) = \left\langle e^{\alpha a^+} e^{-\alpha^* a} \right\rangle = \int \phi_N(z) e^{\alpha z^*} e^{-\alpha^* z} d^2 z; \tag{21a}$$

$$X_A(\alpha) = \left\langle e^{-\alpha^* a} e^{\alpha a^+} \right\rangle = \int \phi_A(z) e^{\alpha z^*} e^{-\alpha^* z} d^2 z; \tag{21b}$$

$$X_W(\alpha) = \left\langle e^{\alpha a^+ - \alpha^* a} \right\rangle = \int \phi_W(z) e^{\alpha z^*} e^{-\alpha^* z} d^2 z. \tag{21c}$$

Hear $\phi_N(z)$, $\phi_A(z)$ and $\phi_W(z)$ are the phase space distribution functions for the normal, anti-normal and Weyl's ordering rules respectively. The distribution $\phi_N(z)$ is essentially the diagonal coherent state representation $\phi(z)$, whereas $\phi_A(z)$, is given by

$$\phi_A(z) = (1/\pi)\langle z|\rho|z\rangle , \tag{22}$$

$$= \frac{1}{\pi}\int \varphi(z') e^{-|z-z'|^2} d^2 z' . \tag{23}$$

It may be readily seen that the characteristic functions X_N, X_A and X_W, are related as follows:

$$X_N(\alpha) = e^{\frac{1}{2}|\alpha|^2} X_W(\alpha) = e^{|\alpha|^2} X_A(\alpha) . \tag{24}$$

Since $\phi_A(z)$ is well behaved bounded positive real function, its Fourier transform X_A is also well behaved and bounded function. We thus find that $\phi(z)$ is the Fourier transform of $X_N(\alpha) = e^{|\alpha|^2} X_A(\alpha)$, and consequently it can be regarded as the limit of a sequence of tempered distributions[6]. In this sense it is a well defined generalized function. It is of interest to mention an alternate expression for $\phi(z)$. One readily obtains from Eq. (16) the relation

$$\langle -\alpha|\rho|\alpha\rangle e^{|\alpha|^2} = \int \phi(z)\exp(-|z|^2 - \alpha^* z + \alpha z^*)\, d^2 z, \qquad (25)$$

which on taking the inverse Fourier transform gives[7]

$$\phi(z) = \frac{\exp(|z|^2)}{\pi^2}\int \langle -\alpha \mid \rho \mid \alpha\rangle \exp(|\alpha|^2 + \alpha^* z - \alpha z^*) d^2\alpha. \qquad (26)$$

This form of $\phi(z)$ is particularly useful to obtain a well behaved (but equivalent) expression for it whenever possible. As an illustration, consider the pure coherent state $|\beta\rangle$ for which the density operator is given by $\rho = |\beta\rangle\langle\beta|$. Making use of Eq. (26) we readily obtain the expression $\phi(z) = \delta(z-\beta)$ for the diagonal coherent state representation for this density operator. It is possible to write a similar expression for the phase space distribution function $\phi_W(z)$ using coherent states[8]:

$$\varphi_W(z) = \frac{2\exp(2|z|^2)}{\pi^2}\int \langle -\alpha|\rho|\alpha\rangle \exp[2(\alpha^* z - \alpha z^*)] d^2\alpha \qquad (27)$$

We observed that coherent states form an over complete set of basis states. It is possible that even a subset of coherent states may suffice to form a complete set. We illustrate this property by observing that the number state $|n\rangle$ may be written in a form[9]

$$|n\rangle = \frac{\sqrt{n!e}}{2\pi}\int_0^{2\pi} |e^{i\theta}\rangle e^{-in\theta} d\theta. \qquad (28)$$

This completeness property may also be expressed in terms of the resolution of the identity

$$\frac{e}{2\pi}(a^+ a)! \int_0^{2\pi} |e^{i\theta}\rangle\langle e^{i\theta}| d\theta = 1. \qquad (29)$$

Relations (28) and (29) clearly demonstrates that even a very restricted sub-set of the coherent states, namely, those on the unit circle ($z = e^{i\theta}$) alone form a complete set. This set is in fact just complete and no more over complete. An arbitrary coherent state $|z\rangle$ may uniquely be expressed in terms of this sub-set as

$$|z\rangle = \frac{\sqrt{e}}{2\pi} e^{-\frac{1}{2}|z|^2} \int \frac{|e^{i\theta'}\rangle}{1 - z e^{-i\theta'}} d\theta'. \qquad (30)$$

It is believed that the set of coherent states on any closed path in a complex plane would form a complete set.

It is felt necessary to comment on the Glauber's work on the so called 'P representation'.

In a later publication[10] Glauber discusses in much detail the representation of the density operator in terms of coherent states. In section 6 of this paper he develops the representation given by [cf. Eq. (15) above]

$$\rho = \int \aleph(\alpha,\alpha')\,|\alpha\rangle\langle\alpha'|\,d^2\alpha\, d^2\alpha' . \tag{31}$$

In section 7, he points out that not all fields require for their description density operators of quite so general a form. He mentions of a '*broad class of radiation fields*' for which it becomes possible to '*reduce*' the density operator to a considerably simpler form which he calls the P representation [Eq. (7.6) of ref. 10]:

$$\rho = \int P(\alpha)|\alpha\rangle\langle\alpha|\,d^2\alpha. \tag{32}$$

If a density operator is specified by means of the P representation, its matrix elements connecting the n-quantum states are given by [Eq. (7.12) of ref. 10]

$$< n\,|\,\rho\,|\,m > = (n!m!)^{-1/2}\int P(\alpha)\alpha^n(\alpha^*)^m e^{-|\alpha|^2} d^2\alpha. \tag{33}$$

However he never gives any method of inverting Eq. (32) even for such special fields which permit his P representation, nor does he mention any way of '*reducing*' his (α,α') representation given by Eq. (31) to that of the 'P representation'. It is thus clear from his discussion that he acknowledges the usefulness of his P representation, but fails to derive an explicit expression for it for any given density matrix. Yet in a discussion of Eq. (32), Glauber states: "In general, however, it is not possible to interpret the function P(α) as a probability distribution in any precise way since the projection operators $|\alpha\rangle\langle\alpha|$ with which it is associated are not orthogonal to one another for different values of α." He thus contradicts himself about the usefulness of P(α) even for cases when it is a well behaved positive definite function.

I wish to emphasize that in ref. 10, Glauber discusses the inversion of his expression [Eq. (5.6) of ref. 10], viz.,

$$T = \frac{1}{\pi^2}\int|\alpha\rangle\Im(\alpha^*,\beta)\langle\beta|\exp\{-\frac{1}{2}|\alpha|^2 - \frac{1}{2}|\beta|^2\}\,d^2\alpha\, d^2\beta \tag{34}$$

to [Eq. (5.7) of ref. 10]

$$\Im(\alpha^*,\beta) = \langle\alpha|T|\beta\rangle\exp\{\frac{1}{2}|\alpha|^2 + \frac{1}{2}|\beta|^2\}, \tag{35}$$

but he never discusses the inversion of his P representation [Eq. (7.6) of ref. 10]

$$\rho = \int P(\alpha)|\alpha\rangle\langle\alpha|\,d^2\alpha\,, \tag{36}$$

even for the simple case of a pure coherent state. He never derives this expression. There is hardly any justification to say that in special cases his representation (31) '*reduces*' to the P representation (36), nor is there any reason whatsoever to dignify it as a '*representation*' at all. The only source from which Glauber gets his so called *'P-representation'* is Sudarshan's published work which he copies, changing ϕ to P and z to α. Despite this, his confusion about the representation is clearly evident between sections 6 and 7 of his Phys. Rev. paper (ref. 10). It is thus incontrovertible that the credit for formulating and discovering the diagonal coherent state representation must go to Sudarshan. It is truly ironic that although these facts are readily accessible, an expression which ought to be called '*Sudarshan's diagonal coherent state representation*' is dubbed '*Glauber's P representation*'. Nor is it correct even as compromise move , to call it the 'Glauber-Sudarshan representation'

Sudarshan's application of the over completeness of coherent states led to a unique expression for $\phi(z)$, the diagonal coherent state representation of the density operator. As such his work is the first and the only formulation of quantum coherence theory.

I will next give a couple of specific examples where the diagonal coherent state representation takes on negative values for some complex z and such a situation corresponds to a true quantum feature of the radiation field. From the Mandel's Poisson transform relation [Eqs. (2a) and (20)] one readily obtains the relation between the fluctuations of number of photons and that of the integrated intensity:

$$\left\langle \Delta n^2 \right\rangle = \left\langle n \right\rangle + \left\langle \Delta W^2 \right\rangle; \tag{37a}$$

$$= \left\langle n \right\rangle + \left\langle \Delta(\left|z\right|^2)^2 \right\rangle. \tag{37b}$$

It is evident that for positive $\phi(z)$ the second term on the right hand side of Eq.(37) is positive and we expect bunching of photons [$\left\langle \Delta n^2 \right\rangle \geq \left\langle n \right\rangle$] for such fields. This for example is the case for blackbody radiation for which $\left\langle \Delta n^2 \right\rangle = \left\langle n \right\rangle + \left\langle n \right\rangle^2$. On the other hand if $\phi(z)$ was negative for some values of z, we *may* expect the second term on the right hand side of Eq. (37) to be negative. In such cases there will be anti-bunching of photons [$\left\langle \Delta n^2 \right\rangle \leq \left\langle n \right\rangle$] and that is a true quantum feature. Such is a case for the number state where there are fixed number of photons and hence $<\Delta n^2>$ is zero. The second term on the right hand side of (37) is $-<n>$. Of course a negative $\phi(z)$ for some values of z will not *always* lead to anti-bunching of photons, but in such a case one should expect other quantum features associated with higher order correlations. A squeezed state (such as the squeezed vacuum) is another example of negative $\phi(z)$ and hence the case of a state having no classical analogue.

Finally I will consider briefly the relationship between operator ordering and coherent state representations. Very often one is interested in writing a given operator in a well-ordered form such as a normal, anti-normal or Weyl (completely symmetric) ordered form. Such ordering plays an important role in phase space representation of quantum mechanics, quantum c-number correspondence etc. In this context we note the following relations which may readily be verified. If $:f_N(a,a^+):$ is the normally ordered form of $F(a,a^+)$ then

$$\left\langle z \,|\, F(a,a^+) \,|\, z \right\rangle = f_N(z,z^*) \tag{38}$$

Similarly if "$f_A(a,a^+)$" is the anti-normal ordered form of $F(a,a^+)$, then

$$F(a,a^+) = \frac{1}{\pi} \int f_A(z,z^*) \left|z\right\rangle\left\langle z\right| d^2 z\,. \tag{39}$$

These relations may effectively be used for operator ordering. As an illustration, let us assume that we want to write $\exp(-\lambda a^+ a)$ in a normally ordered form. We note that

$$\begin{aligned}\langle z|\exp(-\lambda a^+ a)|z\rangle &= \sum \langle z|\exp(-\lambda a^+ a)|n\rangle\langle n|z\rangle \\ &= \sum_n e^{-\lambda n}|\langle z|n\rangle|^2 = \exp(-|z|^2 + |z|^2 e^{-\lambda}) .\end{aligned} \quad (40)$$

Hence on making use of Eq. (38) we obtain the required normal ordered form:

$$\begin{aligned}\exp(-\lambda\, a^+ a) &=: \exp[-a^+ a(1-e^{-\lambda})]: \\ &= \sum \frac{[-(1-e^{-\lambda})]^n}{n!}(a^+)^n a^n .\end{aligned} \quad (41)$$

Alternatively if we know the anti-normal ordered form of $\exp(-\lambda a^+ a)$, viz.,

$$\begin{aligned}\exp(-\lambda a^+ a) &= "\exp[-a^+ a(e^{-\lambda} - 1)]" \\ &= \sum \frac{[-(e^{-\lambda} - 1)]^n}{n!} a^n (a^+)^n ,\end{aligned} \quad (42)$$

we may immediately write down, using Eq. (39), the diagonal coherent state representation for this operator:

$$\exp(-\lambda a^+ a) = \frac{1}{\pi}\int \exp[-|z|^2 (e^{-\lambda} - 1)]|z\rangle\langle z| d^2 z . \quad (43)$$

One may use these techniques to obtain desired ordered forms of other operators in general.

References:

[1] For a review, see Born M and Wolf E 1959 Principles of Optics (Pergamon Press, New York), Chapter X
[2] Mandel L 1958 Proc. Phys. Soc. (London), **72**, 1037, also 1959 **74**, 233
[3] Glauber R J 1963 Phys. Rev. Letters, **10**, 84
[4] Schroedinger E 1962 Naturwissenschaften, 14, 664
[5] Sudarshan E C G 1963 Phys. Rev. Lettes, **10**, 277
[6] Mehta C L and Sudarshan E C G 1965 Phys. Rev., **138**, B 274
[7] Mehta C L 1967 Phys. Rev. Letters, **18**, 752
[8] Agarwal G S and Wolf E 1970 Phys. Rev., D2, 2161
[9] Mehta C L and Sudarshan E C G, to be published
[10] Glauber R J 1963 Phys. Rev., **131**, 2766

Neoclassical Light – An Assessment of the Voyage into Hilbert Space

H. Jeff Kimble

Department of Physics, California Institute of Technology, Pasadena, CA

Email: hjkimble@caltech.edu

1. Introduction

I thank Rodger Walser and his colleagues for organizing this conference to celebrate George Sudarshan's birthday and landmark achievements in science. It is an honor for me to be here and also a great pleasure to be back in Austin to join the chorus of the FOGS.

I would like begin with a brief historical introduction and then turn to topics that we haven't heard so much about thus far in the symposium, namely experiments in Quantum Optics. Let me remind you that blackbody radiation has had a profound impact on physics since the time of Planck. However, even in more modern times after the development of quantum mechanics and quantum electrodynamics, one really did not know how to describe the underlying field fluctuations as required for diverse interference phenomena. Emil Wolf laid critical foundations for this enterprise with his theory of classical coherence developed in collaboration with Max Born in their classic book *Principles of Optics*.

A great challenge to the physics community was the experiment by R. H. Brown and R. Q. Twiss in 1956 [1], which led to a confusing and controversial time for photons and fields. In 1958 Leonard Mandel helped to resolve the controversy by explaining how light fluctuations are converted to photocurrents in the detection process, albeit in with a classical description of the field [2]. The confusion arose from the effort to understand, on the one hand, interference phenomena that are expressed in terms of the complex amplitude of the electromagnetic field, for which only a classical description was available. On the other hand, quantum electrodynamics had been developed based upon energy eigenstates (i.e., Fock states) with photons as quanta. It was really difficult to understand how the electric and magnetic fields behave in terms of these quantum states. For example, the expectation value of any electromagnetic field is zero in a Fock state. The confusion was compounded by the historic invention of the maser and laser and by the pressing need for quantum theories of the dynamical processes of these and other laboratory advances.

2. The Optical Equivalence Theorem

And then in 1963 the fog lifted. Two landmark papers were published in Physical Review Letters, one by George [3] and one by Roy Glauber [4]. Within a few months, we suddenly had Nature's language for the quantum theory of coherence, including for generation, propagation, and detection. Certainly, not all phenomena were understood, but Refs. [3, 4] and publications that followed by Glauber and Sudarshan provided a new foundation for the formulation of quantum theories of the interaction of light and matter.

An extremely important aspect of these developments was the Optical Equivalence Theorem, first stated by George in Ref. [3] and elaborated in his essential book with John Klauder [5]. Over the five decades since then, the Optical Equivalence Theorem has provided the central tool in Quantum Optics for distinguishing between classical and manifestly quantum (or nonclassical) regimes for the electromagnetic field.

As reviewed in the presentation by Professor Mehta, the density operator $\hat{\rho}$ for the electromagnetic field can be expressed in the form [3]

$$\hat{\rho} = \int d^2z\phi(z)\left|z\right\rangle\left\langle z\right|.$$

Here, the function $\phi(z)$ is, in modern terms, called the Glauber-Sudarshan phase-space function [6]. The aspect of this function that George emphasized (and formally established) is its universal character for all states of the electromagnetic field, ranging from blackbody radiation, for which $\phi(z)$ is a Gaussian distribution, to more pathological examples, such as a number (Fock) state $|n\rangle$ of the field, for which

$$\phi(z) = \frac{e^{|z|^2}}{n!}\frac{\partial^{2n}}{\partial z^n \partial z^{*n}}\delta^{(2)}(z).$$

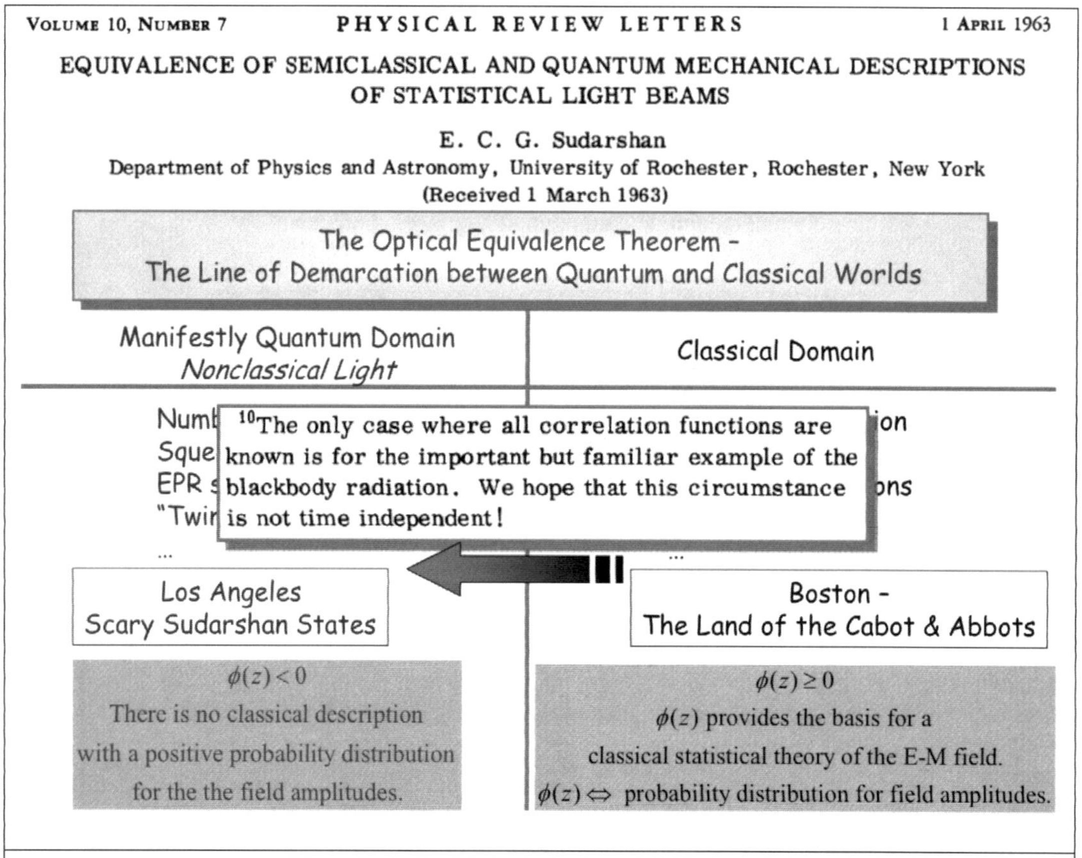

Figure 1. The Optical Equivalence Theorem provides a definitive line of demarcation between classical and quantum worlds [3, 5].

As you can see and as Professor Mehta discussed, this is not an ordinary function. As Dorothy in the Wizard of Oz expressed so appropriately, we're certainly not in Kansas anymore! For an experimentalist, this seems like spooky territory, but in fact, once $\phi(z)$ is clothed by an integral [5], it becomes our best friend.

Indeed as illustrated in Figure 1, the Optical Equivalence Theorem provides a definitive, model independent line of demarcation between the classical and manifestly quantum domains. In this regard, it is critical to know that $\phi(z)$ is everywhere well-defined ("exists") but which becomes negative or more singular than the delta function at a particular boundary. George's Optical Equivalence Theorem provides this foundation. The behavior of $\phi(z)$ leads to operational consequences in the laboratory and has been essential to advances in the creation of nonclassical light.

As expressed qualitatively in Figure 1, the Optical Equivalence Theorem provides a definitive boundary between the classical and quantum domains. But of course, everything is quantum, right? Well, for many phenomena, a manifestly quantum description is not required - such as for the fluctuations of blackbody radiation, radio waves, cell phones, lasers, and so on. It is perfectly fine to treat the field in such cases as a stochastic mixture of appropriate classical electromagnetic fields with $\phi(z)$ serving as the underlying probability. This is what is indicated as the classical domain in the figure.

By contrast, the manifestly quantum or nonclassical domain encompasses the region for which a quantum description is essential and includes number states, squeezed states, photon twins, and even more bizarre states. This is a domain where $\phi(z)$ does not exist in the usual sense of an ordinary function and for which our classical intuition about coherence phenomena fails. The question that has been central to the development of Quantum Optics since the early 1960s is how to move from the classical into the quantum domain depicted in Fig. 1. This is the voyage into Hilbert space that is expressed in the title of my presentation. It is a journey that has taken many decades to travel from the land of classical random processes to manifestly quantum phenomena. Our compass for this journey has been the Optical Equivalence Theorem for providing unambiguous direction into the nonclassical realm.

Certainly, I think that George appreciated much of this in 1963. His paper has the following footnote [3]:

> [10]The only case where all correlation functions are known is for the important but familiar example of the blackbody radiation. We hope that this circumstance is not time independent!

I really like this note, which conveys the sense that light wasn't very interesting in 1963, and that it should be – that we should try to do other, less "familiar" things.

3. The Creation of Nonclassical Light

In the ensuing years after the papers by Glauber and Sudarshan, there were many theoretical papers written in the search for physical processes that could generate quantum effects for light. Some of the theories were quite intimidating, the proposed experiments complicated, and the predicted effects small. However, it turns out there is a really simple way to generate nonclassical light, namely by way of the fluorescence from one atom [7,8]. The idea is to illuminate an atom, collect the scattered field, and direct it to two photodetectors by way of a beam splitter. If a detection event is recorded by one detector, surely a second simultaneous event cannot occur at the other detector since the atom can only scatter photons one by one. More formally, the conditional state of the atom following a detection event is the atomic ground state from which no emission can occur until after a finite interval for re-excitation of the atom.

Simple enough, but it took many years in the 1970s to build this experiment in the laboratory of Professor Leonard Mandel at the University of Rochester. I will not go into the technical problems, which required the marriage of traditional laser spectroscopy with the new field of Quantum Optics.

Results from this effort are described in Ref. [9]. In particular, Figure 2 in Ref. [9] displays the number of coincidences $n(\tau)$ for photoelectric counts as a function of time delay τ between the counts. The tools to record time-resolved detection events from fast photoelectric detectors came from the high energy physics community, with some electronic components borrowed from colleagues at the University of Rochester.

What's interesting about our results in Ref.9? Well, after calibrations and checks and cross-checks, we observed the simple feature $n(0) < n(\tau)$, which is to say that there was a rising slope in coincidences from $\tau = 0$. Anybody who knows Quantum Optics knows that then we're done. Based upon the Optical Equivalence Theorem, if the number of recorded pulse pairs $n(\tau)$ rises from $\tau = 0$, then the field that generated these photoelectric statistics must be nonclassical [6]. This is an absolutely unambiguous signature of a manifestly quantum state of light independent of one's favorite theoretical model.

The experiment in Ref. [9] and subsequent work by Short and Mandel [10] are examples of nonclassical behavior associated with lumps of light - photons. There is another "flavor" of non-classical light that relates more directly to the complex amplitude of the field. Maxwell's equations can be expressed in canonical form to resemble a collection of harmonic oscillators. For each oscillator, the two canonical variables E_1 and E_2 that describe the field are completely analogous to the position x and momentum p for a mechanical oscillator. The qualitative "rules" for states in the phase space are illustrated in Fig. 2 in terms of the Wigner phase-space function $W(x, p)$ for several states of the electromagnetic field.

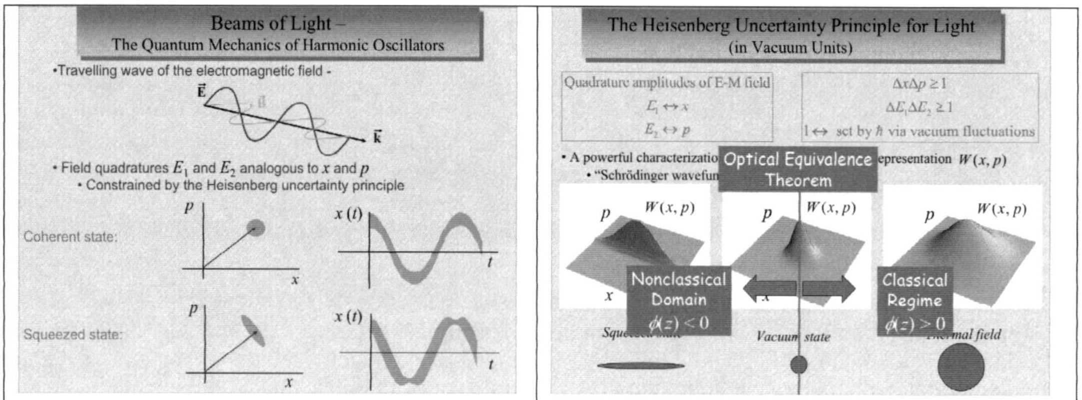

Figure 2. Harmonic oscillators and the uncertainty principle for the electromagnetic field.

The critical question is how to quantify the manifestly quantum character of the fluctuations of E_1, E_2. The Heisenberg uncertainty principle for light is analogous to that for a harmonic oscillator. The uncertainty product for E_1, E_2 in appropriate units is "1" for each oscillator (i.e., each mode of the E-M field), precisely as for (x,p), which I will use in place of (E_1,E_2). For nonclassical states, it is not possible to display the Glauber-Sudarshan phase-space function, which is highly singular. Hence we plot $W(x, p)$ in Fig. 2. The vacuum state has equal fluctuations for x and p set by Planck's constant. A thermal state, which is a particularly important state historically, is a Gaussian just as for the vacuum state, but now with finite average photon number.

Of more interest is a squeezed state, whose nonclassical character follows directly from the Optical Equivalence Theorem [6]. In qualitative terms, a state whose Wigner distribution varies on a scale less than "1" (the dimension of the cells of the quantum phase space) is nonclassical, with the example of a squeezed state shown.

How do we translate these concepts to the laboratory? In 1984 when I was at UT Austin, there was a big race to observe "squeezing," which was won by Richart Slusher and his colleagues at Bell Labs [11]. Subsequently, my group observed squeezing by what has become the world's prototypical system [12]. As before, the key question is: how do we tell if the light coming out of one's favorite black box exhibits quantum squeezing? And again, a model independent criterion is provided by the Optical Equivalence Theorem.

The crux of the matter is to define operationally the line shown in Fig. 2 that sets the level for vacuum fluctuations of light. And believe me, this was a hard part of the experiment in which my group invested considerable effort [Figs. 1 and 3 in Ref. [12]]. Determining this level has nothing to do directly with the physics of squeezing. Rather, it involves great care in understanding whether a particular level of noise current from a specific photodiode and electronic amplifier actually corresponds to the level associated with a vacuum input to a homodyne detector. Different groups accomplished this determination with various levels of "believability". The best that my colleagues and I have done over the years is to define the "1" associated with vacuum fluctuations to an accuracy ~ 0.002.

Once we have that "1" nailed down, we have translated the abstract mathematics of the Optical Equivalence Theorem into a definitive laboratory "meter" for detecting nonclassical light.

Based upon this razor's edge for the determination of squeezing, my group at UT Austin carried out a series of experiments related to squeezed light. Two experiments of which I am particularly proud are (1) the empirical mapping of the Heisenberg Uncertainty Principle for light [12,13] and (2) the first measurement with sensitivity beyond the Standard Quantum Limit (in our case, in optical interferometry) [14].

4. Dual Beam Correlation

The discussion thus far has centered on quantum correlation for a single beam of light, whether in terms of "photons" as for photon antibunching or for fluctuations of the quantum amplitudes as for squeezed light. However, another area of great historic significance relates to the quantum character of correlation between two beams of light. There are a number of nonclassical inequalities that we could discuss, which fall broadly into two classes. The first relates to investigations of Bell inequalities for discrete variables (i.e., polarization correlation for beams of light) that are of profound importance but not so directly related to our story [15]. The second involves correlation in the infinite dimensional Hilbert space of light in the spirit of our previous discussion, about which I'll say more.

An important experiment that is often overlooked was reported by John Clauser in 1974 [16]. As part of his investigations of the inequalities of John Bell, Clauser observed photon correlation for light from an atomic cascade [16], which is roughly the time domain version of the energy correlation studied by Weisskopf and Wigner in 1930 [17]. Clauser's measurements of intensity-intensity correlation for light emitted in the successive steps of an atomic cascade demonstrated a clear violation of a classical Cauchy-Schwarz inequality (roughly, that the magnitude of cross correlation is bounded by the product of autocorrelation for two beams) and hence established the nonclassical character of the beams by way of the Optical Equivalence Theorem.

Leonard Mandel at the University of Rochester was searching for a more powerful means to carry forward his investigations of nonclassical light beyond his early experiments in resonance fluorescence. To circumvent the limitations of atomic radiative processes in free space, in the early 1980s Mandel undertook experiments to utilize parametric down conversion as a source to generate quantum fields. Mandel thus began a landmark set of experiments that continued until his death in 2001 and that opened new vistas into the nature of the quantum world [6,18]. E. Wolf and I summarized the impact of these experiments in 2001 as follows:

> *Beyond what had by then become "the standard model" of manifestly quantum effects, Mandel's genius for probing the quantum realm led to advances throughout the 1990s that challenged and ultimately redefined our understanding of the quantum character of light and, more generally, of quantum measurement.*

There is now a flourishing world community whose work is built upon the foundations laid by Mandel, which in turn, rest upon George's Optical Equivalence Theorem.

Perhaps you will wonder that I have said very little about the "black boxes" employed in these various experiments to generate nonclassical light. It is really important from my perspective that we need not delve into such details to support the case for nonclassical light. Certainly we must understand the characteristics of our detectors [19]. But with this done, the Optical Equivalence Theorem provides an unambiguous means for validating the quantum character of the electromagnetic field for an incredible diversity of experiments.

5. The Paradox of EPR

Another important non-classical phenomenon is related to the well-known paper by Einstein, Podolsky and Rosen [20]. These authors considered the collision of two massive particles that led to entanglement for the positions (x_1, x_2) and momenta (p_1, p_2) for the final state of the particles. Although it has not been possible to realize this *gedanken* experiment with massive particles, there is an optical implementation, as was first analyzed by Margaret Reid in 1989 [21].

The idea is not to collide two particles, but rather two beams of light at a beam splitter. If each incident beam is in a squeezed state and if the relative phase for these two beams is precisely controlled, then the output state is identical to the EPR state in the limit of large squeezing for the inputs. The crux of the EPR paper is that if one measures x_1 then x_2 can be inferred perfectly, and likewise for p_1 and p_2. Hence, the product of uncertainties for the inferences of the variables (x_2, p_2) given measurements of (x_1, p_1) is not constrained by an uncertainty principle with lower bound of "1". What might all this have to do with the Optical Equivalence Theorem? My group at Caltech carried out the initial experiment to realize the EPR paradox in 1992 [22]. We demonstrated that the product of uncertainties for inference of (x_2, p_2) given measurements (x_1, p_1) does indeed lie below the limit set by the uncertainty principle for (x_2, p_2), where the relevant constraint was set by the Optical Equivalence Theorem.

However, it was not until 2000 that Professor Simon [23] and Duan *et al.* [24] were able to show that the line of demarcation set by the Optical Equivalence Theorem also determines the boundary between quantum entanglement and classical correlation for the EPR paradox, including for finite degrees of correlation. I consider this to be quite a remarkable result. Not until the work in Refs. [23,24] did we have the tools to quantify entanglement for states of continuous quantum variables.

6. Angels on a Pinhead?

At this point, one might well wonder "This business about the Optical Equivalence Theorem is all very interesting, but seems somewhat arbitrary in drawing a line of demarcation - so what?" It turns out that there really is an important "so what". Nonclassical states as quantified by the Optical Equivalence Theorem are essential in diverse applications, ranging from the quantum limits to measurement to the teleportation of quantum states.

In fact, in 1987 my group at UT Austin carried out an experiment that first achieved measurement sensitivity beyond the so-called Standard Quantum Limit (SQL) [14]. I will not go into the details except to say that we followed a theoretical proposal by C. Caves [25] and used squeezed states of light to surpass the SQL for optical interferometry. The boundary associated with the SQL was set by the Optical Equivalence Theorem.

Another application of these ideas is the phenomenon called quantum teleportation for transporting quantum states from A to B – Alice to Bob. This protocol was invented by C. Bennett and his colleagues in 1993 [26]. It is a remarkable protocol with diverse applications in Quantum Information Science, including for the realization of quantum networks.

In 1998, my group at Caltech performed an experiment to teleport a beam of light from one side of an optical table to the other [27]. Success in such an experiment is determined by surpassing the limit for the transport of the state without the use of quantum entanglement. Again, it turns out that the demarcation between the classical and quantum domains for teleportation of light beams is based upon the Optical Equivalence Theorem.

7. My First Colloquium

One of my first interactions with George was at a colloquium that I gave at UT Austin in 1979 entitled "Light work with optical forces". George was in the front row (just like he was at the symposium in his honor), and he interrupted me by saying "... that's not right". I remember this very distinctly; I was a new assistant professor just starting my career and "kaboom"! I asked George "why not?" and he said something that I did not understand, but he let me off the hook.

Later, I went to his office to understand his concern. Indeed, the formula that I had showed assumed a classical random walk for an atom's momentum driven by light scattering. In some approximation this is indeed the case, but the basic equations have much more complex and interesting possibilities, including entanglement between atomic momentum steps and the state of the field. George appreciated this in my talk about a subject that he had not previously studied. He encouraged me to go back to the basics and find out where things really come from. This and other interactions with George over the years had a profound impact on me and on the design of my experiments. Although often without success, my goal inspired by George has been always to drill to the core of the matter.

One direction that my research took was that of cavity quantum electrodynamics (cQED). My group and I tried to push not towards conventional lasers and optical processes, but into a new regime to access fundamental radiative processes for single atoms and photons.

It took a couple of decades, but one of the things that we have been able to achieve is the generation of single photons on demand within the setting of cQED [28]. The photons are generated by one atom trapped inside an optical cavity in a deterministic fashion with a Gaussian spatial mode and a "user" controlled wave packet. This was a long voyage from my work in graduate school with photons randomly emitted into 4π.

And how do we know they are single photons? By now the story line is clear; we go back to the Optical Equivalence Theorem, which leads to operational criteria for a nonclassical field.

Further Destinations into Hilbert Space - By now there is a large, worldwide community pursuing experiments in Quantum Optics that rest upon the foundation of the Optical Equivalence Theorem. One example in my group is an experiment that generates the following entangled state [29]:

$$|\Psi_{L,R}\rangle = |0\rangle_L |1\rangle_R \pm e^{i\varphi} |1\rangle_L |0\rangle_R .$$

This state has one quantum of excitation shared collectively between two ensembles of atoms, a left ensemble (*L*) and a right ensemble (*R*) which are separated by 3 meters. Certainly, as written, this is an entangled state, but how do we quantify entanglement in the laboratory? First, we map the purported entangled state of the two material systems (*L*,*R*) to two beams of light (1_L,1_R), and then employ the powerful tools of Quantum Optics to analyze the light. However, for the state $|\Psi_{L,R}\rangle$ and the resulting optical fields (1_L,1_R), the Optical Equivalence Theorem provides a necessary but not sufficient condition to establish entanglement.

To demonstrate entanglement, we must turn to criteria that go beyond the Optical Equivalence Theorem [29,30]. The tool that has served us so well for five decades now proves insufficient for the general task of the classification and quantification of entanglement. However, the Optical Equivalence Theorem does provide the standard for our quest, namely to find model-independent criteria for the classification and understanding of the zoology of states as we journey further into Hilbert space.

Acknowledgements

The work that I have described has been carried out by a remarkable group of young people over the years -- graduate students, postdocs, and visiting scientists, as well undergraduates. We all owe a great debt to Professor Sudarshan for the future that he has so powerfully enabled with his historic contributions to Quantum Optics. We call ourselves DOGS – Disciples of George Sudarshan. And how can we best express our gratitude and honor George's lifetime of landmark contributions? In the best California tradition, all great things are immortalized by a T-shirt, here presented to George with a photograph of two atomic ensembles entangled by his DOGS.

References

[1] Brown R H and Twiss R Q 1956 Nature **177**, 27 (1956)
[2] Mandel L 1958 Proc. Phys. Soc. **72**, 1037
[3] Sudarshan E C G 1963 Phys. Rev. Lett. **10**, 277
[4] Glauber R J 1963 Phys. Rev. Lett. **10**, 84
[5] Klauder J R and Sudarshan E C G 1968 *Fundamentals of Quantum Optics* (W. A. Benjamin, Inc., 1968)
[6] Wolf E and Mandel L 1995 *Optical Coherence and Quantum Optics* (Cambridge University Press)
[7] Kimble H J and Mandel L 1976 Phys. Rev. A **13**, 2123
[8] Carmichael H J and Walls D F 1976 J. Phys. B **9**, L43
[9] Kimble H J, Dagenais M, and Mandel L 1977 Phys. Rev. Lett. **39**, 691
[10] Short R and Mandel L 1983 Phys. Rev. Lett. **51**, 384
[11] Slusher R E *et. al.* 1985 Phys. Rev. Lett. **55**, 2409
[12] Wu L A, Kimble H J, Hall J L, and Wu H 1986 Phys. Rev. Lett. **57**, 2520
[13] Wu L A, Xiao M, and Kimble H J 1987 J. Opt. Soc. B **4**, 1465
[14] Xiao M, Wu L A, and Kimble H J 1987 Phys. Rev. Lett. **59**, 278
[15] Clauser J F and A. Shimony 1978 Reports on Progress in Physics, **41**, 1881
[16] Clauser J F 1974 Phys. Rev. D **9**, 853
[17] Weisskopf V and Wigner E 1930 Z. Phys. **63**, 54 (1930)
[18] Mandel L 1999 Reviews of Modern Physics **71**, S274, 1999 Centenary Issue
[19] Glauber R J 1965 in *Quantum Optics and Electronics* (Les Houches 1964) (Eds.: Cecil de Witt, Blandin A, Cohen-Tannoudji C), Gordon and Breach, New York
[20] Einstein A, Podolsky B, and Rosen N 1935 Phys. Rev. Lett. **47**, 777 (1935)
[21] Reid M D 1989 Phys. Rev. A **40**, 913
[22] Ou Z Y, Pereira S F, Kimble H J, and Peng K C 1992 Phys. Rev. Lett. **68**, 3663
[23] Simon R 2000 Phys. Rev. Lett. **84**, 2726
[24] Duan L M, Giedke G, Cirac J I, and Zoller P 2000 Phys. Rev. Lett. **84**, 2722
[25] Caves C M 1981 Phys. Rev. D **23**, 1693
[26] Bennett C H, *et. al.* 1993 Phys. Rev. Lett. **70**, 1895
[27] Furusawa A, Sorensen J L, Braunstein S L, Fuchs C A, Kimble H J, and Polzik E S 1998 Science **282**, 706
[28] McKeever J, Boca A, Boozer A D, Miller R, Buck J R, Kuzmich A, and Kimble H J 2004 Science **303**, 1992
[29] Chou C W, de Riedmatten H, Felinto D, Polyakov S V, van Enk S J, and Kimble H J 2005 Nature **438**, 828
[30] Wooters W K 1998 Phys. Rev. Lett. **80**, 2245

Sudarshan: Seven Science Quests IOP Publishing
Journal of Physics: Conference Series **196** (2009) 012016 doi:10.1088/1742-6596/196/1/012016

Sudarshan's diagonal representation: the ecstasy and agony of another major discovery in science

R. Simon
The Institute of Mathematical Sciences, Tharamani, Chennai 600 113
E-mail: simon@imsc.res.in

M.D. Srinivas
Centre for Policy Studies, No. 6, Balaiah Avenue, Luz, Mylapore, Chennai 600 004
E-mail: mdsrinivas50@gmail.com

1. Introduction

If the V–A theory represents Professor Sudarshan's mid-century adventure in Particle Physics, the Diagonal Representation or Optical Equivalence Theorem represents his mid-century adventure in Non-classical Optics. Particle physics already existed as a discipline, but the latter discipline was heralded by this very adventure. The diagonal representation too seems to have brought to its discoverer brought lot more agony than ecstasy.

The ecstasy part is best described by quoting from the abstract of Professor Jeff Kimble to this conference: *In 1963 Professor E.C.G. Sudarshan presented the Optical Equivalence Theorem and thereby provided a quantitative, model independent boundary between the classical and manifestly quantum domains of light. More than a decade then passed before nonclassical light first flickered dimly in the laboratory. By now, a zoology of nonclassical sources shine in laboratories around the world, with applications ranging from quantum measurement to quantum computation and communication. The Optical Equivalence Theorem of Sudarshan has been the critical, guiding light in these quests.*

While referring to the Born and Wolf classic, the unusual book of J.W. Simmons and M.J. Guttmann [States, Waves and Photons: A Modern Introduction to Light, Addison-Wesley, 1970] takes the liberty to add the subtitle *The Bible of Non-quantum Optics*, to emphasise the fact that on any issue of optics, which is not *specifically* quantum, this work has some authoritative insight to offer. The point is that *non-quantum* subsumes *classical*. Similarly, *nonclassical* is more specific than *quantum*. The question of when exactly nonclassical optics was born is thus a sharp question with a clear answer. Sudarshan's paper may be called *the corner-stone of non-classical optics*, notwithstanding the fact that this paper did not occupy anything much more than two printed pages. After all, the poems of *Thiruvalluvar* occupied not even two full lines each; yet they remain ranked among the greatest works mankind has ever seen!

The agony part is expressed most effectively by Sudarshan himself, with his characteristic brevity in eloquence: *Give unto Glauber only what is his.* We wanted to understand, perhaps even experience, the nature and magnitude of the agony capsuled in this Biblical quote; and this paper is the result.

The context of the quote is clearly the documents released by the Nobel Foundation to accompany the announcement of Physics Nobel 2005 to Professor Roy Glauber. These documents assert, in unambiguous terms, that all action took place in the year 1963. Since we are not old enough to have been around doing science in 1963 and have the benefit of first hand experience of what happened, our assessment relies much on the writings of Glauber and Sudarshan during that year. Before we set out in detail our impressions and findings, however, we wish to clarify that understanding the agony is our aim, and not criticising the individuals involved or the authors of the Nobel documents. We believe our description below is how posterity will judge 1963.

2. Before Sudarshan

The semiclassical theory of optical coherence and photoelectric detection was formulated by the Rochester School in the fifties and early sixties. The key element emphasised in these formulations was the central role of the *positive frequency* part of the *real* optical field; this is the *analytical signal*. It was Glauber who showed in his seminal paper (PRL 1963) that quantum correlations could be, and had to be, formulated by simply substituting the positive frequency part of the *hermitian* field operators in place of the analytic signals in the corresponding classical expressions, leading to normally ordered expressions. He thus brought the *coherent states*, the eigenstates of these positive frequency operators, into the centre stage. He observed that the density operator of the thermal state is diagonal in the coherent state basis. Finally, Glauber related the photoionization probability of a pair of atoms, atom 1 at t_1 and atom 2 at t_2, to the quantum correlation function of order four in the field operators (order two in the intensity).

It seems Glauber, at this stage, had little reason to even remotely suspect that this diagonal form for thermal states could be part of a more general result or, equivalently, that the coherent state matrix elements of a density operator [his Eq.(3)] are not all independent of each other. In any case, there is nothing in the text of his paper, or in the list of references cited by him, to suggest that Glauber was, at the time of writing this paper of his, aware of the *overcompleteness* of the coherent states or its consequences. And the literature he cites reinforces this impression.

Glauber's PRL paper appeared in print on February 01, 1963. It was received on December 27, 1962.

3. Enter Sudarshan

Sudarshan's paper appeared in the April 01, 1963 issue of PRL. It was received on March 01, 1963.

Referring to Glauber's PRL paper, Sudarshan notes: *statistical states of a quantized electromagnetic field have been considered recently, and a quantum mechanical definition of coherence function of arbitrary order presented.* Sudarshan exhibits little interest in hair-splitting on the fact that Glauber seems to have considered only correlation functions of order four.

Sudarshan refers to the papers of Bargmann, Segal, Klauder, and Schweber on the analytic function representation of operators. None of these had been referred to by Glauber.

Sudarshan draws the reader's attention to the non-orthogonality and over-completeness of the coherent states, and immediately gets to the heart of the central problem: *We can make use of the overcompleteness of the states to represent every density matrix in the "diagonal" form . . .*

To complete the diagonal representation

$$\rho = \int d^2z\, \phi(z)\, |z\rangle\langle z|,$$

he presents the *inversion formula* expressing the phase-space 'density', quasi-probability, or diagonal weight function $\phi(z)$ in terms of the density operator ρ. *Note that the diagonal representation becomes a representation only with the presentation of the inversion formula!*

Sudarshan notes that $\phi(z)$, the 'density' function in the diagonal representation, is real in view of the hermiticity of ρ, *but not necessarily positive definite.* He shows that, in taking expectation values, every operator should be written in the normally-ordered form to go hand in hand with the diagonal representation, so that the quantum-mechanical expression for all correlation functions will have the same form as their classical counterparts, with the quasi-probability $\phi(\{z_k\})$ now playing the role played in the classical case by the true probability density of the classical ensemble. This is the essence of the *Optical Equivalence Theorem.* All nonclassicalities, if any, of a given state ρ are *fully* captured in the departure of the corresponding $\phi(\{z_k\})$ from being a genuine classical probability, Sudarshan notes.

This departure continues to remain the very definition of *nonclassicality* or *quantumness*, and the guiding light for nonclassical optics. Finally, in Eq. (7) Sudarshan shows how the density operator can be determined from the measured correlation functions, if all the correlation functions are known.

Note: At the time when Sudarshan's paper is sent to PRL (March 1, 1963), and in fact even at the time when it appeared in print on April 1, 1963, there seems to be no statement or hint of the diagonal representation as a general theorem from anyone else, even as a preprint.

4. Interlude

Glauber's next paper appeared in the Physical Review of June 15, 1963. It was received on February 11, 1963. It says: *The present paper, which is the first of a series of fundamental papers on optics, is devoted largely to defining the concept of coherence.* Glauber also gives some hint as to what is to be expected in the next instalment of this series. For, after introducing the coherent states in Eqs. (4.9), (4.10) he states: *We shall discuss the properties of such states at length in the paper to follow.* There is absolutely no indication of anything remotely resembling the diagonal representation as a general theorem! The citation is similar to his PRL paper: no Bargmann, Segal, Klauder, or Schweber.

5. Glauber's second Phys. Rev. paper, or Sudarshan's Diagonal representation in Glauber's own words

This paper appeared in the Physical Review of September 15, 1963. It was received on April 29, 1963, four weeks after Sudarshan's paper appeared in print. Though Glauber promised a *series* in his first Phys. Rev. paper, the series seems to have been aborted or abandoned with this one.

Towards the end of the abstract Glauber notes: *A particular form is exhibited for the density operator which makes it possible to carry out many quantum mechanical calculations by methods resembling those of classical theory. This representation presents clear insights into the essential distinction between the quantum and classical descriptions of the field.* An unsuspecting reader can easily get this wrong, and so a warning could help: this is intended to be part Glauber's abstract of his own work, and is not to be misread as his eloquent summary of Sudarshan's work!

Again, while summarizing the contents of the paper, Glauber states: *The application of the formalism to physical problems is begun in section VII, where we introduce a particular form for the density operator which seems especially suited to the treatment of radiation by macroscopic sources.* Glauber makes here an important point, namely that no useful application of the formalism will emerge without this *particular form* or representation. At this point Glauber makes no reference to Sudarshan, thereby tempting the reader to approach Section VII with great reverence and even awe, in anticipation of a genuinely new representation which would form the basis for *all* applications.

Writing Section VII in general, and page 2776 in particular, appears to have been a painful task of great agony, even for Glauber who is one of the all time great expositors. First, Glauber had to avert the risk of this Section creating an impression in the reader's mind that the previous Sections V and VI are, perhaps, just remnants from an earlier draft of Glauber before Sudarshan's paper appeared in print. Thus Glauber begins by presenting his view of why the (unnecessary) discussion in those two Sections had to be gone through: *In the preceding Sections we have demonstrated the generality of the use of coherent states as a basis. Not all fields require for their description density operators of quite so general a form. Indeed, for a broad class of radiation fields which includes, as we shall see, virtually all of those studied in optics, it becomes possible to reduce the density operator to a considerably simpler form. This form is one which brings to light many similarities between quantum electro-dynamical calculations and the corresponding classical ones.* And he goes on to add: *Its use offers deep insights into the reasons why some of the fundamental laws of optics, such as those for the superposition of fields and calculation of the resulting intensities, are the same as in classical theory, even when very few quanta are involved.* This is precisely what Sudarshan had asserted in respect of his diagonal representation, but Glauber does not seem to care, for he has chosen not to cite Sudarshan as yet!

The promised particular form or representation is presented in Eq.(7.6):

$$\rho = \int P(\alpha)\,|\alpha\rangle\langle\alpha|\,d^2\alpha.$$

Glauber anticipates the risk that at this point the reader could wonder thus: But is this not simply Sudarshan's diagonal representation, with ϕ replaced by P and z by α? So this time he adds, with manifest reluctance, a footnote: *The existence of this form for the density operator has also been observed by E.C.G. Sudarshan, Phys. Rev. Lett.* **10**, *277 (1963). His 'note' is discussed briefly at the end of Sec. X.* Note that the phrase used is *observed*, and not even *independently discovered*, whereas the normal way of acknowledging an earlier fundamental discovery forming the very soul of the discipline would have been to say: *discovered earlier by E.C.G. Sudarshan.*

So the curious reader moves on to this 'brief discussion' in Section X. Indeed there is a second and last reference to the work of Sudarshan: *During the completion of the present paper a note by Sudarshan has appeared which deals with some of the problems of photon statistics that have been treated here. Sudarshan has observed the existence of what we have called the P representation of the density operator and has stated its connection with the representation based on the n-quantum states. To that extent, his work agrees with ours in Section VII and IX.*

The above way of citing an earlier published paper, and that too the single most important discovery in the field, is extraordinary, to say the least. Note that the same phrase 'observed' strikes again. How does the poor reader know that Glauber expects 'observed' to be read as 'formulated and proved earlier', and 'stated' to be read as 'established'? Note also that it is not 'To that extent, our work of Section VII and IX is along the same lines as implied in Sudarshan's work', but it is 'To that extent, his work agrees with ours in Section VII and IX". How does Glauber know, with certainty, that Sudarshan's *gift of prophesy* was so strong as to ensure that what Sudarshan writes in February would exactly agree with what Glauber will choose to claim and publish a few months later? Finally, since Sudarshan had observed 'the existence of' such an all important representation much before Glauber, even by Glauber's own admission, how did it become lawful for Glauber to name Sudarshan's baby?

Back to p.2776. Glauber goes on to specify the class of states for which 'his' P-representation exists or, more precisely, states he would be willing to *allow*. On this one he is categorical: *The function $P(\alpha)$ need not be subject to any regularity conditions, but its singularities must be integrable ones. It is convenient to allow $P(\alpha)$ to have delta function singularities so that we may think of a pure coherent state as represented by a special case of equation (7.3).* There is no question of Glauber allowing $P(\alpha)$ to assume negative values, for he says: *It is the kind of* *operator we might be naturally led to if we were given knowledge that the oscillator is in a coherent state, but one which corresponds to an unknown eigenvalue* α. That is, only *incoherent mixtures* or *convex sums* of coherent states are acceptable for 'his' P-representation. [One wonders if Glauber was really aware of the fact that these states will have nothing nonclassical in them. Probably not, as our next consideration shows.]

This lands Glauber on his next difficulty: All the states Glauber has just allowed for 'his' representation are already classical according to Sudarshan's diagonal representation theorem. Being apparently unaware of this fact, and with much zeal, Glauber proceeds to propose towards the end of the page a restricted or fine-tuned definition of classicality/non-classicality divide; this one is original and independent of Sudarshan. He alerts the reader to his observation earlier in the paper that coherent states $|\alpha\rangle$ and $|\alpha'\rangle$ are approximately orthogonal to one another only when $|\alpha - \alpha'| >> 1$, and goes on to add: *When the function* $P(\alpha)$ *tends to vary little over such large ranges of the parameter* α*, the non-orthogonality of the coherent states will make little difference, and* $P(\alpha)$ *will then be interpretable approximately as a probability density. The functions* $P(\alpha)$ *which vary this slowly will, in general, be associated with strong fields, ones which may be described approximately in classical terms.* Glauber does not seem to care that this class which he is willing to allow is nearly empty. Nor does he seem to care that if this decree of his is accepted, even the mixture of a discrete set of coherent states will fail to be classical! In any case, we are not aware of anyone who has taken Glauber's attempt to modify Sudarshan's definition of the classicality/nonclassicality divide seriously.

As for his final difficulty in respect of writing this page, Glauber should have been acutely aware of the fact that he has no explicit expression of his own for 'his' P-representation. He even goes to the extent of omitting Sudarshan's formula for the diagonal representation density function, even if it meant that in his entire paper there will be no explicit formula for the function $P(\alpha)$ that plays such a crucial and central role in his theory.

To a beginning student it may appear that Eq.(7.6) is 'derived' through Eq.(7.3). This is not true, appearance notwithstanding; indeed, the latter is just the coherent state matrix element of the former.

Glauber knows very well that no integral representation is defined until the inversion formula is established. That is why in the case of his $R(\alpha^*, \beta)$ representation wherein he knows the inversion formula he presents it not just once, but four times: Eqs.(5.4), (5.7), (6.1), and (6.2). That he is not able to define his P-representation in the first place, however, does not seem to deter him from referring in Section IX to some 'generalized P distribution', whatever that might mean.

We have always admired Glauber for the clarity of his writings. Thus, we wonder how the half-hearted and self-contradicting incoherent hotchpotch discussions in the second half of this paper, dealing with the diagonal representation, or the P-representation as Glauber would like to name it, could ever have flown out of Glauber's own pen.

There is little evidence to suggest that Glauber had appreciated in any significant manner the depth, breadth, or universality of Sudarshan's fundamental theorem. His reluctance to credit Sudarshan fully for the diagonal representation, the soul of nonclassical optics is obvious; may be he was not impressed by the length of Sudarshan's 'note'. It is also clear that this representation cannot be perturbatively assimilated into Glauber's scheme of things, with much conviction.

It should be clear that there was and is only one diagonal representation, and Sudarshan had discovered it in its totality. Glauber's apparent position that he had a P representation whose details he will not divulge in verifiable terms is bound to be a sure recipe for agony.

6. The Nobel Documents

As noted earlier, the Nobel documents form the immediate provocation for the Biblical 'Give unto Glauber only what is his'. This part of the agony turns out to be much easier to understand.

Advanced information on the Nobel Prize in Physics 2005 makes several statements which can prove potentially misleading, but we have space for only a few. We begin by quoting from page 4: *On the other hand, there lingered an impression that quantum noise only supplied ripples on the field amplitudes of the classical fields. Thus random function theory would account for the observed effects. As an example we quote [8]: "In the conditions under which light fluctuations are usually measured by photoelectric detectors, the semi-classical treatment applies as readily to light of non-thermal origin as to thermal light, and to non-stationary as well as stationary fields." The subsequent experimental progress rapidly proved the short-comings of the semi-classical approach. The correct theory was published by Roy Glauber in 1963 [9], and this has been the basis for all subsequent theoretical considerations.*

To an unsuspecting reader, this quote will seem to clearly identify three phases or time-periods. The first phase is that of die-hard classical or semi-classical physicists of which the authors of [8] are alleged to form a sample. The second is a phase of rapid experimental progress discrediting the alleged die-hards. The third phase is the work of [9], which was to form the final theory for all time to come. And the naive reader will naturally take it that these phases are causally ordered.

Now to the truth: [9] is Glauber's PRL paper, and we are not aware of Glauber explaining or predicting anything non-classical. [8] is by Mandel, Sudarshan, and Wolf (the Rochester group), the parents of optical coherence at all levels, and this paper was submitted in June 1964, long after Glauber's PRL and Phys. Rev. papers of 1963. Theirs was a criticism of Glauber's Nobel winning work in that Glauber had no non-classical aspect on offer, notwithstanding his often repeated anti-classical refrain. Indeed, and as we have seen, Glauber's view of the diagonal P-representation was too restrictive to allow for anything non-classical, even potentially.

Later part of the page goes on to add: *Glauber introduced the concept of quasiprobabilities into Quantum Optics.* But we have already seen that it was Sudarshan who showed that the diagonal weight function exists as a quasiprobability for all states. Glauber in his later work, renamed it as the P-distribution but claimed that it does not exist for all states. As we will see, this is a position he held even during his Nobel lecture. The real agony is that, to the best of our knowledge, Glauber has never stated clearly and in verifiable terms what 'exists' meant in his scheme of things.

Page 5 reads: *E.C.G. Sudarshan [11] drew attention to the use of coherent state representations for the approach of classical physics.* This is an astonishing way of characterizing Sudarshan's pivotal contribution. Without this work we will be left simply with anti-classical optics with no non-classical component. No wonder, Sudarshan was so provoked!

Another seriously, but more subtly, flawed statement in this document is the one concerning Professor Mandel: *L. Mandel has used the theory to design many ingenious experiments illuminating the quantum nature of light signals.* While seemingly very appreciative of Mandel's contributions, this statement it terribly patronizing and unjust to the memory of one of the greatest physicists of the second half of the twentieth century. In fact, Mandel was one of the main contributors to the development of coherence theory at all levels: Classical, Semi-classical, and Quantum. It was not Glauber who developed coherence theory to arrive at predictions which could be put to experimental test. It was mainly Mandel who developed quantum coherence theory into a testable theory. Were he alive, it would have been impossible to overlook his contributions while awarding a Nobel prize for the development of quantum theory of coherence.

7. The Nobel Lecture

Commenting on *Information Theory* which came into being, all in one stroke, through the remarkable 1948 work of Shannon, A. I. Khinchin writes in his Mathematical Foundations of Information Theory (Dover, 1957): *Rarely does it happen in mathematics that a new discipline achieves the character of a mature and developed scientific theory in the first investigation devoted to it. Such in its time was the case with the theory of integral equations, after the fundamental work of Fredholm; so it was with information theory after the work of Shannon.*

The Nobel documents had upped the ante by painting as if a similar thing happened with quantum optics in 1963, with Glauber doing a Shannon. Since this was far from being true, the world was keenly looking forward to his Nobel lecture and to his unmatched eloquence to see how he will bridge the gap between these documents and reality. But the lecture proved to be a disappointing anticlimax in this regard. Like floor 13, there is no mention of 1963 in his entire lecture, except for the, by now familiar, negative reference to Sudarshan's work. None of Glauber's own three papers of 1963 is cited. Rather, 1951 paper of his, which he himself did not cite in his first two papers of 1963, is cited! Regarding the diagonal weight or 'his' P-distribution he says: *It is a member, as we shall see, of a broader class of quasiprobability densities.* But we could not find anything like that in the rest of the lecture, and so his quasiprobability remains to be defined. [English not being our native language, we seem to have misunderstood 'later' to mean 'later in this lecture']. Then he goes on to add: *The P-representation, unfortunately, is not always available.* And this is where he cites Sudarshan! This is astonishing even by the standards Glauber had heralded in his second Phys. Rev. paper.

8. Final Remarks

Having traced in some detail the source of the agony, it seems appropriate to raise some questions with colleagues in the quantum optics and quantum information community. In raising these questions we refuse to be discouraged by the proverb in *Tamil*, our mother tongue, which liberally translated reads:*you can wake up the one who is sleeping, but how will you wake up the one who pretends to be sleeping.*

- How long should an individual be subjected to this kind of agony? And for whose fault?
- If Glauber has indeed introduced an integral representation for the density operator, and if it is different from Sudarshan's diagonal representation, how does one construct this representation for a given density operator?
- If Glauber says his representation does not exist for all density operators, then where does one find the characterization of the division of the set of all density operators into two subsets – the subset for which the representation exists and the one for which it does not?
- Can 'exist' in the sense of Glauber be defined in such a way that we can run the test on a given density operator and conclude 'exists' or 'does not exist', without having to run to Glauber for his impressionistic verdict?
- If Glauber has introduced a quasiprobability distinct from Sudarshan's diagonal weight, where does one find a characterization of the same?
- Who discovered Glauber-Sudarshan representation? When? As far as we could investigate, they never entered into a joint venture. And when Sudarshan discovered his diagonal representation, he discovered it completely, with no room for anyone to get a piece of the cake.

If 40 plus years of waiting and a Nobel will not encourage Professor Glauber to divulge a precise definition of his P-representation and his 'broader class of quasiprobability densities', it should certainly be a matter of agony, not only for Professor Glauber, but also for the entire community.

Journal of Physics: Conference Series **196** (2009) 011010 doi:10.1088/1742-6596/196/1/011010

Introduction to "Quantum Zeno Effect" - the fifth quest

Iwo Bialynicki-Birula

Center for Theoretical Physics, Warsaw, Poland

E-mail: birula@cft.edu.pl

"An unstable particle observed continuously whether it has decayed or not will never be found to decay!" With these words Baidyanath Mishra and George Sudarshan described thirty years ago in their Journal of Mathematical Physics paper a striking phenomenon that they named the Zeno's paradox in quantum theory. This paper started an avalanche. The literature on the Zeno effect grew to hundreds of papers --- both experimental and theoretical but the subject is still alive and well. To answer all the pertinent questions about the Zeno effect we will probably have to wait until the final resolution of the measurement problem in quantum theory.

$$\lim_{N\to\infty}\left(1-\frac{itV}{N}\right)^{N}\to 1$$

E C G Sudarshan at the reception party

Speakers at Quantum Zeno: Sudarshan's fifth quest

Iwo Bialynicki-Birula

Giuseppe Marmo

Saverio Pascazio

Wayne Itano

Charles Chiu

$$\lim_{N\to\infty}\left(1-\frac{itV}{N}\right)^{N}\to 1$$

Quantum Zeno dynamics and quantum Zeno subspaces

Paolo Facchi[1], Giuseppe Marmo[2], Saverio Pascazio[3]

[1] Dipartimento di Matematica, Università di Bari
and Istituto Nazionale di Fisica Nucleare, Sezione di Bari, I-70125 Bari, Italy
[2] Dipartimento di Scienze Fisiche, Università di Napoli "Federico II"
and Istituto Nazionale di Fisica Nucleare, Sezione di Napoli, I-80126 Napoli, Italy
[3] Dipartimento di Fisica, Università di Bari
and Istituto Nazionale di Fisica Nucleare, Sezione di Bari, I-70126 Bari, Italy

E-mail: saverio.pascazio@ba.infn.it

Abstract. A quantum Zeno dynamics can be obtained by means of frequent measurements, frequent unitary kicks or a strong continuous coupling and yields a partition of the total Hilbert space into quantum Zeno subspaces, among which any transition is hindered. We focus on the "continuous" version of the quantum Zeno effect and look at several interesting examples. We first analyze these examples in practical terms, towards applications, then propose a novel experiment.

1. Introduction

The quantum Zeno effect is intertwined with the name of George Sudarshan, who cast the problem in a rigorous mathematical framework and proposed the classical allusion to the sophist philosopher in a seminal article written in collaboration with Baidyanath Misra in 1977 [1]. Somewhat curiously, the most remarkable practical application of the quantum Zeno effect consists in reducing (and eventually suppressing) decoherence and dissipation, which are other issues in which George Sudarshan was a protagonist, by establishing a firm physical and mathematical description that is known nowadays as the Gorini-Kossakowski-Sudarshan-Lindblad equation [2]. The aim of this article is to illustrate the main features of the quantum Zeno effect (QZE), clarify what is a quantum Zeno dynamics, discuss examples and recent experiments and propose applications.

The quantum Zeno effect has an interesting history. It was first understood by von Neumann [3], who proved that any given quantum state ϕ can be "steered" into any other state ψ, by applying a suitable series of measurements. If ϕ and ψ coincide (modulo a phase factor), the evolution yields, in modern language, a quantum Zeno effect. After 35 years (!) Beskow and Nilsson [4] considered a particle in a bubble chamber (thought of as an apparatus that "continuously checks" whether the particle has decayed) and wondered whether this "measurement" mechanism can hinder decay. This interesting intuition was then reconsidered by other authors, both from a physical [5, 6] and more genuinely mathematical perspective [7]. Notice that a rigorous formulation hinges upon difficult mathematical issues [8], most of which are yet unsolved [9].

It was not until 1988 that Cook realized that the quantum Zeno *effect* (not a "paradox" as people tended to regard it) could be *tested* on oscillating (two- or three-level) systems [10]. This proposal deviated from the original framework, that dealt with *bona fide* unstable systems [4, 1], but was nonetheless an interesting and concrete idea, that led to a beautiful experimental test a few years later, performed by Itano and collaborators [11]. The debate that followed [12] motivated novel experimental tests, involving diverse physical systems, such as photon polarization, chiral molecules and ions [13] and very recently Bose-Einstein condensates [14] and cavity QED [15]. New experiments have been proposed with neutron spin [16].

The presence of a short-time quadratic (Zeno) region for a *bona fide* unstable quantum mechanical system (particle tunnelling out of a confining potential) was experimentally confirmed by Raizen and collaborators in 1997 [17]. A few years later, the same group demonstrated the Zeno effect (hindered evolution by frequent measurements) [18]. This experiment also proved the occurrence of the curious inverse (or anti) Zeno effect (IZE) [19, 20], first suggested in 1983 (!), according to which the evolution can be *accelerated* if the measurements are frequent, but not *too* frequent. The IZE has remarkable links with chaos and dynamical localization [21].

The QZE arises as a straightforward consequence of general features of the Schrödinger equation, that yield quadratic behavior of the survival probability at short times [22, 23]. It is usually understood as the hindrance of the dynamics out of the intial state due to frequent von Neumann measurements. Nowadays, in view of possible applications, this picture appears too restrictive in two respects. First, the QZE does not necessarily freeze the dynamics. On the contrary, for frequent projections onto a *multi*dimensional subspace, the system can evolve away from its initial state, although it remains in the "Zeno subspace" defined by the measurement [24]. The resulting constrained evolution is called "quantum Zeno dynamics" [25]. Second, the QZE can be reformulated in terms of a "continuous coupling" [23], without making use of projection operators and non-unitary dynamics, obtaining the same physical effects. We emphasize that the idea of a continuous formulation of the QZE is not new [26, 6], but has appeared in the literature in different contexts and at different times. These two observations will enable us to focus on possible interesting applications and novel experimental proposals.

This article is organized as follows. We start by reviewing some notions related to the (familiar) "pulsed" formulation of the Zeno effect in Sec. 2. We then summarize the celebrated Misra and Sudarshan theorem in Sec. 3. This theorem is generalized in Sec. 4, in order to accommodate nonselective measurements. We briefly introduce the "kicked" version of the QZE in Sec. 5 and the continuous formulation in Sec. 6. The three procedures (pulsed, kicked and continuous QZE) are compared in Sec. 7, where their physical equivalence is discussed. We then look at several elementary but relevant examples in Secs. 8-10. This is a central part of the article, where a new experiment is proposed. We conclude with a few comments in Sec. 11.

2. Preliminaries: the "pulsed" formulation

We start off by giving an elementary introduction to the QZE, in its most familiar formulation, in terms of von Neumann measurements represented by one-dimensional projections. Let H be the total Hamiltonian of a quantum system Q and $|\psi_0\rangle$ its initial state at $t=0$. The survival amplitude and probability in state $|\psi_0\rangle$ at time time t read ($\hbar=1$)

$$\mathcal{A}(t) = \langle\psi_0|e^{-iHt}|\psi_0\rangle, \tag{1}$$

$$p(t) = |\mathcal{A}(t)|^2 = |\langle\psi_0|e^{-iHt}|\psi_0\rangle|^2 \tag{2}$$

and a short-time expansion yields a quadratic behavior

$$p(t) \sim 1-t^2/\tau_Z^2, \tag{3}$$

$$\tau_Z^{-2} \equiv \langle\psi_0|H^2|\psi_0\rangle - \langle\psi_0|H|\psi_0\rangle^2, \tag{4}$$

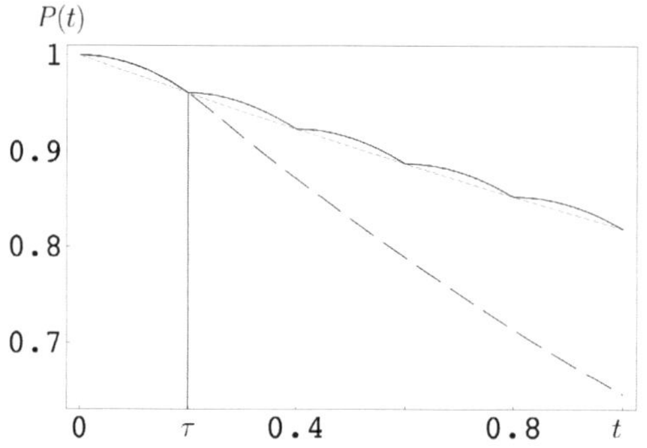

Figure 1. Quantum Zeno effect in the "pulsed" formulation. The dashed (full) line is the survival probability without (with) measurements. The gray line is the interpolating exponential (8). The units on the abscissae are arbitrarily chosen for illustrative purposes.

where τ_Z is the Zeno time [27]. Observe that if the Hamiltonian is divided into a free and an (off-diagonal) interaction parts

$$\begin{aligned} H &= H_0 + H_{\rm int}, \\ H_0|\psi_0\rangle &= \omega_0|\psi_0\rangle, \\ \langle\psi_0|H_{\rm int}|\psi_0\rangle &= 0, \end{aligned} \tag{5}$$

the Zeno time reads

$$\tau_Z^{-2} = \langle\psi_0|H_{\rm int}^2|\psi_0\rangle \tag{6}$$

and depends only on the interaction Hamiltonian.

Let us now show how frequent projective measurements can hinder evolution away from the initial state. Perform N measurements at time intervals $\tau = t/N$, in order to check whether the system is still in its initial state $|\psi_0\rangle$. The survival probability $p^{(N)}(t)$ at time t reads

$$p^{(N)}(t) = p(\tau)^N = p\left(t/N\right)^N \sim \exp\left(-t^2/\tau_Z^2 N\right) \stackrel{N\to\infty}{\longrightarrow} 1. \tag{7}$$

For large N the evolution is slowed down and in the $N \to \infty$ limit the evolution is completely hindered. Notice that the survival probability after N pulsed measurements ($t = N\tau$) is interpolated by an exponential law [20]

$$p^{(N)}(t) = p(\tau)^N = \exp(N \log p(\tau)) = \exp(-\gamma_{\rm eff}(\tau)t), \tag{8}$$

with an effective decay rate

$$\gamma_{\rm eff}(\tau) \equiv -\frac{1}{\tau}\log p(\tau) = -\frac{2}{\tau}\log|\mathcal{A}(\tau)| = -\frac{2}{\tau}\,{\rm Re}\left[\log \mathcal{A}(\tau)\right] \geq 0\ . \tag{9}$$

For $\tau \to 0$ ($N \to \infty$) one gets $p(\tau) \sim \exp(-\tau^2/\tau_Z^2)$, whence

$$\gamma_{\rm eff}(\tau) \sim \tau/\tau_Z^2, \qquad \tau \to 0. \tag{10}$$

The Zeno evolution for "pulsed" measurements is pictorially represented in Figure 1.

3. Misra and Sudarshan's theorem

The formulation of the previous section is very intuitive and does not rest on a firm mathematical ground. The existence of the moments of the Hamiltonian and the convergence of the expansions are all taken for granted and subtle field-theoretical issues related to the Fermi golden rule [28] are not considered [29, 27].

The formulation by Misra and Sudarshan consists in a theorem that requires a few preliminary notions. We first introduce incomplete measurements, represented by multidimensional projections, by applying the von Neumann-Lüders [3, 30] formulation in terms of projection operators and adopting some definitions given by Schwinger [31] (but see also Peres [32]). We will say that a measurement is "incomplete" if some outcomes are lumped together. This happens, for example, if the experimental equipment has insufficient resolution (and in this sense the information on the measured observable is incomplete). The projection operator P, that selects a particular lump, is therefore multidimensional.

Let $\mathcal{H}$ be the Hilbert space of system Q and let the evolution be described by the unitary operator $U(t) = \exp(-iHt)$, where H is a time-independent lower-bounded Hamiltonian. Let P be a projection operator and $\text{Ran}P = \mathcal{H}_P$ its range, with $\dim\mathcal{H}_P = \text{Tr}P = s \leq \infty$ (not necessarily finite dimensional). We assume that the initial density matrix ρ_0 of system Q belongs to $\mathcal{H}_P$:

$$\rho_0 = P\rho_0 P, \qquad \text{Tr}[\rho_0 P] = 1. \tag{11}$$

Under the sole action of the Hamiltonian H (no measurements), the state at time t reads

$$\rho(t) = U(t)\rho_0 U^\dagger(t) \tag{12}$$

and the survival probability (namely the probability that the system is still in $\mathcal{H}_P$) at time t reads

$$p(t) = \text{Tr}\left[U(t)\rho_0 U^\dagger(t)P\right]. \tag{13}$$

No distinction is made between one- and multi-dimensional projections.

The above evolution is "undisturbed:" Q evolves under the sole action of its Hamiltonian for a time t, without undergoing any measurement process. We now perform a *selective* measurement at time τ, in order to check whether Q has "survived" inside $\mathcal{H}_P$. By "selective", we mean that we filter out the survived component and stop the other ones. (Think for instance of decomposing a spin in a Stern-Gerlach setup and absorbing away the unwanted components.) The state of Q changes into

$$\rho_0 \to \rho(\tau) = \frac{PU(\tau)\rho_0 U^\dagger(\tau)P}{p(\tau)}, \tag{14}$$

where

$$p(\tau) = \text{Tr}\left[U(\tau)\rho_0 U^\dagger(\tau)P\right] = \text{Tr}\left[V(\tau)\rho_0 V^\dagger(\tau)\right], \qquad V(\tau) \equiv PU(\tau)P \tag{15}$$

is the survival probability in $\mathcal{H}_P$. The QZE is the following. We prepare Q in the initial state ρ_0 at time 0 and perform a series of (selective) P-observations at time intervals $\tau = t/N$ (by this we mean that Q is found in $\mathcal{H}_P$ at every step). The state at time t reads

$$\rho^{(N)}(t) = \frac{V_N(t)\rho_0 V_N^\dagger(t)}{p^{(N)}(t)}, \qquad V_N(t) = \left[V\left(\frac{t}{N}\right)\right]^N, \tag{16}$$

where

$$p^{(N)}(t) = \text{Tr}\left[V_N(t)\rho_0 V_N^\dagger(t)\right] \tag{17}$$

is the survival probability in $\mathcal{H}_P$. Equations (16)-(17) are the formal statement of the QZE, according to which very frequent observations modify the dynamics of the quantum system: under general conditions, if N is sufficiently large, transitions outside $\mathcal{H}_P$ are inhibited.

The $N \to \infty$ limit requires some technical hypotheses: assume that the strong limit

$$\mathcal{V}(t) \equiv \lim_{N\to\infty} V_N(t) \tag{18}$$

exists $\forall t > 0$. The final state of Q is then

$$\rho(t) = \lim_{N\to\infty} \rho^{(N)}(t) = \mathcal{V}(t)\rho_0\mathcal{V}^\dagger(t) \tag{19}$$

and the survival probability in $\mathcal{H}_P$ is

$$\mathcal{P}(t) \equiv \lim_{N\to\infty} p^{(N)}(t) = \mathrm{Tr}\left[\mathcal{V}(t)\rho_0\mathcal{V}^\dagger(t)\right]. \tag{20}$$

By assuming the strong continuity of $\mathcal{V}(t)$ at $t = 0$

$$\lim_{t\to 0^+} \mathcal{V}(t) = P, \tag{21}$$

Misra and Sudarshan proved that under general conditions the operators $\mathcal{V}(t)$ exist $\forall t \in \mathbb{R}$ and form a semigroup. Moreover, by time-reversal invariance

$$\mathcal{V}^\dagger(t) = \mathcal{V}(-t), \tag{22}$$

one gets $\mathcal{V}^\dagger(t)\mathcal{V}(t) = P$. This implies, by (11), that

$$\mathcal{P}(t) = \mathrm{Tr}\left[\rho_0\mathcal{V}^\dagger(t)\mathcal{V}(t)\right] = \mathrm{Tr}\left[\rho_0 P\right] = 1. \tag{23}$$

If the particle is very frequently observed, in order to check whether it has survived inside $\mathcal{H}_P$, it will never make a transition to $\mathcal{H}_P^\perp$. In general, if N is sufficiently large in (16)-(17), transitions outside $\mathcal{H}_P$ are inhibited. However, if N is not *too* large the system can display an inverse Zeno effect [19, 20], by which decay is accelerated. Both effects have been demonstrated [18]. We will not elaborate on this here.

A few comments are in order. Notice that the dynamics (16)-(17) is not reversible. On the other hand, the dynamics in the $N \to \infty$ limit is often time reversible [24] (although, in general, the operators $\mathcal{V}(t)$ form a *semi*group). Observe also that the Misra and Sudarshan theorem *does not* state that the system *remains* in its initial state, after the series of very frequent measurements. Rather, the system *evolves* in the subspace $\mathcal{H}_P$, instead of evolving "undisturbed" in the total Hilbert space $\mathcal{H}$. The limiting *Zeno dynamics* within $\mathcal{H}_P$ is governed by (18), which can be a *unitary group*. Therefore, starting from the dynamics (16), which is irreversible and probability-nonconserving, one may end up with a fully unitary evolution. Unitarity is recovered in the limit. The features of this evolution in some simple cases will be the object study of the following sections. We anticipate that if $\mathcal{H}_P \subset D(H)$, the domain of the Hamiltonian H, the limiting time evolution has the explicit form

$$\mathcal{V}(t) = P\exp(-iPHPt), \tag{24}$$

namely $\mathcal{V}(t)$ is unitary within $\mathcal{H}_P$ and is generated by the self-adjoint Hamiltonian PHP: reversibility is recovered in the $N \to \infty$ limit. In particular, Eq. (24) is always valid for bounded Hamiltonian, since $D(H) = \mathcal{H}$. In more general cases (infinite dimensional projectors, $s = \dim\mathcal{H}_P = \infty$, unbounded H) one can always formally write the limiting evolution in the form (24), but has to define the meaning of PHP. In such a case one has to study the self-adjointness of the formal limiting Hamiltonian PHP [7, 24, 8, 9].

4. Extension of the Misra and Sudarshan theorem: The quantum Zeno subspaces

We now consider more general measurements and extend the Misra and Sudarshan theorem in order to accommodate multiple projectors. We will say that a measurement is "nonselective" [31] if the measuring apparatus does not "select" the different outcomes, so that all the "branch waves" undergo the whole Zeno dynamics. In other words, a nonselective measurement destroys the phase correlations between different branch waves, provoking the transition from a pure state to a mixture. Let

$$\{P_n\}_n, \qquad P_n P_m = \delta_{mn} P_n, \qquad \sum_n P_n = 1, \tag{25}$$

be a (countable) collection of projection operators and $\mathrm{Ran}P_n = \mathcal{H}_{P_n}$ the relative subspaces. This induces a partition on the total Hilbert space

$$\mathcal{H} = \bigoplus_n \mathcal{H}_{P_n}. \tag{26}$$

Consider the associated nonselective measurement described by the superoperator [3, 30]

$$\hat{P}\rho = \sum_n P_n \rho P_n. \tag{27}$$

The free evolution reads

$$\hat{U}_t \rho_0 = U(t) \rho_0 U^\dagger(t), \qquad U(t) = \exp(-iHt) \tag{28}$$

and the Zeno evolution after N measurements in a time t is governed by the superoperator

$$\hat{V}_t^{(N)} = \hat{P} \left(\hat{U}\left(t/N\right) \hat{P} \right)^{N-1}. \tag{29}$$

This yields the evolution

$$\rho(t) = \hat{V}_t^{(N)} \rho_0 = \sum_{n_1,\ldots,n_N} V_{n_1\ldots n_N}^{(N)}(t)\; \rho_0 \; V_{n_1\ldots n_N}^{(N)\dagger}(t), \tag{30}$$

where

$$V_{n_1\ldots n_N}^{(N)}(t) = P_{n_N} U\left(t/N\right) P_{n_{N-1}} \cdots P_{n_2} U\left(t/N\right) P_{n_1}, \tag{31}$$

which should be compared to Eq. (16). We follow Misra and Sudarshan [1] and assume, as in Sec. 3, the existence of the strong limits ($t \in \mathbb{R}$)

$$\mathcal{V}_n(t) = \lim_{N\to\infty} V_{n\ldots n}^{(N)}(t), \qquad \lim_{t\to 0} \mathcal{V}_n(t) = P_n, \quad \forall n\ . \tag{32}$$

Then $\mathcal{V}_n(t)$ form a semigroup for every n and

$$\mathcal{V}_n^\dagger(t) \mathcal{V}_n(t) = P_n. \tag{33}$$

Moreover, one can show that

$$\lim_{N\to\infty} V_{n\ldots n'\ldots}^{(N)}(t) = 0, \qquad \text{for} \quad n' \neq n, \tag{34}$$

strongly. Notice that, for any *finite* N, the off-diagonal operators (31) are in general nonvanishing, i.e. $V^{(N)}_{n\ldots n'\ldots}(t) \neq 0$ for $n' \neq n$. It is only in the limit (34) that these operators vanish. This is because $U(t/N)$ provokes transitions among different subspaces $\mathcal{H}_{P_n}$. By Eqs. (32)-(34) the final state is

$$\rho(t) = \hat{\mathcal{V}}_t\rho_0 = \sum_n \mathcal{V}_n(t)\rho_0\mathcal{V}_n^\dagger(t), \quad \text{with} \quad \sum_n \mathcal{V}_n^\dagger(t)\mathcal{V}_n(t) = \sum_n P_n = 1. \tag{35}$$

The components $\mathcal{V}_n(t)\rho_0\mathcal{V}_n^\dagger(t)$ make up a block diagonal matrix: the initial density matrix is reduced to a mixture and any interference between different subspaces $\mathcal{H}_{P_n}$ is destroyed (complete decoherence). In conclusion,

$$p_n(t) = \mathrm{Tr}\left[\rho(t)P_n\right] = \mathrm{Tr}\left[\rho_0 P_n\right] = p_n(0), \quad \forall n. \tag{36}$$

Probability is conserved in each subspace and no probability "leakage" between any two subspaces is possible: the total Hilbert space splits into invariant subspaces and the different components of the wave function (or density matrix) evolve independently within each sector. One can think of the total Hilbert space as the shell of a tortoise, each invariant subspace being one of the scales. Motion among different scales is impossible. (See Fig. 2 in the following.)

If $\mathcal{H}_{P_n} \subset D(H)$, then the limiting evolution operator $\mathcal{V}_n(t)$ (32) within the subspace $\mathcal{H}_{P_n}$ has the form (24),

$$\mathcal{V}_n(t) = P_n \exp(-iP_n H P_n t), \tag{37}$$

$P_n H P_n$ is bounded and self-adjoint and $\mathcal{V}_n(t)$ is unitary in $\mathcal{H}_{P_n}$. Precautions are necessary in more general cases, as mentioned after Eq. (24).

The result (23) is reobtained when $p_n(0) = 1$ for some n, in (36): the initial state is then in one of the invariant subspaces and the survival probability in that subspace remains unity. In conclusion, when the necessary care is taken for mathematical rigor and subtleties, the Zeno evolution can be written

$$\rho(t) = \hat{\mathcal{V}}_t\rho_0 = \sum_n \mathcal{V}_n(t)\rho_0\mathcal{V}_n^\dagger(t), \tag{38}$$

$$\mathcal{V}_n(t) = P_n \exp(-iH_{\mathrm{Z}}t) \tag{39}$$

where

$$H_{\mathrm{Z}} = \hat{P}H = \sum_n P_n H P_n \tag{40}$$

is the "Zeno" Hamiltonian.

5. Unitary kicks

The formulation of the preceding section hinges upon projections *à la* von Neumann. Projections are instantaneous processes, yielding the collapse of the wave function (an ultimately nonunitary process). However, one can obtain the QZE without making use of nonunitary evolutions. In this section we further elaborate on this issue, obtaining the QZE by means of a *generic* sequence of frequent *instantaneous unitary* processes. We will only give the main results, as additional details and a complete proof, which is related to von Neumann's ergodic theorem [33], can be found in [34].

Consider the dynamics of a quantum system Q undergoing N "kicks" U_{kick} (instantaneous unitary transformations) in a time interval t

$$U_N(t) = \left[U_{\mathrm{kick}} U\left(\frac{t}{N}\right)\right]^N. \tag{41}$$

In the large N limit, since the dominant contribution is U_{kick}^N, one considers the sequence of unitary operators

$$V_N(t) = U_{\text{kick}}^{\dagger N} U_N(t) = U_{\text{kick}}^{\dagger N} \left[U_{\text{kick}} U\left(\frac{t}{N}\right)\right]^N \tag{42}$$

and its limit

$$\mathcal{U}(t) \equiv \lim_{N\to\infty} V_N(t). \tag{43}$$

One can show that

$$\mathcal{U}(t) = \exp(-iH_Z t), \tag{44}$$

where

$$H_Z = \hat{P}H = \sum_n P_n H P_n \tag{45}$$

is the Zeno Hamiltonian, P_n being the spectral projections of U_{kick}

$$U_{\text{kick}} = \sum_n e^{-i\lambda_n} P_n. \quad (e^{-i\lambda_n} \neq e^{-i\lambda_l}, \text{ for } n \neq l). \tag{46}$$

In conclusion

$$U_N(t) \sim U_{\text{kick}}^N \mathcal{U}(t) = U_{\text{kick}}^N \exp(-iH_Z t) \tag{47}$$

$$= \exp\left(-i\sum_n N\lambda_n P_n + P_n H P_n t\right). \tag{48}$$

The unitary evolution (41) yields therefore a Zeno effect and a partition of the Hilbert space into Zeno subspaces, like in the case of repeated projective measurements discussed in the previous section. The appearance of the Zeno subspaces is a direct consequence of the wildly oscillating phases between different eigenspaces of the kick.

6. Continuous coupling

The formulation of the preceding sections hinges upon instantaneous processes, that can be unitary or nonunitary. However, the main features of the QZE can be obtained by making use of a continuous coupling, when the external system takes a sort of steady "gaze" at the system of interest. The mathematical formulation of this idea is contained in a theorem [25] on the (large-K) dynamical evolution governed by a *generic* Hamiltonian of the type

$$H_K = H + KH_c, \tag{49}$$

which again need not describe a *bona fide* measurement process. H is the Hamiltonian of the quantum system investigated and H_c can be viewed as an "additional" interaction Hamiltonian performing the "measurement." K is a coupling constant.

Consider the time evolution operator

$$U_K(t) = \exp(-iH_K t). \tag{50}$$

In the "infinitely strong coupling" ("infinitely quick detector") limit $K \to \infty$, the dominant contribution is $\exp(-iKH_c t)$. One therefore considers the limiting evolution operator

$$\mathcal{U}(t) = \lim_{K\to\infty} \exp(iKH_c t)\, U_K(t), \tag{51}$$

that can be shown to have the form

$$\mathcal{U}(t) = \exp(-iH_Z t), \tag{52}$$

where

$$H_Z = \hat{P}H = \sum_n P_n H P_n \tag{53}$$

is the Zeno Hamiltonian, P_n being the eigenprojection of H_c belonging to the eigenvalue η_n

$$H_c = \sum_n \eta_n P_n, \qquad (\eta_n \neq \eta_m, \quad \text{for } n \neq m) \ . \tag{54}$$

This is formally identical to (40) and (45). In conclusion, the limiting evolution operator is

$$U_K(t) \sim \exp(-iKH_c t)\mathcal{U}(t) \tag{55}$$

$$= \exp\left(-i\sum_n Kt\eta_n P_n + P_n H P_n t\right), \tag{56}$$

whose block-diagonal structure is explicit. Compare with (48). The above statements can be proved by making use of the adiabatic theorem [35].

7. A few comments on the three formulations

The three different formulations of QZE summarized in the previous sections are physically equivalent and the analogy among them can be pushed very far [34].

After all, a projection *à la* von Neumann is a handy way to "summarize" the complicated physical processes that take place during a quantum measurement. A measurement process is performed by an external (macroscopic) apparatus and involves dissipative effects (an interaction and an exchange of energy with and often a flow of probability towards the environment). The external system performing the observation need not be a *bona fide* detection system, namely a system that "clicks" or is endowed with a pointer. For instance, a spontaneous emission process is often a very effective measurement process, for it is irreversible and leads to an entanglement of the state of the system (the emitting atom or molecule) with the state of the apparatus (the electromagnetic field). The von Neumann rules arise when one traces away the photonic state and is left with an incoherent superposition of atomic states. However, it is clear that the main features of the Zeno effects are still present if one formulates the measurement process in more realistic terms, introducing a physical apparatus, a Hamiltonian and a suitable interaction with the system undergoing the measurement. It goes without saying that one can still make use of projection operators, if such a description turns out to be simpler and more economic (Occam's razor).

A comparison of the three Zeno procedures outlined in the three preceding sections suggests that the continuous version might be the most effective strategy against decoherence [36]. We shall therefore look at some simple examples that yield a quantum Zeno subspace by making use of a strong continuous coupling. As already emphasized, the idea that a strong continuous coupling to an external apparatus can yield a Zeno-like behavior has often been proposed in the literature of the last two decades [26, 23]. The novelty in this context is the appearance of effective superselection rules that are *de facto* equivalent to the celebrated "W^3" ones [37], but turn out to be a mere consequence of the Zeno dynamics. This remarkable aspect is reminiscent of some results on "classical" observables [38], the semiclassical limit [39] and quantum measurement theory [40, 31]. The appearance of a superselection rule in the Hilbert space of the system is pictorially represented in Fig. 2. We conclude by observing that the algebra of observables in the Zeno subspaces is in itself an open problem and a topic worth investigating [41].

Figure 2. The Hilbert space of the system: a dynamical superselection rule appears as the coupling K to the apparatus or the frequency N of measurements is increased.

8. Example: three level system

Let us now look at some simple examples, involving only finite-dimensional Hilbert spaces. Let us first consider the experiment schematically shown in Fig. 3. Two levels 1 and 2 are Rabi coupled and frequent pulses rapidly sweep the population from level 2 to level 3. If the decay rate Γ out of level 3 is fast enough, this can be viewed as a very efficient measurement procedure and QZE takes place. This is the scheme adopted by Itano and collaborators in their celebrated experiment [11]. Observe that in Fig. 3 one implicitly assumes $\pi/2$-pulses; on the other hand, if one makes the pulses extremely frequent and accordingly less intense (technically, one rescales the Hamiltonian [34]), one gets a situation similar to that shown in Fig. 4. In this sense, both "pulsed" and "continuous" measurements are physically equivalent in that they both yield QZE. This observation motivated a recent experiment performed at MIT with a BEC [14].

Consider now the scheme shown in Fig. 4, that describes a three-level system undergoing two different Rabi oscillations and coupled to the photon field. Photons drain the population away from level 3 (towards some other level not shown in the figure), so that if one restricts one's attention to the three atomic levels 1, 2 and 3, the effective Hamiltonian in the Markovian approximation reads ($\hbar = 1$)

$$H = \frac{\omega}{2}(|1\rangle\langle 2| + |2\rangle\langle 1|) + \frac{\omega'}{2}(|2\rangle\langle 3| + |3\rangle\langle 2|) - i\frac{\Gamma}{2}|3\rangle\langle 3| = \frac{1}{2}\begin{pmatrix} 0 & \omega & 0 \\ \omega & 0 & \omega' \\ 0 & \omega' & -i\Gamma \end{pmatrix}, \tag{57}$$

where

$$\langle 1| = (1,0,0), \quad \langle 2| = (0,1,0), \quad \langle 3| = (0,0,1). \tag{58}$$

This is the effective Hamiltonian that better describes the MIT experiment [14]. Once a Γ-photon has been emitted, the atom recoils and leaves the condensate. Here and in the following we will oversimplify the analysis in order to focus on the main ideas.

The Hamiltonian (57) is not Hermitian and probability is not conserved (the photon field has been traced out). Let us prepare the system in state $|1\rangle$ and let

$$p_1(t) = |\langle 1|e^{-iHt}|1\rangle|^2 \tag{59}$$

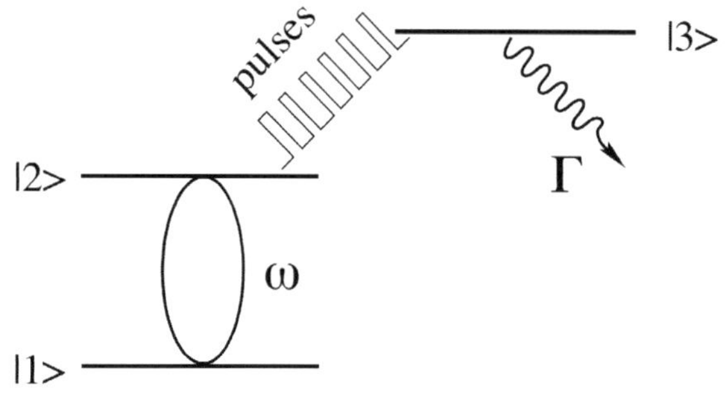

Figure 3. "Pulsed" QZE, as in the experiment by Itano *et al* [11].

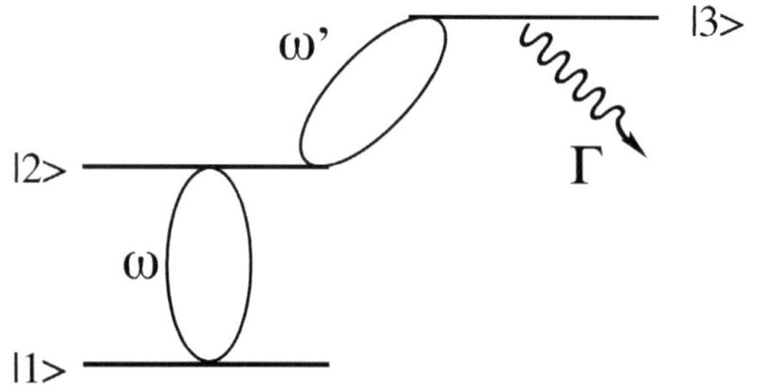

Figure 4. "Continuous" observation, as in the experiment performed at MIT with a BEC [14].

be the survival probability. This is our key quantity, that can be studied in different regimes.

The main idea is that a larger coupling to the photon field (via ω' and Γ) yields a more effective "continuous" observation of the population of level 2 (and therefore of level 1), due to a quicker response of the environment/detection system. As a consequence, the evolution is slower and one obtains QZE. We assume [14]

$$\Gamma, \omega' \gg \omega, \tag{60}$$

which enables one to adopt the effective description of Fig. 5: Consider only the two levels

$$\langle 1| = (1,0), \quad \langle 2| = (0,1), \tag{61}$$

with the effective Hamiltonian

$$H_{\rm eff} = \frac{1}{2}\begin{pmatrix} 0 & \omega \\ \omega & -i\gamma \end{pmatrix} = \frac{\omega}{2}(|1\rangle\langle 2| + |2\rangle\langle 1|) - i\frac{\gamma}{2}|2\rangle\langle 2|, \tag{62}$$

with

$$\gamma = \omega'^2/\Gamma. \tag{63}$$

This yields Rabi oscillations of frequency ω, but at the same time absorbs away the population of level $|2\rangle$, performing in this way a "measurement." Again, probabilities are not conserved.

An elementary calculation [23] yields the survival probability

$$p_1(t) = \left|\langle 1|e^{-iH_{\rm eff}t}|1\rangle\right|^2 \quad = \quad \left|\frac{1}{2}\left(1 + \frac{\gamma}{\sqrt{\gamma^2 - 4\omega^2}}\right) e^{-(\gamma - \sqrt{\gamma^2 - 4\omega^2})t/4}\right.$$

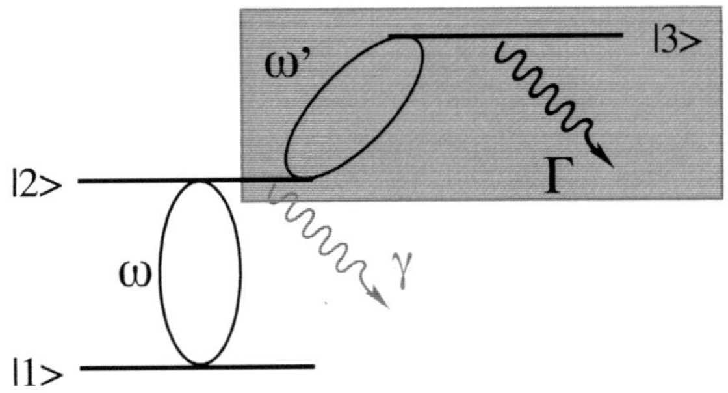

Figure 5. Effective description of the system of Fig. 4 in terms of only two levels: $\gamma = \omega'^2/\Gamma$.

$$+\frac{1}{2}\left(1-\frac{\gamma}{\sqrt{\gamma^2-4\omega^2}}\right)e^{-(\gamma-\sqrt{\gamma^2+4\omega^2})t/4}\Bigg|^2. \tag{64}$$

As expected, probability is (exponentially) absorbed away as $t \to \infty$. When $\gamma \gg \omega$ the above expression reads

$$p_1(t) = \left|\langle 1|e^{-iH_{\rm eff}t}|1\rangle\right|^2 \sim e^{-t\omega^2/\gamma} = e^{-t\omega^2\Gamma/\omega'^2}. \tag{65}$$

As expected, probability is (exponentially) absorbed away as $t \to \infty$. However, as ω' (and hence γ) increases, the effective decay rate ω^2/γ becomes smaller, eventually halting the "decay" (and consequent absorption) of the initial state and yielding an interesting example of QZE: a larger γ entails a more "effective" measurement of the initial state. Additional details on the derivation of the effective decay rate ω^2/γ can be found in [42]. Notice that the expansion (65) is not valid at very short times (where there is a quadratic Zeno region), but becomes valid very quickly, on a time scale of order ω^{-1} (the duration of the Zeno region [23, 27]).

The (non-Hermitian) Hamiltonian (62) can be obtained by considering the evolution engendered by a Hermitian Hamiltonian acting on a larger Hilbert space and then restricting the attention to the subspace spanned by $\{|1\rangle, |2\rangle\}$: consider the Hamiltonian

$$\tilde{H}_\gamma = \Omega(|1\rangle\langle 2| + |2\rangle\langle 1|) + \int dk\; k|k\rangle\langle k| + \sqrt{\frac{\gamma}{2\pi}}\int dk\; (|2\rangle\langle k| + |k\rangle\langle 2|), \tag{66}$$

which describes a two-level system coupled to the photon field $\{|k\rangle\}$ in the rotating-wave approximation. It is not difficult to show [23] that, if only state $|1\rangle$ is initially populated, this Hamiltonian is equivalent to (62), in that they both yield the *same* equations of motion in the subspace spanned by $|1\rangle$ and $|2\rangle$. QZE is obtained by increasing γ: a larger coupling to the environment leads to a more effective continuous observation on the system (quicker response of the apparatus), and as a consequence to a slower decay (QZE). The quantity $1/\gamma$ is the response time of the apparatus.

We now discuss a different regime, yielding a different Zeno picture. Rather than the situation (60), consider

$$\Gamma \ll \omega'. \tag{67}$$

In this regime, the "control" of the evolution, performed by ω', is made with no dissipation/measurement. Let us first discuss in which sense one gets a Zeno effect. Look at the idealized case $\Gamma = 0$ (Fig. 6). The (Hermitian) Hamiltonian reads

$$H_{\rm ideal} = \frac{\omega}{2}(|1\rangle\langle 2| + |2\rangle\langle 1|) + \frac{\omega'}{2}(|2\rangle\langle 3| + |3\rangle\langle 2|) = \frac{1}{2}\begin{pmatrix} 0 & \omega & 0 \\ \omega & 0 & \omega' \\ 0 & \omega' & 0 \end{pmatrix}: \tag{68}$$

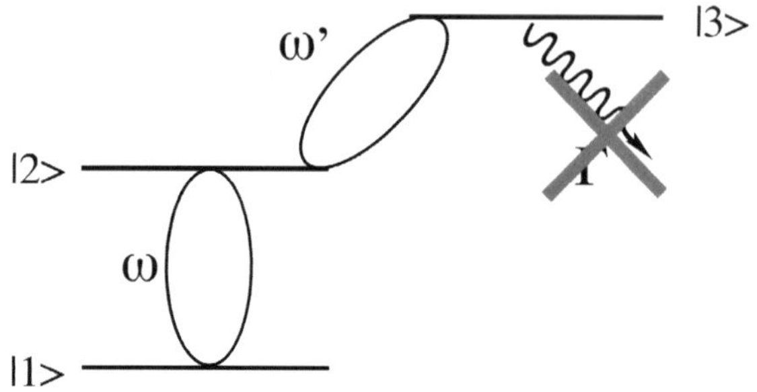

Figure 6. Three stable levels, no dissipation.

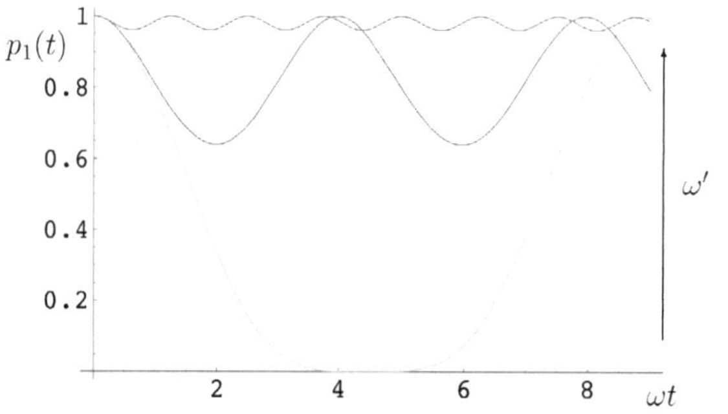

Figure 7. Survival probability for a continuous Rabi coupling with $\omega' = \omega$ (green), $\omega' = 3\,\omega$ (blue), $\omega' = 9\,\omega$ (red).

we shall argue that level 3 plays here the effective role of measuring apparatus, although there is no *bona fide* measurement involved: as soon as the system is in $|2\rangle$ it undergoes Rabi oscillations to $|3\rangle$. We expect level $|3\rangle$ to perform better as a measuring apparatus when the strength ω' of the coupling becomes larger.

The survival probability in the initial state $|1\rangle$ reads

$$p_1(t) = \left|\langle 1|e^{-iH_{\rm ideal}t}|1\rangle\right|^2 = \frac{1}{(\omega'^2+\omega^2)^2}\left[\omega'^2 + \omega^2\cos\left(\sqrt{\omega'^2+\omega^2}t/2\right)\right]^2 \qquad (69)$$

and is shown in Fig. 7 for $\omega' = 1, 3, 9\,\omega$. We notice that for large ω' the state of the system does not change much: as ω' is increased, level $|3\rangle$ performs a better "observation" of the state of the system, hindering transitions from $|1\rangle$ to $|2\rangle$. This can be viewed as a QZE due to a continuous, Hermitian coupling to level $|3\rangle$. This is the situation displayed in Fig. 6.

Note that one can also interpret the above results as an experiment of electromagnetically induced transparency [43]. Indeed, the strong coupling ω' between levels 2 and 3 splits state $|2\rangle$ into two dressed states $|\pm\rangle = (|2\rangle \pm |3\rangle)/\sqrt{2}$ of energy $\pm\omega'$. For large values of ω' these states run out of resonance and the system becomes transparent to the laser resonant with the transition $1 \leftrightarrow 2$.

But what happens for small Γ? An expansion yields

$$p_1(t) \stackrel{\Gamma\ \rm small}{\simeq} \frac{1}{(\omega'^2+\omega^2)^2}\left[\omega'^2 e^{-\Gamma\omega^2 t/2(\omega^2+\omega'^2)} + \omega^2 e^{-\Gamma\omega'^2 t/4(\omega^2+\omega'^2)}\cos(\sqrt{\omega'^2+\omega^2}t/2)\right]^2. \qquad (70)$$

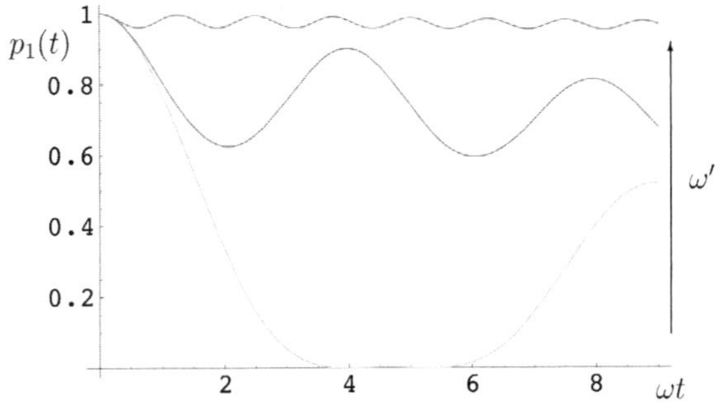

Figure 8. Survival probability for a continuous Rabi coupling with a small dissipation $\Gamma = 0.2\,\omega$ on level #3. Like before, $\omega' = \omega$ (green), $\omega' = 3\,\omega$ (blue), $\omega' = 9\,\omega$ (red).

We stress that this is a very accurate formula (technically, the expansion is uniform), essentially valid even for $\Gamma = O(\omega)$ or $\Gamma = O(\omega')$ (or both!). The results are shown Fig. 8 for $\omega' = 1, 3, 9\,\omega$ and $\Gamma = 0.2\,\omega$.

In Figs. 7 and 8 we used arbitrary values of the parameters for the sake of illustration. It is remarkable how efficient the procedure turns out to be for more physical values of ω, ω', Γ: one can get a quantitative idea by plotting (with care!) Eq. (70).

Clearly, in the very same regime (67), one can use a pulsed technique, as in Fig. 3. This is sometimes called "bang-bang." In fact, it is nothing but pulsed Zeno.

9. Some comments on interpretation

Let us reinterpret the results of the previous section in the light of the theorem proved in Sec. 6. Reconsider Hamiltonian (68) (here ω' plays the role of K in Sec. 6)

$$H_{\text{ideal}} = \frac{1}{2}\begin{pmatrix} 0 & \omega & 0 \\ \omega & 0 & \omega' \\ 0 & \omega' & 0 \end{pmatrix} = H + \omega' H_{\text{c}}, \tag{71}$$

where

$$H = \frac{\omega}{2}(|1\rangle\langle 2| + |2\rangle\langle 1|) = \frac{\omega}{2}\begin{pmatrix} 0 & 1 & 0 \\ 1 & 0 & 0 \\ 0 & 0 & 0 \end{pmatrix}, \tag{72}$$

$$H_{\text{c}} = \frac{1}{2}|2\rangle\langle 3| + |3\rangle\langle 2| = \frac{1}{2}\begin{pmatrix} 0 & 0 & 0 \\ 0 & 0 & 1 \\ 0 & 1 & 0 \end{pmatrix}. \tag{73}$$

As ω' is increased, the Hilbert space is split into three invariant subspaces (the eigenspaces of H_{c}), $\mathcal{H} = \bigoplus \mathcal{H}_{P_n}$, with

$$\mathcal{H}_{P_0} = \text{span}\{|1\rangle\}, \quad \mathcal{H}_{P_1} = \text{span}\{|2\rangle + |3\rangle\}, \quad \mathcal{H}_{P_{-1}} = \text{span}\{|2\rangle - |3\rangle\}, \tag{74}$$

corresponding to the projections

$$P_0 = \begin{pmatrix} 1 & 0 & 0 \\ 0 & 0 & 0 \\ 0 & 0 & 0 \end{pmatrix}, \quad P_1 = \frac{1}{2}\begin{pmatrix} 0 & 0 & 0 \\ 0 & 1 & 1 \\ 0 & 1 & 1 \end{pmatrix}, \quad P_{-1} = \frac{1}{2}\begin{pmatrix} 0 & 0 & 0 \\ 0 & 1 & -1 \\ 0 & -1 & 1 \end{pmatrix}, \tag{75}$$

with eigenvalues $\eta_0 = 0$ and $\eta_{\pm 1} = \pm 1$. The diagonal part of the system Hamiltonian H vanishes, $H_{\rm diag} = \sum P_n H P_n = 0$, and the Zeno evolution is governed by

$$H_Z = \omega' H_c = \frac{1}{2}\begin{pmatrix} 0 & 0 & 0 \\ 0 & 0 & \omega' \\ 0 & \omega' & 0 \end{pmatrix}. \tag{76}$$

Any transition between $|1\rangle$ and $|2\rangle$ is inhibited: a watched pot never boils.

10. A watched cook can freely watch a boiling pot

In the previous discussion we endeavored to make QZE on level 1. Essentially, a Rabi transition from $|1\rangle$ to $|2\rangle$ was inhibited by making use of different procedures. It would be interesting if one could hinder transitions out of a Zeno subspace, say a two-dimensional one, such as a qubit. Let us therefore modify the level scheme, in order to understand whether one can "protect" the evolution of a qubit. Let

$$\langle 0| = (1,0,0,0), \quad \langle 1| = (0,1,0,0), \quad \langle 2| = (0,0,1,0), \quad \langle 3| = (0,0,0,1) \tag{77}$$

and consider the four level Hamiltonian

$$H_{\rm 4lev} = \begin{pmatrix} 0 & \Omega & 0 & 0 \\ \Omega & 0 & \omega & 0 \\ 0 & \omega & 0 & \omega' \\ 0 & 0 & \omega' & 0 \end{pmatrix}, \tag{78}$$

where states $|0\rangle$ and $|1\rangle$ undergo Rabi oscillations,

$$\Omega(|0\rangle\langle 1| + |1\rangle\langle 0|) = \begin{pmatrix} 0 & \Omega & 0 & 0 \\ \Omega & 0 & 0 & 0 \\ 0 & 0 & 0 & 0 \\ 0 & 0 & 0 & 0 \end{pmatrix}, \tag{79}$$

while state $|2\rangle$ "observes" $|1\rangle$

$$\omega(|1\rangle\langle 2| + |2\rangle\langle 1|) = \begin{pmatrix} 0 & 0 & 0 & 0 \\ 0 & 0 & \omega & 0 \\ 0 & \omega & 0 & 0 \\ 0 & 0 & 0 & 0 \end{pmatrix} \tag{80}$$

and state $|3\rangle$ "observes" $|2\rangle$

$$\omega'(|2\rangle\langle 3| + |3\rangle\langle 2|) = \begin{pmatrix} 0 & 0 & 0 & 0 \\ 0 & 0 & 0 & 0 \\ 0 & 0 & 0 & \omega' \\ 0 & 0 & \omega' & 0 \end{pmatrix}. \tag{81}$$

For the sake of simplicity, we assumed no dissipation and redefined the Rabi frequencies in order to get rid of factors 2.

One can amuse oneself by deciding which "observation" is more effective [25, 44]. In the following, it is helpful to think of $|0\rangle$ and $|1\rangle$ as the states of a qubit (Ω being its dynamics), while ω is a caricature of a "decoherence" process and ω' represents the "control" (whose objective is to suppress decoherence). See Fig. 9.

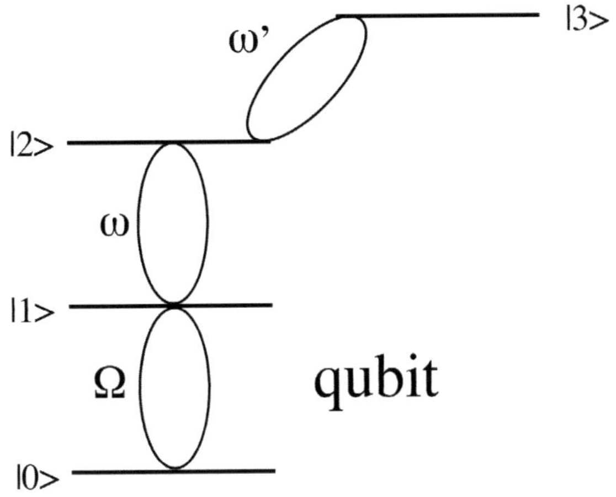

Figure 9. "Control" of the dynamics of a qubit and suppression of "decoherence."

Let us consider the worst situation, where $\omega \gg \Omega$ and $\omega \gg \omega'$. Then one easily shows that the total Hilbert space $\mathcal{H}$ splits into the three "Zeno" subspaces:

$$\mathcal{H}_0 = \mathrm{span}\{|0\rangle, |3\rangle\}, \quad \mathcal{H}_+ = \mathrm{span}\{|1\rangle + |2\rangle\}, \quad \mathcal{H}_- = \mathrm{span}\{|1\rangle - |2\rangle\} \tag{82}$$

and the Zeno evolution is governed by

$$H_{4\mathrm{lev}} = \begin{pmatrix} 0 & \Omega & 0 & 0 \\ \Omega & 0 & \omega & 0 \\ 0 & \omega & 0 & \omega' \\ 0 & 0 & \omega' & 0 \end{pmatrix} \xrightarrow{\omega \text{ large}} H_{\mathrm{Z}}^{(\omega)} = \begin{pmatrix} 0 & 0 & 0 & 0 \\ 0 & 0 & \omega & 0 \\ 0 & \omega & 0 & 0 \\ 0 & 0 & 0 & 0 \end{pmatrix}. \tag{83}$$

The Rabi oscillations between states $|0\rangle$ and $|1\rangle$ are hindered (as well as those between $|2\rangle$ and $|3\rangle$). The dynamics Ω of the qubit $\{|0\rangle, |1\rangle\}$ is destroyed by decoherence ω. This is detrimental.

On the other hand, if $\omega' \gg \omega$, and even if $\omega \gg \Omega$, the total Hilbert space splits into the three subspaces:

$$\mathcal{H}'_0 = \mathrm{span}\{|0\rangle, |1\rangle\}, \quad \mathcal{H}'_+ = \mathrm{span}\{|2\rangle + |3\rangle\}, \quad \mathcal{H}'_- = \mathrm{span}\{|2\rangle - |3\rangle\} \tag{84}$$

[notice the differences with (82)] and the Zeno Hamiltonian reads

$$H_{4\mathrm{lev}} = \begin{pmatrix} 0 & \Omega & 0 & 0 \\ \Omega & 0 & \omega & 0 \\ 0 & \omega & 0 & \omega' \\ 0 & 0 & \omega' & 0 \end{pmatrix} \xrightarrow{\omega' \text{ large}} H_{\mathrm{Z}}^{(\omega')} = \begin{pmatrix} 0 & \Omega & 0 & 0 \\ \Omega & 0 & 0 & 0 \\ 0 & 0 & 0 & \omega' \\ 0 & 0 & \omega' & 0 \end{pmatrix}. \tag{85}$$

The conclusions are interesting and beautiful: the Rabi oscillations between states $|0\rangle$ and $|1\rangle$ are fully *preserved* (*even if and in spite of* $\omega \gg \Omega$). This is the control of a qubit: its dynamics Ω is restored, even in presence of strong decoherence $\omega >$ (or even $\gg$) Ω.

In conclusion, if $\omega' \gg \omega, \Omega$, so that ω'^{-1} is the shortest timescale in the problem, *a watched cook can freely watch a boiling pot.* The suppression of the transition out of level 1 can be of great help in protecting the dynamics of the qubit $\{|0\rangle, |1\rangle\}$. Unlike the one-dimensional case, an experiment on the multidimensional QZE has a twofold objective: i) prevent probability leakage *out* of the Zeno subspace; ii) preserve the Zeno dynamics *in* the subspace. An experiment on the preservation of the Zeno dynamics of a multidimensional system (such as a qubit) has never been done.

11. Conclusions

We considered three physically equivalent formulations of the QZE and focused on the "continuous coupling" version, in order to analyze some simple examples involving two, three and four level systems. These examples are of interest in the light of possible applications. Among these, we considered the possibility of tailoring the interaction so as to obtain robust subspaces against decoherence, useful for applications in quantum computation. The experimental scheme proposed in Sec. 10 would enable one to control the dynamics of a qubit. It is remarkable that a problem like the quantum Zeno effect, that was considered academic until no more than 20 years ago, might lead in the next future to novel experiments and very practical applications, motivating the invention of techniques to counter decoherence and dissipation.

Acknowledgments

We would like to thank Wolfgang Ketterle for an interesting email exchange. This work is partially supported by the European Union through the Integrated Project EuroSQIP.

References

[1] Misra B and Sudarshan E C G 1977 *J. Math. Phys.* **18** 756
[2] Gorini V, Kossakowski A and Sudarshan E C G 1976 *J. Math. Phys.* **17** 821; Lindblad G 1976 *Comm. Math. Phys.* **48** 119
[3] von Neumann J 1932 *Die Mathematische Grundlagen der Quantenmechanik* (Springer, Berlin). [English translation by E. T. Beyer 1955 *Mathematical Foundation of Quantum Mechanics* (Princeton University Press, Princeton)]
[4] Beskow A and Nilsson J 1967 *Arkiv für Fysik* **34** 561
[5] Khalfin L A 1968 *JETP Letters* **8** 65
[6] Peres A 1980 *Am. J. Phys.* **48** 931; Kraus K 1981 *Found. Phys.* **11** 547; Sudbery A 1984 *Ann. Phys.* **157** 512
[7] Friedman C N 1972 *Indiana Univ. Math. J.* **21** 1001; Gustafson K and Misra B 1976 *Lett. Math. Phys.* **1** 275; Gustafson K 1983 *Irreversibility questions in chemistry, quantum-counting, and time-delay*, in *Energy storage and redistribution in molecules*, edited by Hinze J (Plenum, New York), and refs. [10,12] therein; Gustafson K 2002 *A Zeno story* quant-ph/0203032
[8] Exner P and Ichinose T 2005 *Ann. H. Poincare* **6** 195; Exner P, Ichinose T, Neidhardt H, Zagrebnov V 2007 **57** 67
[9] Facchi P and Pascazio S 2008 *J. Phys. A: Math. Theor.* **41** 493001
[10] Cook, R J 1988 *Phys. Scr.* T **21** 49
[11] Itano W M, Heinzen D J, Bollinger J J and Wineland D J 1990 *Phys. Rev.* A **41** 2295
[12] Petrosky T, Tasaki S, and Prigogine I 1990 *Phys. Lett.* A **151** 109; Petrosky T, Tasaki S, and Prigogine I 1991 *Physica* A **170** 306; Peres A and Ron A 1990 *Phys. Rev.* A **42** 5720; Pascazio S, Namiki M, Badurek G and Rauch H 1993; *Phys. Lett.* A **179** 155 Altenmüller T P andSchenzle A 1994 *Phys. Rev.* A **49** 2016; Cirac J I, Schenzle A and Zoller P 1994 *Europhys. Lett.* **27** 123; Pascazio S and Namiki M 1994 *Phys. Rev.* A **50**, 4582; Berry M V 1995 *Two-State Quantum Asymptotics*, in *Fundamental Problems in Quantum Theory*, eds.; Greenberger D M and Zeilinger A (*Ann. N.Y. Acad. Sci.* **755** 303; Beige A and Hegerfeldt G 1996 *Phys. Rev.* A **53** 53; Luis A and Peřina J 1996 *Phys. Rev. Lett.* **76**, 4340; Beige A, Braun D, Tregenna B, and Knight P L 2000 *Phys. Rev. Lett.* **85** 1762; Agarwal G S, Scully M O and Walther H 2001 *Phys. Rev. Lett.* **86** 4271
[13] Kwiat P, Weinfurter H, Herzog T, Zeilinger A, and Kasevich M 1995 *Phys. Rev. Lett.* **74** 4763; Nagels B, Hermans L J F and Chapovsky P L 1997 *Phys. Rev. Lett.* **79** 3097; Balzer C, Huesmann R, Neuhauser W and Toschek P E 2000 *Opt. Comm.* **180** 115; Toschek P E and Wunderlich C 2001 *Eur. Phys. J.* D **14** 387; Wunderlich C, Balzer C, and Toschek P E 2001 *Z. Naturforsch.* **56a** 160
[14] Streed E W, Mun J, Boyd M, Campbell G K, Medley P, Ketterle W and Pritchard D E 2006 *Phys. Rev. Lett.* **97** 260402
[15] Bernu J, Deléglise S, Sayrin C, Kuhr S, Dotsenko I, Brune M, Raimond J M and Haroche S 2008 *Phys. Rev. Lett.* **101** 180402
[16] Jericha E, Schwab D E, Jäkel M R, Carlile C J, and Rauch H 200 *Physica* B **283**, 414; Rauch H 2001 *Physica* B **297** 299
[17] Wilkinson S R, Bharucha C F, Fischer M C, Madison K W, Morrow P R, Niu Q, Sundaram B and Raizen M G 1997 *Nature* **387** 575

[18] Fischer M C, Gutiérrez-Medina B and Raizen M G 2001 *Phys. Rev. Lett.* **87** 040402
[19] Lane A M 1983 *Phys. Lett.* A **99** 359; Schieve W C, Horwitz L P and Levitan J 1989 *Phys. Lett.* A **136** 264; Kofman A G and Kurizki G 2000 *Nature* **405** 546; Elattari B and Gurvitz S A 2000 *Phys. Rev.* A **62** 032102
[20] Facchi P, Nakazato H and Pascazio S 2001 *Phys. Rev. Lett.* **86** 2699
[21] Kaulakys B and Gontis V 1997 *Phys. Rev.* A **56** 1131; Facchi P, Pascazio S and Scardicchio A 1999 *Phys. Rev. Lett.* **83** 61; Flores J C 1999 *Phys. Rev.* B **60** 30) Flores J C 2000 *Phys. Rev.* B **62** R16291; Gurvitz A 2000 *Phys. Rev. Lett.* **85** 812; Gong J and Brumer P 2001 *Phys. Rev. Lett.* **86** 1741; Luis A 2001 *J. Opt.* B **3** 238
[22] Nakazato H, Namiki M and Pascazio S 1996 *Int. J. Mod. Phys.* B **10** 247; Home D and Whitaker M A B 1997 *Ann. Phys.* **258** 237
[23] Facchi P and Pascazio S 2001 *Progress in Optics*, edited by E. Wolf (Elsevier, Amsterdam) **42** 147
[24] Facchi P, Gorini V, Marmo G, Pascazio S and Sudarshan E C G 2000 *Phys. Lett.* A **275** 12; Facchi P, Pascazio S, Scardicchio A and Schulman L S 2002 *Phys. Rev.* A **65** 012108
[25] Facchi P and Pascazio S 2002 *Phys. Rev. Lett.* **89** 080401
[26] Simonius M 1978 *Phys. Rev. Lett.* **40** 980; Harris R A and Stodolsky L 1978 *Phys. Lett. B* **78** 313; Harris R A and Stodolsky L 1981 *J. Chem. Phys.* **74** 2145; Harris R A and Stodolsky L 1982 *Phys. Lett.* B **116** 464; Cina J A and Harris R A 1995 *Science* **267** 832; Venugopalan A and Ghosh R 1995 *Phys. Lett.* A **204** 11; Plenio M P, Knight P L and Thompso R C 1996; *Opt. Comm.* **123** 278; Berry M V and Klein S 1996 *J. Mod. Opt.* **43** 165; Panov A D 1996 *Ann. Phys.* (NY) **249** 1; Mihokova E, Pascazio S and Schulman L S 1997 *Phys. Rev.* A **56** 25; Luis A and Sánchez–Soto L L 1998 *Phys. Rev.* A **57** 781; Thun K and Peřina J 1998 *Phys. Lett.* A **249** 363; Schulman L S 1998 *Phys. Rev.* A **57** 1509; Panov A D 1999 *Phys. Lett.* A **260** 441; Řeháček J, Peřina J, Facchi P, Pascazio S and Mišta L 2000 *Phys. Rev.* A **62** 013804; Facchi P and Pascazio S 2000 *Phys. Rev.* A **62** 023804; Militello B, Messina A and Napoli A 2001 *Phys. Lett.* A **286** 369; Luis A 2001 *Phys. Rev.* A **64** 032104
[27] Facchi P and Pascazio S 1998 *Phys. Lett.* A **241** 139; Facchi P and Pascazio S 1999 *Physica* A **271** 133; Antoniou I, Karpov E, Pronko G and Yarevsky E 2001 *Phys. Rev.* A **63** 062110
[28] Fermi E 1932 *Rev. Mod. Phys.* **4** 87; Fermi E 1950 *Nuclear Physics* (University of Chicago, Chicago) 136, 148; Fermi E 1954 *Notes on Quantum Mechanics. A Course Given at the University of Chicago in 1954*, edited by Segré E (University of Chicago, Chicago) Lec. 23
[29] Bernardini C, Maiani L and Testa M 1993 *Phys. Rev. Lett.* **71** 2687; Maiani L and Testa M 1998 *Ann. Phys.* (NY) **263** 353; Joichi I, Matsumoto Sh and Yoshimura M 1998 *Phys. Rev.* D **58** 045004; Alvarez-Estrada R F and Sánchez-Gómez J L 1999 *Phys. Lett.* A **253** 252; Panov A D 2000 *Physica* A **287** 193
[30] Lüders G 1951 *Ann. Phys.* (Leipzig) **8** 322
[31] Schwinger J 1959 *Proc. Nat. Acad. Sc.* **45** 1552; Schwinger J 1991 *Quantum kinematics and dynamics* (Perseus Publishing, New York) p 26
[32] Peres A 1998 *Quantum Theory: Concepts and Methods* (Kluwer Academic Publishers, Dordrecht)
[33] Reed M and Simon B 1980 *Functional Analysis I* (Academic Press, San Diego) p 57
[34] Facchi P, Lidar D A and Pascazio S 2004 *Phys. Rev.* A **69** 032314
[35] Messiah A 1961 *Quantum mechanics* (Interscience, New York); Born M and Fock V 1928 *Z. Phys.* **51** 165; Kato T 1950 *J. Phys. Soc. Jap.* **5** 435; Avron J E and Elgart A 1999 *Comm. Math. Phys.* **203** 445 and references therein
[36] Facchi P, Tasaki S, Pascazio S, Nakazato H, Tokuse A and Lidar D A 2005 *Phys. Rev.* A **71** 022302
[37] Wick G C , Wightman A S and Wigner E P 1952 *Phys. Rev.* **88** 101 Wick G C , Wightman A S and Wigner E P 1970 *Phys. Rev.* D **1** 3267
[38] Jauch J M 1964 *Helv. Phys. Acta* **37** 293
[39] Berry M V *Chaos and the semiclassical limit of quantum mechanics (is the moon there when somebody looks?)*, in *Quantum Mechanics: Scientific perspectives on divine action* (eds Robert John Russell, Philip Clayton, Kirk Wegter-McNelly and John Polkinghorne), Vatican Observatory CTNS publications p 41
[40] Machida S and Namiki M 1980 *Prog. Theor. Phys.* **63** 1457; Machida S and Namiki M 1980 *Prog. Theor. Phys.* **63** 1833; Araki H 1962 *Einführung in die Axiomatische Quantenfeldtheorie I, II* (ETH-Lecture, ETH, Zurich); Araki H 1980 *Prog. Theor. Phys.* **64** 719
[41] Facchi P, Marmo G, Pascazio S, Scardicchio A and Sudarshan E C G 2004 *J. Opt. B: Quantum Semicl. Opt.* **6** S492
[42] Cohen-Tannoudji C, Dupont-Roc J and Grynberg G 1998 *Atom-Photon Interactions: Basic Processes and Applications* (Wiley, New York)
[43] Boller K J, Imamoglu A and Harris S E 1991 *Phys. Rev. Lett.* **66** 2593
[44] Militello B, Messina A and Napoli A 2001 *Fortschr. Phys.* **49** 1041

Perspectives on the quantum Zeno paradox[1]

Wayne M. Itano

Time and Frequency Division, National Institute of Standards and Technology, Boulder, CO 80305 USA

E-mail: `itano@boulder.nist.gov`

Abstract. As of October 2006, there were approximately 535 citations to the seminal 1977 paper of Misra and Sudarshan that pointed out the quantum Zeno paradox (more often called the quantum Zeno effect). In simple terms, the quantum Zeno effect refers to a slowing down of the evolution of a quantum state in the limit that the state is observed continuously. There has been much disagreement as to how the quantum Zeno effect should be defined and as to whether it is really a paradox, requiring new physics, or merely a consequence of "ordinary" quantum mechanics. The experiment of Itano, Heinzen, Bollinger, and Wineland, published in 1990, has been cited around 347 times and seems to be the one most often called a demonstration of the quantum Zeno effect. Given that there is disagreement as to what the quantum Zeno effect *is*, there naturally is disagreement as to whether that experiment demonstrated the quantum Zeno effect. Some differing perspectives regarding the quantum Zeno effect and what would constitute an experimental demonstration are discussed.

1. Introduction

A recent entry in Wikipedia, an Internet-based encyclopedia, defines the quantum Zeno effect as follows:

> The quantum Zeno effect is a quantum mechanical phenomenon first described by George Sudarshan and Baidyanaith Misra of the University of Texas in 1977. It describes the situation that an unstable particle, if observed continuously, will never decay. This occurs because every measurement causes the wavefunction to "collapse" to a pure eigenstate of the measurement basis [1].

This definition is close to the original language of Misra and Sudarshan [2], but is not sufficiently general to describe the many situations that are considered to be examples of the quantum Zeno effect. It is true that the quantum Zeno effect describes the situation in which the decay of a particle can be prevented by observations on a sufficiently short time scale. However, the quantum Zeno effect is much more general, since it describes the situation in which the time evolution of *any* quantum system can be slowed by sufficiently frequent "observations." The references to observations and to wavefunction collapse tend to raise unnecessary questions related to the interpretation of quantum mechanics. Actually, all that is required is that some interaction with an external system disturb the unitary evolution of the quantum system in a way that is effectively like a projection operator. Finally, the word "never" describes a limiting case. A slowing of the time evolution, as opposed to a complete freezing, is generally regarded as a demonstration of the quantum Zeno effect.

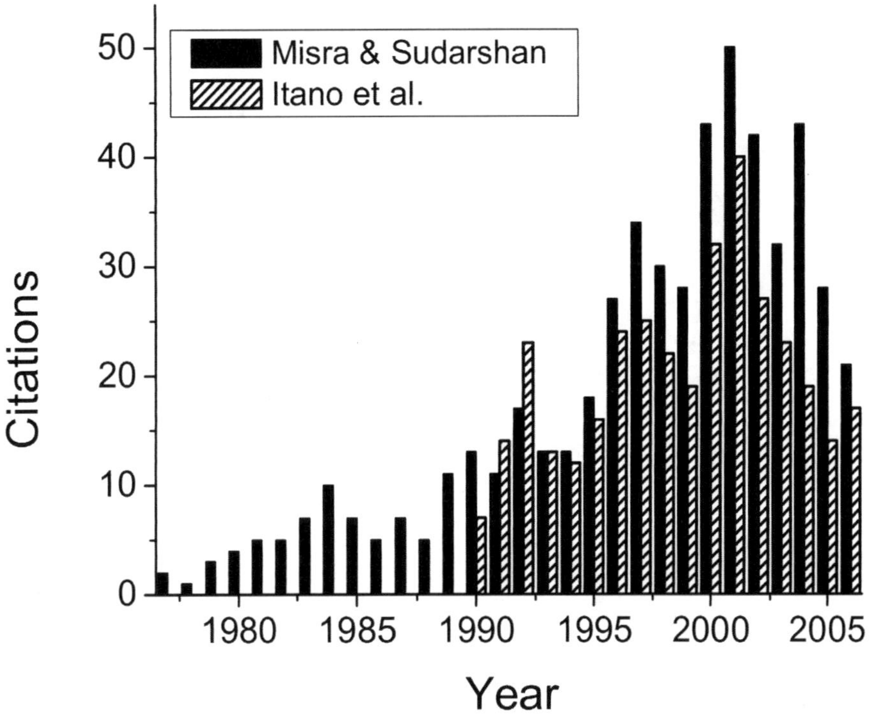

Figure 1. Graph of the number of citations per year to Misra and Sudarshan [2] and to Itano *et al* [3].

2. The Misra and Sudarshan Paper

The 1977 article "The Zeno's paradox in quantum theory" by Misra and Sudarshan [2] studied the evolution of a quantum system subjected to frequent ideal measurements. They showed that, in the limit of infinitely frequent measurements, a quantum system would remain in its initial state. Applied to the case of an unstable particle whose trajectory is observed in a bubble chamber or film emulsion, this result seemed to imply that such a particle would not decay, in contradiction to experiment. In this case, the resolution to the apparent paradox lies in the fact that the interactions between the particle and its environment that lead to the observed track are not sufficiently frequent to modify the particle's lifetime.

The time distribution of literature citations to Misra and Sudarshan [2] is shown in Fig. 1. The total number of citations listed in the Web of Science database in October 2006 was 535. The graph shows a relatively low but steady number of citations per year for about a decade, followed by a large increase that continues for over a decade, possibly peaking about 25 years after the original publication date. The great increase in the rate of citations in recent years is partially due to the increased interest in quantum information processing, where the quantum Zeno effect may find practical applications.

3. Simple derivation of the quantum Zeno effect

The quantum Zeno effect can be derived in an elementary way by considering the short-time behavior of the state vector [4]. (The treatment of Misra and Sudarshan [2] is more general since it involves the density matrix.) Let $|\phi\rangle$ be the state vector at time $t = 0$. If H is the Hamiltonian, in units where $\hbar = 1$, then the state vector at time t is $e^{-iHt}|\phi\rangle$, and the survival

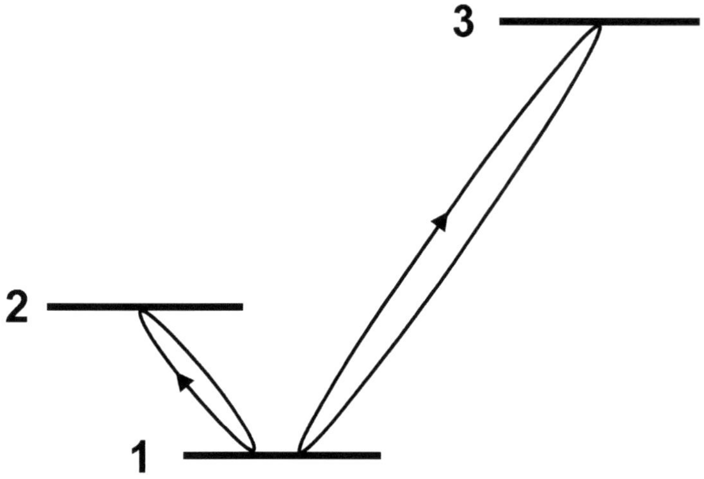

Figure 2. Energy-level diagram for the IHBW demonstration of the quantum Zeno effect.

probability is $S(t) = |\langle\phi|e^{-iHt}|\phi\rangle|^2$. If t is small enough, it should be possible to make a power series expansion:

$$e^{-iHt} \approx I - iHt - \frac{1}{2}H^2t^2\ldots, \tag{1}$$

so that the survival probability is

$$S(t) = |\langle\phi|e^{-iHt}|\phi\rangle|^2 \approx [1 - (\Delta H)^2t^2], \tag{2}$$

where

$$(\Delta H)^2 \equiv \langle\phi|H^2|\phi\rangle - \langle\phi|H|\phi\rangle^2. \tag{3}$$

Many quantum systems have states whose survival probability appears on ordinary time scales to be a decreasing exponential in time. This is inconsistent with the quadratic time dependence of Eq. (1) and implies that in such cases Eq. (1) holds only for very short times. Consider the survival probability $S(T)$, where the interval $[0, T]$ is interrupted by n measurements at times $T/n, 2T/n, \ldots, T$. Ideally, these measurements are instantaneous projections and the initial state $|\phi\rangle$ is an eigenstate of the measurement operator. In that case, the survival probability is

$$S(T) \approx [1 - (\Delta H)^2(T/n)^2]^n, \tag{4}$$

which approaches 1 as $n \to \infty$.

It is important to note that at this level there should be nothing controversial or problematic about the existence of the quantum Zeno effect. The quantum Zeno effect should be observed as long as the physical system can be made to display the behavior shown in Eq. (4). For a given system, it may be difficult or impossible to make measurements quickly enough for the quadratic time dependence of the survival probability to be observed, so that, as a practical matter, the quantum Zeno effect cannot be observed. It should be noted that the semantic arguments over terms such as "measurement" or "observation" can be avoided if we accept that a "measurement" is an operation that interrupts the unitary time evolution governed by H in such a way as to yield Eq. (4) as a good approximation. That is, the "measurement" should effectively act as a projection operator. According to this view, it is not necessary that the "measurements" be recorded by a macroscopic apparatus or that they be instantaneous.

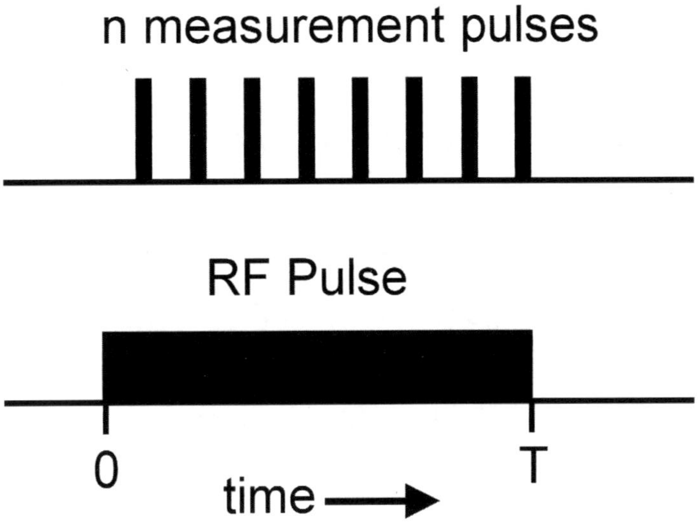

Figure 3. Timing of the radio frequency and optical fields applied to the beryllium ions in the IHBW experiment

4. The IHBW Experiment

The experiment of Itano, Heinzen, Bollinger, and Wineland (IHBW) [3] was based on a proposal of Cook [5] for observing the quantum Zeno effect in a three-level atom (see Fig. 2). Levels 1 and 2 are stable on the time scale of the experiment. Level 3 decays to level 1 with the emission of a photon. In the experiment of IHBW, levels 1 and 2 were two of the hyperfine sublevels of the ground $^2S_{1/2}$ state of the Be^+ ion. Level 3 was a sublevel of the $^2P_{3/2}$ excited state that decayed only to level 1.

The experiment was carried out with a sample of about 5000 Be^+ ions confined by electric and magnetic fields in a Penning trap. The steps in the experiment were as follows:

(i) The ions were prepared in level 1 by optical pumping with the laser beam.

(ii) A resonant radio frequency (RF) magnetic field was applied for the interval required to drive the ions to level 2.

(iii) During the time that the RF pulse was applied, a variable number n of equally spaced short laser pulses was applied to the ions (see Fig. 3).

(iv) The laser (resonant with the 1-to-3 transition) was turned on, and the induced fluorescence was recorded.

The intensity of the laser-induced fluorescence at the end of the experiment was proportional to the population of level 1. If there are no optical pulses during the long RF pulse, the population of level 2 as a function of the time t that the RF pulse is applied is

$$P_2(t) = \sin^2(\Omega t/2) = \frac{1}{2}[1 - \cos(\Omega t)], \tag{5}$$

where Ω is proportional to the amplitude of the RF field. If the duration of the RF pulse is chosen to be $T = \pi/\Omega$ (a pi-pulse), then all of the population is transferred from level 1 to level 2. If n equally-spaced laser pulses of negligible duration are applied during the RF pi-pulse, the population of level 2 at time T is

$$P_2(T) = \frac{1}{2}\left[1 - \cos^n(\pi/n)\right], \tag{6}$$

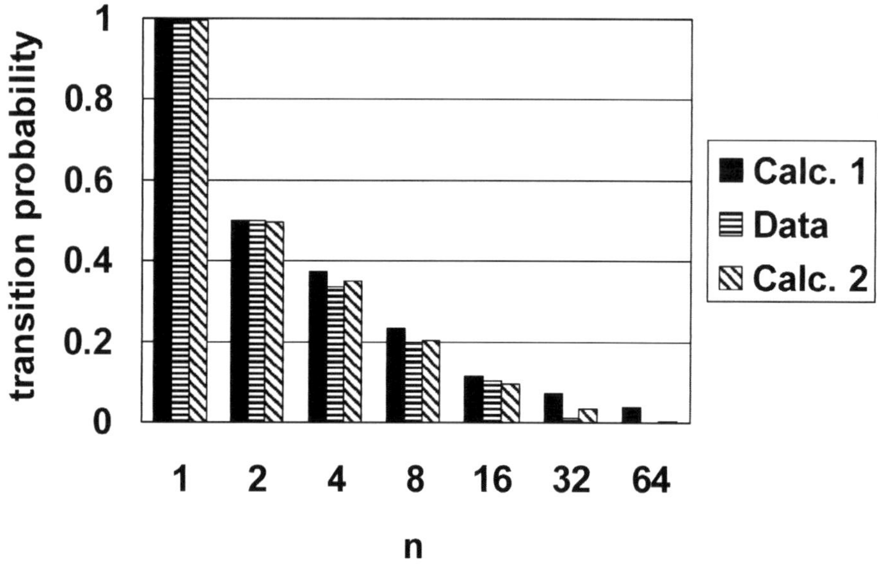

Figure 4. Probability of making the 1-to-2 transition as a function of the number n of optical "measurement" pulses.

which approaches 0 as n goes to infinity.

Figure 4 compares the data to theory. The solid bars represent the transition probability as a function of n according to the simplified calculation of Eq. (6). The bars with horizontal stripes represent the data. The bars with diagonal stripes represent a calculation that takes into account the finite duration of the laser pulses and optical pumping effects. The data are in reasonably good agreement with the simplified calculation and in better agreement with the improved calculation. The decrease in $P_2(T)$ as n increases demonstrates the quantum Zeno effect.

A variation of the experiment was carried out by initializing the ion in level 2 and then applying the RF field and the laser pulses. In this case, the transition from level 2 to level 1 was inhibited as n increased. This is another example of the quantum Zeno effect. In this case, the inhibition of the transition is accompanied by the *absence* of laser-induced fluorescence.

Recently, the quantum Zeno effect was observed for an unstable quantum system by Fischer *et al* [6]. The quantum Zeno effect for induced transitions and for unstable systems are not fundamentally different, since they both follow from the general arguments of Misra and Sudarshan [2], but it has been difficult to observe in the latter case, because of the short times over which the decay is nonexponential. Fischer *et al* were able to create an artificial system (atoms tunneling from a standing-wave light field) in which the interactions could be controlled so as to observe the desired effects.

5. Responses to the IHBW Experiment

As can be seen by the history of citations (Fig. 1), the publication of the IHBW experiment [3] generated considerable interest. Initially, some of the responses were critical in one way or another. Some (e. g., Ref. [7]) objected to the use of the term "wavefunction collapse" in describing the experiment. The authors responded that the concept of wavefunction collapse was not essential, and that any interpretation of quantum mechanics that yielded the same prediction of the experimental results should be regarded as valid [8]. Some objected to the fact that photons were not actually observed during the intermediate "measurements," in the sense of having the scattered photons registered by a detector, so that the experiment did not actually demonstrate the quantum Zeno effect [7, 9, 10, 11]. However, the results are predicted to be the same whether or not the intermediate measurements are made. It is enough that the measurements *could* have been made. As long as the laser interactions act effectively as projection operators, so that the algebra of Eqs. (1)–(4) is followed, the experiment should be regarded as a demonstration of the quantum Zeno effect. It should be noted that none of the criticisms were directed at the execution of the experiment itself, only at the interpretation. For the most part, the citations to Ref. [3] simply accept it as a demonstration of the quantum Zeno effect. In fact, it is cited in quantum mechanics textbooks [12, 13, 14, 15] and popular science books [16, 17, 18].

6. Distinctions

While Misra and Sudarshan originally used the term "quantum Zeno paradox," as did Peres [4] and others, the more recent work usually uses the term "quantum Zeno effect," perhaps because the effect no longer seems paradoxical. Some authors distinguish between the quantum Zeno *paradox* and the quantum Zeno *effect*, but they do so in differing ways. Pascazio and Namiki [19] call the situation in which the frequency of measurements is finite and the evolution is slowed the quantum Zeno *effect*, and the limiting case in which the frequency of measurements is infinite and the evolution is frozen the quantum Zeno *paradox*. Block and Berman [20] call the inhibition of spontaneous decay the quantum Zeno *paradox* and the inhibition of induced transitions (as in the IHBW experiment) the quantum Zeno *effect*. In Ref. [21], Home and Whitaker reserve the term quantum Zeno *paradox* for a negative-result experiment involving observations with a macroscopic apparatus. This definition of the quantum Zeno paradox seems to exclude most, if not all, feasible experiments. In this context, the IHBW experiment is not regarded as an example of the quantum Zeno paradox because a local interaction is present between the laser field and the atoms, and also because the electromagnetic field, containing zero or a few scattered photons, is not regarded as a macroscopic observation apparatus. They regard the type of experiments where the time evolution of a quantum system is affected by a direct interaction, for example with an external field, as examples of the quantum Zeno *effect*. However, in a later publication [22] the same authors treat the terms quantum Zeno *paradox* and quantum Zeno *effect* as synonymous and restrict both to nonlocal negative-result experiments involving a macroscopic observation apparatus. Experiments that do not meet these criteria would not be examples of *either* the quantum Zeno paradox *or* the quantum Zeno effect, according to their later definition.

7. Extensions

Several variations on the general theme of quantum Zeno effects have been described. Soon after the IHBW experiment was carried out, Peres and Ron [23] showed that a partial quantum Zeno effect results if the measurements are too weak to completely destroy the coherence of the state of the measured system. A modification of the IHBW experiment was proposed in which the measurement laser pulses are weakened. Jordan *et al* [24] showed that a related effect, damped oscillations of the state populations, can occur if the duration of the experiment is extended, while weak measurements are made.

Some, including Kofman and Kurizki [25] and Facchi *et al* [26] have shown that the decay of an unstable quantum system can be accelerated by frequent observations. This is called the quantum anti-Zeno effect or the inverse quantum Zeno effect. As is the case for the quantum Zeno effect, the observations must take place before the decay becomes exponential. Unlike the quantum Zeno effect, which follows from rather general arguments, e. g. Eqs. (1)–(4), the possibility of observing a quantum anti-Zeno effect depends on the details of the system.

The experiment of Fischer *et al* [6] demonstrated the quantum anti-Zeno effect as well as the quantum Zeno effect.

An interesting generalization of the quantum Zeno effect is the concept of quantum Zeno dynamics [27, 28]. Frequent measurements can confine the evolution of a quantum system to a subspace of the Hilbert space rather than simply to the initial state. Compared to the ordinary quantum Zeno effect, the difference is that the measurements distinguish not between the initial *state* and all other states but between a *subspace* and the rest of the Hilbert space. This form of quantum Zeno effect may find application in quantum information processing.

8. Applications

As already noted, the recent increase in the rate of citations to the articles of Misra and Sudarshan [2] and IHBW [3] is partially related to increased interest in quantum information processing. In this context, there have been various proposals to use the quantum Zeno effect to preserve quantum systems from decoherence.

Beige *et al* [29] have proposed an arrangement of atoms inside an optical cavity capable of carrying out quantum logic operations with low error rates within a decoherence-free subspace of the Hilbert space. States outside the decoherence-free subspace are coupled strongly to the environment. The quantum Zeno effect then leads to effective confinement of the system to the decoherence-free subspace, which is an example of a quantum Zeno subspace.

Franson *et al* [30] have proposed use of the quantum Zeno effect to suppress errors in a linear optics implementation of quantum computation. In this implementation, the presence of two photons in the same mode indicates an error. The presence of a strong two-photon absorber in an optical fiber takes the role of the "observer" and suppresses the errors. Other proposed applications of the quantum Zeno effect to error prevention in quantum computation are discussed in Refs. [31, 32, 33, 34].

Quantum "bang-bang" control and related dynamical decoupling techniques [35, 36] utilize frequent, pulsed interactions to effectively prevent decoherence of a quantum system by confining the dynamics to a subspace. This is not exactly the quantum Zeno effect, since the interactions are unitary, but the results are mathematically similar to those for the quantum Zeno effect.

Dhar *et al* [37] have discussed the "super-Zeno effect," which preserves a state (or more generally, keeps a quantum system within a subspace of the Hilbert space) with a set of pulsed interactions unequally spaced in time. The timing of these interactions can be arranged so as to be more efficient than can be done with the same number of equally spaced interactions (ordinary quantum Zeno effect). Also, it should be noted that the pulsed interactions are unitary kicks, as in the so-called "bang-bang control" [35], and not observations in the usual sense.

9. Conclusion

The 1977 publication of Misra and Sudarshan stimulated a great deal of theoretical and experimental work that has enhanced our understanding of the time development of quantum systems, such as the short-time nonexponential decay of unstable quantum systems. The results of the IHBW experiment, published in 1990, was a clear confirmation of the existence of the quantum Zeno effect for the case of the inhibition of an induced transition. Interest in the quantum Zeno effect continues to be high, partially due to the possibility of practical applications in quantum information processing.

References

[1] `URL:http://en.wikipedia.org/wiki/Quantum_Zeno_effect`
[2] Misra B and Sudarshan E C G 1977 *J. Math. Phys.* **18** 756
[3] Itano W M, Heinzen D J, Bollinger J J, and Wineland D J 1990 *Phys. Rev.* A **41** 2295
[4] Peres A 1980 *Am. J. Phys.* **48** 931

[5] Cook R J 1988 *Physica Scripta* **T21** 49
[6] Fischer M C, Gutiérrez-Medina B, and Raizen M G 2001 *Phys. Rev. Lett.* **87** 040402
[7] Ballentine L E 1991 *Phys. Rev.* A **43** 5165
[8] Itano W M, Heinzen D J, Bollinger J J, and Wineland D J 1990 *Phys. Rev.* A **43** 5168
[9] Petrosky T, Tasaki S, Prigogine I 1990 *Phys. Lett.* A **151** 109
[10] Inagaki S, Namiki M, and Tajiri T 1992 *Phys. Lett.* A **166** 5
[11] Pascazio S, Namiki M, Badurek G, and Rauch H 1993 *Phys. Lett.* A **179** 155
[12] Griffiths D J 1995 *Introduction to Quantum Mechanics* (Englewood Cliffs, NJ: Prentice-Hall) p 385
[13] Sakurai J J 1994 *Modern Quantum Mechanics* Rev. Ed., ed S F Tuan (Reading, MA: Addison-Wesley) p 485
[14] Peres A 1993 *Quantum Theory: Concepts and Methods* (Dordrecht: Kluwer Academic) p 394
[15] Braginsky V B and Khalili F Ya 1992 *Quantum Measurement* ed K S Thorne (Cambridge: Cambridge) p 94
[16] Whitaker A 1996 *Einstein, Bohr and the Quantum Dilemma* (Cambridge: Cambridge) p 313
[17] Coveney P and Highfield R 1990 *The Arrow of Time* (New York: Fawcett Columbine) p 137
[18] Schulman L S 1997 *Time's Arrows and Quantum Measurement* (Cambridge: Cambridge) p 64
[19] Pascazio S and Namiki M 1994 *Phys. Rev.* A **50** 4582
[20] Block E and Berman P R 1991 *Phys. Rev.* A **44** 1466
[21] Home D and Whitaker M A B 1992 *J. Phys.* A **25** 657
[22] Home D and Whitaker M A B 1997 *Annals of Physics* **258** 237
[23] Peres A and Ron A 1990 *Phys. Rev.* A **42** 5720
[24] Jordan T F, Sudarshan E C G, and Valanju P 1991 *Phys. Rev.* A **44** 3340
[25] Kofman A G and Kurizki G 2000 *Nature* **405** 546
[26] Facchi P, Nakazato H, and Pascazio S 2001 *Phys. Rev. Lett.* **86** 2699
[27] Facchi P, Gorini V, Marmo G, Pascazio S, and Sudarshan E C G 2000 *Phys. Lett.* A **275** 12
[28] Facchi P and Pascazio S 2002 *Phys. Rev. Lett.* **89** 080401
[29] Beige A, Braun D, Tregenna B, and Knight P L 2000 *Phys. Rev. Lett.* **85** 1762
[30] Franson J D, Jacobs B C, and Pittman T B 2004 *Phys. Rev.* A **70** 062302
[31] Zurek W H 1984 *Phys. Rev. Lett.* **53** 391
[32] Vaidman L, Goldenberg L, and Wiesner S *Phys. Rev.* A **54** R1745
[33] Pachos J and Walther H 2002 *Phys. Rev. Lett.* **89** 187903
[34] Ralph T C, Gilchrist A, Milburn G J, Munro W J, and Glancy S 2003 *Phys. Rev.* A **68** 042319
[35] Viola L and Lloyd S 1998 *Phys. Rev.* A **58** 2733
[36] Viola L, Knill E, and Lloyd S 1999 *Phys. Rev. Lett.* **82** 2417
[37] Dhar D, Grover L K, and Roy S M 2006 *Phys. Rev. Lett.* **96** 100405

Journal of Physics: Conference Series **196** (2009) 012019 doi:10.1088/1742-6596/196/1/012019

Quantum Zeno paradox: survival and decay

Charles B. Chiu

Department of Physics, University of Texas at Austin, Austin, Texas

Abstract. Quantum mechanics enables the calculation of the time evolution of a dynamical system provided the Hamiltonian is defined and the initial state is specified. Due to the time reversal invariance, the time evolution of a given initial state at t=0 does not have a time-arrow. These problems have been extensively studied by E. C. G. Sudarshan and collaborators, including myself. Here the survival probability, which is the square of the survival amplitude, is an even function of t. Using the survival amplitude as a common framework, we discuss Zeno effects in several processes: the decay of an unstable particle, pion productions in proton-nucleus collisions and the Aharonov-Vardi effect which is another manifestation of the Zeno effect.

Introduction: The quantum Zeno effect is the slowing down of the time evolution of a quantum system due to repeated measurements. I would like to discuss three examples in connection with this effect. My talk is based on several pieces of work of E. C. G. Sudarshan in collaboration with Misra, Valanju and myself.

Example 1. Unstable quantum system Consider an unstable quantum system where the Hamiltonian is defined. We prepare the unstable quantum state $|M\rangle$, at $t = 0$. The survival amplitude at t is given by the overlap between this state and the time evolved state, i.e.

$$a(t) = \langle M|e^{-iHt}|M\rangle = \int_{\lambda_{th}}^{\infty} d\lambda \rho(\lambda) e^{-i\lambda t} \tag{1}$$

In the last step, this amplitude is expressed as the Fourier transform of the energy spectrum continuum of H, i.e.

$$\rho(\lambda) = \langle M|\lambda\rangle\langle\lambda|M\rangle. \tag{2}$$

The survival probability is by definition the square of the survival amplitude,

$$Q(t) = |a(t)|^2. \tag{3}$$

Its time derivative at $t = 0$ is given by

$$\dot{Q}(0) = \dot{a}^*(0)a(0) + \dot{a}(0)a^*(0) = 0. \tag{4}$$

where the identity

$$\dot{a}^*(0) = -\dot{a}(0), \tag{5}$$

which is a consequence of the Fourier representation of (1), together $a(0) = 1$ (with by definition,)and thus $a^*(0) = 1$ were used.

In ref. [1], Misra and Sudarshan show that eq. (4), along with some general assumptions and a rigorous analysis, leads to the quantum Zeno paradox. In particular, if the unstable quantum system is monitored over time T in the N equal time intervals, in large N limit the survival probability at T is given by

$$Q(T) = Q(T/N)^N \rightarrow 1. \tag{6}$$

So repeated measurement will prolong the survival. And in the large N limit there will be 100% survival, or nondecay.

Lee model: In ref. [2], we investigated the survival amplitude based on a Lee model. In the Lee model we considered, the basis states contain a discrete state $|V\rangle$, and the continuum states $|N\theta(E)\rangle$, where the energy E, ranges from the threshold energy to infinity. The interaction Hamiltonian specifies that V is coupled to $N\theta$ through a coupling function

$$f(E) =< N\theta(E)|H|V > . \tag{7}$$

In this model the survival amplitude of V at t is given by

$$a(t) \equiv < V|e^{-iHt}|V >= \frac{i}{2\pi}\int_C \frac{e^{-izt}}{\gamma(z)}dz. \tag{8}$$

For this talk we will be mainly interested in the situation near $t = 0$. The last step of eq.(8) gives the contour integral representation of the survival amplitude where the contour C wraps around the right-hand cut of the denominator function $\gamma(z)$ in a clockwise manner in the complex z-plane, where $z = E - E_{th}$, with z being the displaced complex energy variable E with a displacement $-E_{th}$ (see Fig. 1). It is helpful to discuss contour deformation using the new conformal variable

$$k = [ze^{i\pi/2}]^{1/2}. \tag{9}$$

The complex k-plane is shown in Fig. 2. Here the right hand cut contour C of Fig. 1 corresponds to a line passing through the origin O, and runs along the 45°-line. It is labeled as the contour S. The region above the line corresponds to the physical sheet and below the line, the second sheet. We associate an unstable state with a pole on the second sheet. Hermiticity of the Hamiltonian requires that the second sheet poles are present in pairs. For our case, there is a pole in the first quadrant and the companion pole in the third quadrant. They are indicated in Fig. 2. We are interested in the temporal evolution in the positive time domain, $t \geq 0$. The appropriate deformed contours for this time domain are shown in Fig. 2. The contours of concern satisfy the relationship:

$$S = S_+ + S_1 + S_2 = S_+ + S_1. \tag{10}$$

In the last step, the fact that for $t \geq 0$, $S_2 = 0$ at $|k| = \infty$, was used. Thus the survival amplitude can be written as

$$a(t) = a_{pole}(t) + a_{bg}(t), \text{ with } a_{pole}(t) \sim e^{-i(E_0 - i\Gamma/2)t}. \tag{11}$$

<u>Cancellation near $t = 0$:</u> Notice that the pole term gives

$$\text{Re}\, \dot{a}_{pole}(0) = -\Gamma/2, \tag{12}$$

which is the characteristic slope of the survival amplitude at $t = 0$ associated with an exponential decay. On the other hand, eq. (5) implies that the real part of the survival amplitude at $t = 0$ vanishes, since

$$\text{Re}\dot{a}(0) = \frac{1}{2}[\dot{a}^*(0) + \dot{a}(0)] = 0. \tag{13}$$

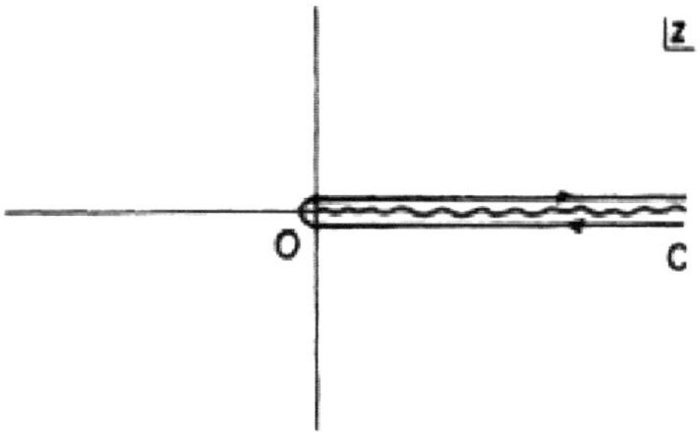

Figure 1. The right hand cut in the complex energy plane.

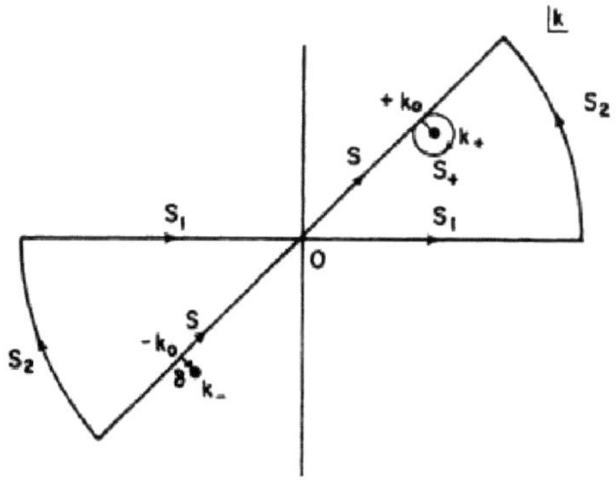

Figure 2. The k-plane: $S = S_+ + S_1 + S_2$. $S_2 = 0$, for $t \geq 0$.

For a sufficiently small region in t, say $t \leq T_1$, eqs. (11), (12) and (13) together imply that there must be a cancellation between the pole-term and the background term. So eq. (13) is satisfied for $t \leq T_1$.

In ref. [2], we have shown that this cancellation does take place in the region $0 < t < T_1$, where

$$T_1 \sim 1/(E_0 - E_{th}), \tag{14}$$

Beyond $t = T_1$, the background term falls off with $t^{-3/2}$ behavior and the pole term begins to dominate. We also found that in the small t region, the survival amplitude takes on the general form

$$Q(t) \rightarrow 1 - const \times t^{\beta}/\beta, \text{ thus } \dot{Q}(t) \rightarrow -const \times t^{\beta-1}. \tag{15}$$

For present case of interest, to ensure $\dot{Q}(0) = 0$, $\beta > 1$. In large n limit, with $t/n << T_1$,

$$Q_n(t) \rightarrow \left[1 - \frac{const}{\beta} \times \left(\frac{t}{n}\right)^{\beta}\right]^n \rightarrow 1. \tag{16}$$

So the $\beta > 1$ case, demonstrates the Zeno paradox.

Zeno time: We see that there is a delay near $t = 0$ before the pole contribution, which is characterized by an exponential decay, becomes dominate. One may understand qualitatively the value of T_1 given in eq. (14) in the following way. From eq. (1) one sees that in the survival amplitude at $t = 0$ the pole spectrum is integrated over. The presence of the specific pole contribution becomes important only when the exponent in the fourier integral, $\Delta\lambda t$, is of the order of unity, where $\Delta\lambda$ represents some significant energy scale in the spectrum. Eq. (14) indicates that this scale should be associated with the distance from the pole to the threshold energy. Since the energy scale given here is general, we suggest that the presence of Zeno time should be a common feature among unstable quantum systems.

In the small time domain, there are other physics considerations that may prevent Zeno effect from manifesting. Take for example the decay of a charged pion $\pi \to \mu\nu$. $\Gamma \sim (3 \times 10^{-8} sec)^{-1}$ and $T^{-1} = m_\pi - m_\mu = 34MeV = (2 \times 10^{-23} sec)^{-1}$. Here the natural check points are spaced at the inter-atomic distance $10^{-8}/(3 \times 10^{10}) = 3 \times 10^{-19} sec$. There is no way to monitor the natural evolution of a system having time scale of 10^{-23} sec by inter-atomic time scale of 10^{-19} sec. On the other hand inter-nucleonic distance within a nucleus may provide a more promising arena for Zeno to perform. This leads us to consider the second example.

Example 2: Evidence of "Equilibrium time" in hadron-nucleus collisions. Our first example suggests that when a quantum state is created, it is a non-equilibrium state. There is a delay time for the state to reach the equilibrium state. We refer to this delay time as the equilibrium time or the Zeno time. We mentioned that the smallness of this time prevents the direct demonstration of the Zeno effect in unstable particle decay if the inter-atomic distance is to be used as the distance between the adjacent checkpoints. In ref. [9], which is based on Dr. Valanju′s dissertation work, we followed the approach of Feinberg[6] in estimating the equilibrium time. We found that inter-nucleon collisions provide a promising arena in demonstrating the Zeno effect.

Bremsstrahlung process: In ref. [6], Feinberg considered the traversal of an electron through a matter medium. Here a bremsstrahlung process is typically preceded by a precursor collision where there is a large momentum change. There is a time delay between the precursor collision and the radiation. In particular, he observed that after the collision, the electron becomes "bare". It takes the order of the equilibrium time before the electron can radiate again. This time is given by

$$T_k = \gamma/k_0, \tag{17}$$

where k_0 is the photon energy defined in the rest frame of the electron and γ the Lorentz factor of the electron in the medium. So T_k is the equilibrium time of the electron in the medium for the emission of the next k_0 quantum.

Equilibrium time is Zeno time. We suggest that this equilibrium time is a Zeno time, since this is the time scale for the bare particle to continue to "survive" in the bare state. We explain below how one may look for the effect of the equilibrium time in pion production from pp and p-nucleus collisions at high energy.

Equilibrium time effect in p-nucleus collisions. Fig. 3 are schematic plots illustrating the "equilibrium time" effect for an ideal scenario.

pp-plot. Fig. 3a shows an ideal pion rapidity spectrum in pp collision. The spectrum is symmetric about the "midpoint" which is indicated in the plot. Pions to the right of the midpoint, which will be referred to as being in the forward range, are to be associated with pion-production due to the projectile, and those to the left, the backward range, associated with pion-production due to the target-proton.

pA-plot. Now we turn to the corresponding pion rapidity spectrum in proton-nucleus production. The situation is schematically depicted in Fig. 3b. It involves successive collisions between the projectile nucleon and ν target-nucleons.

- Forward range: We envision that after the first collision, the projectile nucleon turns into a

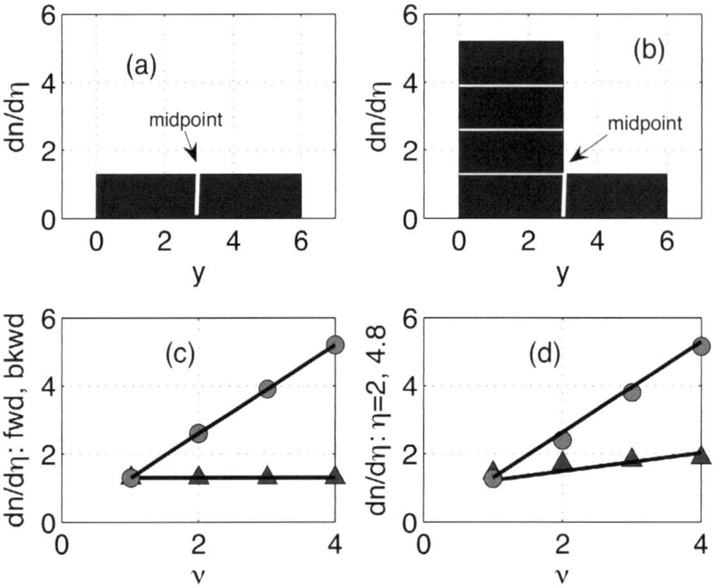

Figure 3. Schematic illustrations of pion production spectra (a) Pion inclusive distribution for pp collision. Forward pions produced due to projectile p, and backward pions due to target p. (b) pA case. Forward production independent of collision number ν, and backward $\propto \nu$. (c) ν-dependence in pA collision for the case (b) shown. Forward case-flat, and backward case $\propto \nu$. (d) Comparison between ν dependence of our model prediction (lines) and the data (points, triangles) at a typical y, i.e. 4.8, in the forward range, and the circles at $y = 2$, in the backward range.

bare state. For an ideal scenario, we assume that the equilibrium time of the bare projectile is much longer than the time taken for it to traverse through the nucleus. Consequently during the time when the bare-projectile traverses through the entire nucleus, no pions are produced. The pion production due to the projectile takes place only after the projectile has reached its equilibrium state, which takes place outside of the nucleus. The pion spectrum produced by this projectile proton is assumed to be comparable to the pion-spectrum produced by the projectile proton in pp collision.

- Backward range: On the other hand, at the target-nucleus side, for each collided target-nucleon after its equilibrium time is reached there will be the production of a pion-spectrum, which is assumed to be comparable to that produced by the target-proton in a pp collision. Thus below the midpoint, i.e. in the backward region, for the ideal scenario the pion-spectrum from a p-nucleus collision should be ν times the pion-spectrum of the pp case. Fig. 3b illustrates the ideal spectra for ν=1, 2, 3 and 4 both for the forward range as well as the backward range.

Signal of equilibrium time effect. We now proceed to display the effect of equilibrium time by comparing the ν dependence of the heights of the forward spectra to that of the backward spectra. Fig. 3c illustrates the situation. For the heights in the forward range of the present ideal case there is no ν dependence, so the plot of height vs ν is flat. On the other hand, for the spectra in the backward region, the height is proportional to ν. This would be the characteristic signal which would demonstrates the presence of the equilibrium time effect.

A quantitative model. So far we have considered an ideal scenario. In the work of Valanju et al[9], a quantitative model was constructed, which is an extension of our earlier phenomenological model for p-A multiple scattering[8]. For the quantitative model, analogous to the electron bremsstrahlung process, the equilibrium time of pion production is taken to be

$$T_\pi \sim \frac{\gamma}{M_{eff}}, \text{ with } M_{eff} = \sqrt{M_c^2 + p_T^2} \tag{18}$$

where M_{eff} is the average transverse mass of a mean hadronic cluster, which has typically ~ 3 pions. We have also taken into account the internal motion of pions within the cluster. The experimental shape of pion rapidity distribution in pp collision is used as an input. Also we use the mean free path in the description of the multiple scattering process and take into account the effect of conservation of energy.

Comparison with the data.

- The comparison between the data (histograms) and our model predictions (lines) is shown in fig. 4. The agreement is reasonable.
- The ν-dependence of the pion-spectrum at a typical forward rapidity, $y = 4.8$, and that at a typical backward rapidity, $y = 2$ are shown in Fig. 3d. The data are indicated by points, and our model predictions by lines. Notice there is indeed the expected pattern from the equilibrium time signal shown in the plot, i.e. the proportional-to-ν dependence associated with the backward region and the relatively flat ν dependence associated with the forward region.

Thus our quantitative model supports the equilibrium-time interpretation of the data and contributes to the evidence of the Zeno effect in proton-nucleus collision.

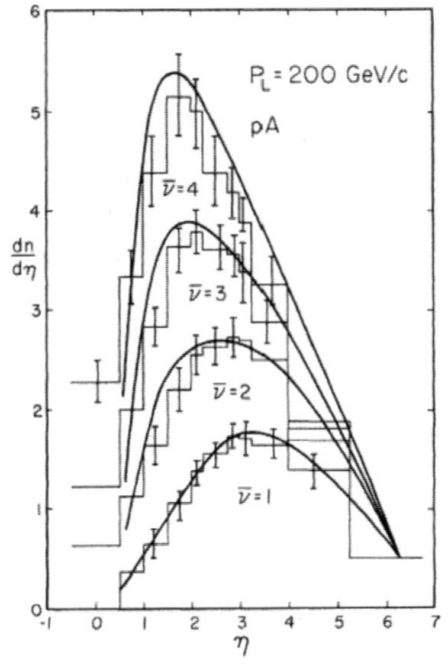

Figure 4. A comparison between a quantitative model considered by Valanju et al[9] (lines) and the data (histogram). Here pseudorapidity η-distributions are compared. Each pseudorapidity η-distribution here is approximately the same as the corresponding rapidity y-distribution.

Example 3: Zeno's effect & moving observer.

This example is based on recent discussions between E. C. G. Sudarshan and myself on the quantum measurement related effect discussed by Aharonov and Vardi[10] over two decades ago.

AV effect: Let me first recall their observation. Consider the following 3 sets of spinor states:

- A stationary spinor state: $|\psi_0>=|\sigma_x=+1\rangle$,
- A rotating spinor state: $|\psi(t)>=R_z(\theta(t))|\psi_0\rangle$, where $\theta(t)=\omega t$, with ω being some constant frequency.
- A set of fiducial spinor states $|\mu_i>\equiv|\psi(t_i)\rangle$. In words, the $|\mu_i\rangle$'s are a set of discrete states, judiciously chosen to follow the rotating spinor state $|\psi(t)\rangle$. At the discrete points, the fiducial states are identical to the corresponding rotated states.

Now consider the effect due to invoking N equal-time consecutive measurements on the successive measured quantum-states, beginning with the initial state $|\psi_0\rangle$. Here measurements are carried out in the time sequence $\{\ t=t_i\}$, with $i=1,\ 2,\ \ldots,\ N$. The corresponding measuring fiducial states are $\mu_i's$. In the large N limit, the "survival probability" at T is given by,

$$Q(T)=\left|<\mu_1|\psi_0\rangle^2\right|\cdot\left|<\mu_2|\psi_0\rangle^2\right|\cdots\left|<\mu_N|\psi_0\rangle^2\right|\to 1. \tag{19}$$

In other words, quantum measurements on $|\psi_0\rangle$, using a judiciously chosen set of fiducial states which follows the rotating state, causes $|\psi_0\rangle$ to rotate in the same manner as the rotating state. We will refer to quantum measurements leading the motion of a stationary state as the AV effect.

Our objective and approach: We would like to show that the AV effect may be regarded as a Zeno effect as seen by a moving observer. To avoid considerations involving a rotating frame (which is a non-initial frame), we will first construct a toy model, which involves only inertial frames. Using this toy model, we will simulate the AV effect in an inertial frame. We will then re-interpret the simulated effect in terms of the Zeno effect.

A toy model

- A box-state:

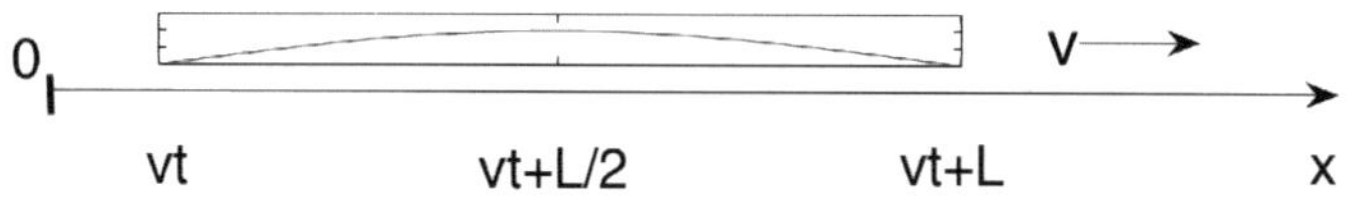

Figure 5. Schematic illustration of the box-state.

Consider a one-dimensional box state as shown in Fig. 5. The box is moving in the x-direction with a velocity v. The wave function inside of the box is given by

$$\psi(x,vt)=f(x,vt),\ \text{where}\ f(x,vt)=\sqrt{\frac{2}{L}}sin\frac{\pi(x-vt)}{L}. \tag{20}$$

Outside of the box, the wave function is zero.

- Fiducial states: The measuring device box is moving in the x-direction with a velocity u. The wave function of the ith fiducial state of the device is defined in terms of the corresponding wave function of the box-state at $t=t_i$, i.e.

$$\mu(x,ut_i)=f(x,ut_i). \tag{21}$$

where $t_i=iT/N$, with $i=1,2,\cdots$ N.

- Survival amplitude: The survival amplitude at $t = t_i$ is given as the overlap between the ith fiducial state and the box-state evaluated at $t = t_i$, which is given by

$$a(t_i) = \int_{ut_i}^{L+ut_i} \mu_i(x, ut_i)\psi(x, vt_i)dx = cos\left[\frac{\pi(v-u)}{L}t_i\right] + R(u, v, t_i). \qquad (22)$$

  The wave function of the box-state is constrained to be non-vanishing only inside of the box. The same constraint is also applicable for the fiducial state. It follows that, for the integrand of the x-integral in eq. (22), when $v > u$, a step function $\theta(x - vt_i)$ is present, and for $v < u$, $\theta(L + vt_i - x)$ is present. On the right hand side of eq. (22), the first term is the contribution, when the step function constraint is not included, while the second term is the correction term after the inclusion of the step-function. In the large N limit, it turned out that $R(u, v, T/N) \propto (T/N)^2$.
- Quantum measurement: We invoke the orthodox theory of quantum measurement, that at the i-th measurement the box-state is collapsed to the $|\mu_i(x, ut_i)\rangle$ state. Then with this new state-identity, the box-state further evolves.

This completes our definition of the toy model.

Zeno effect case & AV effect case

- Zeno effect case: $v > 0$ and $u = 0$. Here the box state is moving with a velocity v and the measuring device is stationary. Now invoke the standard N equal-time-interval measurement over the time T. From eq. (22) the survival amplitude in the large N limit at $t = T$ is given by

$$Q(T) = \left[cos\left(\frac{\pi v}{L}\cdot\frac{T}{N}\right) + R(0, v, T/N)\right]^N \sim \left[1 - \frac{const}{N^2}\right]^N \rightarrow exp\left[-\frac{const}{N}\right] \rightarrow 1. \qquad (23)$$

  In other words, in the large N limit, the box state is stationary with respect to the device which is stationary. In other words, due to large-N quantum measurements, the original moving box-state, is now stationary. This is the Zeno effect.
- AV effect case: $v = 0$ and $u > 0$. Here the box-state is stationary and the measuring device is moving with a velocity u. Again perform the standard set of N equal-time measurements on the box state over a time interval T. From eq.(22), in the large N limit the survival probability at T is given by,

$$Q(T) = \left[cos\left(\frac{\pi u}{L}\cdot\frac{T}{N}\right) + R(u, 0, T/N)\right]^N \sim \left[1 - \frac{const}{N^2}\right]^N \rightarrow exp\left[-\frac{const}{N}\right] \rightarrow 1. \qquad (24)$$

  So with the measurements, the original stationary box-state is now moving along with the measuring device, which is the AV effect.
- Reinterpretation: Now consider an observer co-moving with the measuring device. Without measurements, the observer sees that the box-state is moving backward with a velocity $-u$. On the other hand, with measurements, in large N limit the survival probability is again 1. This implies that the box-state is now stationary relative to the measuring device, or to the co-moving observer. Thus the large-N measurements interrupt the observed backward motion of the box-state. This is Zeno effect, more specifically the Zeno effect as seen by the moving observer and the wave moves with the same velocity.

Summary: We have looked at 3 examples of Zeno effect. We would like to make some comments in a somewhat broader context.

- Unstable quantum system. Our Lee model study suggests that in general there should be a Zeno-time T_1 associated with the decay of an unstable particle. In all cases we have considered, the time scale turned out to be very small.
- The equilibrium time: Our investigation of solvable models agrees with Feinberg′s conclusion that when a quantum system is created, it is in a non-equilibrium state. Some finite time is required for the system to reach equilibrium. We identify this equilibrium time as the Zeno time. In this quantum Zeno paradox session of this symposium, you have already learned about experimental efforts to demonstrate the Zeno effect in the talks [12, 13, 14]. We would like to add to the list of evidence for Zeno effect the pion-production spectrum in proton-nucleus collisions at high energies.
- AV effect: According to the orthodox theory of quantum measurement, each measurement leads to the collapse of the wave function of the measured state into that of the fiducial state. Thus when successive fiducial states are varying slowly, the measured quantum-state is expected to be guided along by the successive fiducial states. This is the AV effect, which is the Zeno effect seen by an observer co-moving with the measuring device. In this context one may regard the Zeno effect hitherto considered as a special case of our example 3. since in earlier cases considered, the observer is co-moving with measuring device with a zero velocity,

To conclude, let us reiterate that the Quantum Zeno effect is a manifestation of general properties of nature (i.e quantum systems where the Hamiltonian is defined). We have already seen that Zeno effect shows up quite convincingly in proton-nucleus collision. We anticipate Zeno effect may also play an important role in relativistic heavy ion production.[16]

Acknowledgement

Special thanks to Professor Rodger Walser and the organizing committee for inviting me to give this talk. George has been an important mentor in my research career. I am indebted to him for his guidance and invaluable discussions throughout these years, including during my preparation of this talk. As he continues to be active in physics, Zeno effect should continue to be operative. The survival probability of his brilliant physics mind should remain at unity. My best wishes to George′s health, and to many active and productive years to come.

Thanks to Dr. Preshant Valanju and my students Man-Fung Cheung and Matt Haley for discussions.

References

[1] Misra B and Sudarshan E C G 1977 J. Math. Phys. 10, 756
[2] Chiu C B, Misra B, and Sudarshan E C G 1977 Phys. Rev. D16, 520
[3] Lee T D 1954 Phys. Rev. 95, 1329
[4] Fredrichs K O 1948 Commun. Pure Appl. Math. 1, 361
[5] Dirac P A M 1935 The Principles of quantum Mechanics (clarendon Press, London), 2nd edition, Chap. 9
[6] Feinberg E L 1966 Sov. Phys. JETP 23, 132
[7] Elias J *et. al.* 1977 Phys. Rev. Lett. 39, 1499
[8] Bialkowski G, Chiu C B and Tow D 1978 Phys. Rev. D 17, 862
[9] Valanju P, Sudarshan E C G, and Chiu C B 1980 Phys. Rev. D, 21, 1304
[10] Aharonov Y and Vardi M 1980 Phys. Rev. D, 21, 2235
[11] Jenkins F A and White H E 1957 Fundamentals of Optics, McGraw-Hill, Chapters 27, 28
[12] Marmo G 2006 Zeno Dynamics, talk at this symposium
[13] Pascazio S 2006 Quantum Zeno Dynamics, talk at this symposium
[14] Itano W M 2006 Perspectives on the Quantum Zeno Paradox, talk at this symposium
[15] RHIC data: See for example Adler S S *et. al.* 2006 (PHENIX collaboration), Phys. Rev. Lett. **97**, 052301
[16] At the end of my talk, Professor Sid Meshkov commented that it is worthwhile to investigate the implication of Zeno effect in relativistic heavy ion collisions (RHIC)[15]. His comment is very appropriate. Indeed, we have just begun our investigation on this problem.

Journal of Physics: Conference Series **196** (2009) 011011 doi:10.1088/1742-6596/196/1/011011

Introduction to "Tachyons" : the sixth quest

David Freedman

Address: 45 Rivard Rd. Needham, MA 02492

Email: david@freedman.com

Thank you for inviting me to chair this session.

I am not sure whether George knows this or not - the article I wrote about Tachyons was my first major magazine article and it started my career. In fact, I may not have been able to make it as a writer if it had not been for that first article which was very well received. And of course I credit George entirely for that. On top of all the wonderful things that have been said about George yesterday and today, George is a fantastic interviewee. He was the first major scientist I interviewed, and I assumed that all scientists would be this accessible, this interesting, this provocative, and I have spent the rest of career in vain trying to duplicate that fir terrific experience with George, It may have been the most interesting article I have ever written - I just happen to hit the lottery my very first time out.

The other thing I would like to say about Tachyons - it is a subject that has captured the public's fascination - perhaps more than any other subject in Physics. In fact, it has become a staple of science fiction. A lot of physicists look askance when things get into public mind like this - thinking maybe it makes the subject silly, wrong, or distorts it. But I will argue that it is a very very good thing because it makes the connection between physics and the public - especially the young people who are drifting towards physics are drawn to it because of things like this. I think it great when people make these connections - and yet another thing for which George deserves credit.

Journal of Physics: Conference Series **196** (2009) 011011 doi:10.1088/1742-6596/196/1/011011

Speakers at Tachyons: Sudarshan's Sixth quest

David Freedman

Erasmo Recami

Oleksa-Myron Bilaniuk

Samir Bose

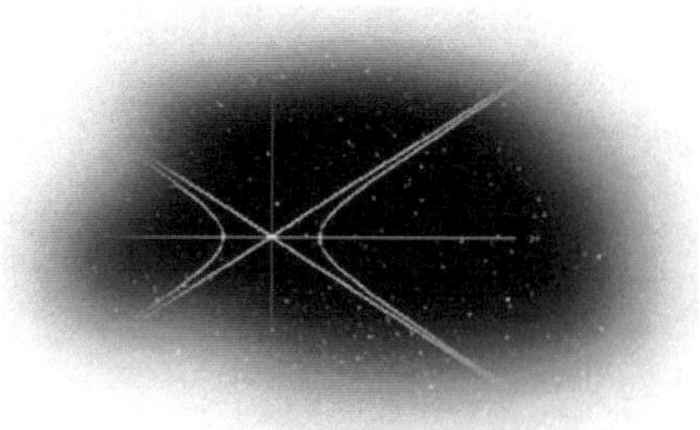

Journal of Physics: Conference Series **196** (2009) 012020 doi:10.1088/1742-6596/196/1/012020

Superluminal waves and objects: an overview of the relevant experiments

Erasmo Recami

Facoltà di Ingegneria, Università statale di Bergamo, Dalmine (BG), Italy; and INFN—Sezione di Milano, Milan, Italy.

E-mail: recami@mi.infn.it

1. Introduction.

It is a great privilege and pleasure to be able to present (after a brief theoretical introduction) a panoramic view of the experiments that —revealing the apparent existence of Superluminal motions— seem to confirm the pioneering works published by E.C.George Sudarshan, already in the sixties, about tachyons.

The question of Superluminal ($V^2 > c^2$) objects or waves has a long story, starting perhaps in 50 B.C. with Lucretius' *De Rerum Natura pervolgant* lumina *coelum?"*. Still in pre-relativistic times, one meets various related works, from those by J.J. Thomson to A. Sommerfeld.

With Special Relativity, however, since 1905 the conviction spread over that the speed c of light in vacuum was the *upper* limit of any possible speed.

For instance, R.C.Tolman in 1917 believed to have shown by his "paradox" that the existence of particles endowed with speeds larger than c would have allowed sending information into the past. Such a conviction blocked for more than half a century – aside from an isolated paper (1922) by the Italian mathematician G.Somigliana – any research about Superluminal speeds. Our problem started to be tackled again essentially in the fifties and sixties, in particular after the mentioned, epoch-making papers by George Sudarshan *et al.*[1], which provoked much further work, in particular by E.Recami and coworkers[2], as well as by H.C.Corben and others (to confine ourselves to the *theoretical* researches). The first experiments looking for faster-than-light objects were performed by T.Alväger *et al.*[2].

Superluminal objects were called tachyons, T, by G.Feinberg, from the Greek word $\tau\alpha\chi\acute{\upsilon}\varsigma$, quick (and this induced the present author in 1970 to coin the term bradyon, for ordinary subluminal ($v^2 < c^2$) objects, from the Greek word $\beta\rho\alpha\delta\acute{\upsilon}\varsigma$, slow). Finally, objects travelling exactly at the speed of light are called "luxons".

In recent years, terms as "tachyon" and "superluminal" fell unhappily into the (cunning, rather than crazy) hands of pranotherapists and mere cheats, who started squeezing money out of simple-minded people; for instance by selling plasters (!) that should cure various illnesses by "emitting tachyons"... We are dealing with tachyons here, however, since at least four different experimental sectors of physics seem to indicate the actual existence of Superluminal motions (thus confirming long-standing theoretical predictions [1,3]).

In Sec. 2 (after a non-technical theoretical introduction) we mention the various experimental sectors of physics in which Superluminal motions seem to appear. In particular, a bird's-eye view is presented of the experiments with evanescent waves (and/or tunnelling photons), and with the "localized Superluminal solutions" (SLS) to the wave equations, like the so-called X-shaped waves; the shortness of this review is compensated by a number of references, sufficient in some cases to provide the interested readers with reasonable bibliographical information.

2. General concepts

As far as classical tachyons are concerned, let us insert Sudarshan's original contributions within the picture provided by Special Relativity (SR), once one *does not restrict* it[2,3] to subluminal motions.

Let us premise that SR, abundantly confirmed by experience, can be built on the two simple, natural Postulates:

1) that the laws (of electromagnetism and mechanics) are valid not only for a particular observer, but for the whole class of the "inertial" observers;

2) that space and time are homogeneous and space is moreover isotropic.

From these Postulates one can theoretically *infer* that one, and only one, *invariant* speed exists: and experience tells us such a speed to be the one, c, of light in vacuum (namely, 299.792458 km/s). Indeed, ordinary light possesses the peculiar feature of presenting always the same speed in vacuum, even when we run towards or away from it. It is just that feature, of being invariant, that makes the speed c quite exceptional: no bradyons, and no tachyons, can enjoy the same property.

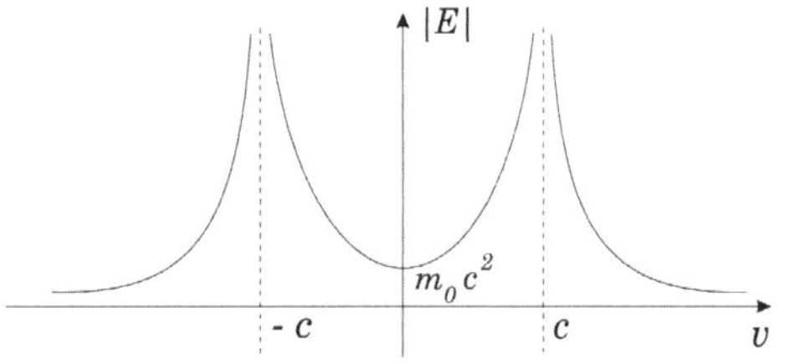

Figure 1. Energy of a free object as a function of its speed.

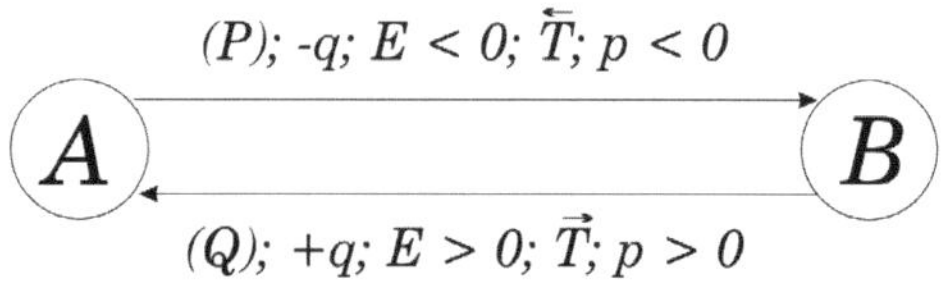

Figure 2. The "switching rule" (or reinterpretation principle) by Stueckelberg-Feynman-Sudarshan.

Another (known) consequence of our Postulates is that the total energy of an ordinary particle increases when its speed v increases, tending to infinity when v tends to c. Therefore, infinite forces would be needed for a bradyon to reach the speed c. This fact generated the popular opinion that speed c can be neither achieved nor overcome.

However, as speed c photons exist which are born, live, and die always at the speed of light[1] (without any need of accelerating from rest to the light speed), so objects can exist[4] always endowed with speeds V larger than c (see Fig.1). This circumstance has been picturesquely illustrated by Sudarshan (1972) with reference to an imaginary demographer studying the population patterns of the Indian subcontinent:

> Suppose a demographer calmly asserts that there are no people North of the Himalayas, since none could climb over the mountain ranges! That would be an absurd conclusion. People of central Asia are born there and live there: they did not have to be born in India and cross the mountain range. So with faster-than-light particles.

Let us add that, still starting from the above two Postulates (besides a third postulate, even more obvious, the theory of relativity can be generalized[1,3] in such a way as to accommodate also Superluminal objects; such a *non-restricted* version of SR is sometimes called "extended relativity". Also within extended relativity[3] the speed c, besides being invariant, is a limiting velocity: but every limiting value has two sides, and one can a priori approach it both from the left and from the right.

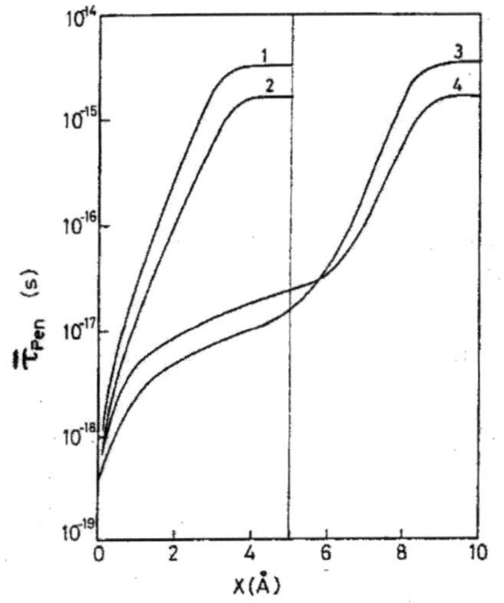

Figure 3. Behaviour of the average "penetration time" spent by a tunnelling wavepacket, as a function of the penetration depth down a potential barrier.

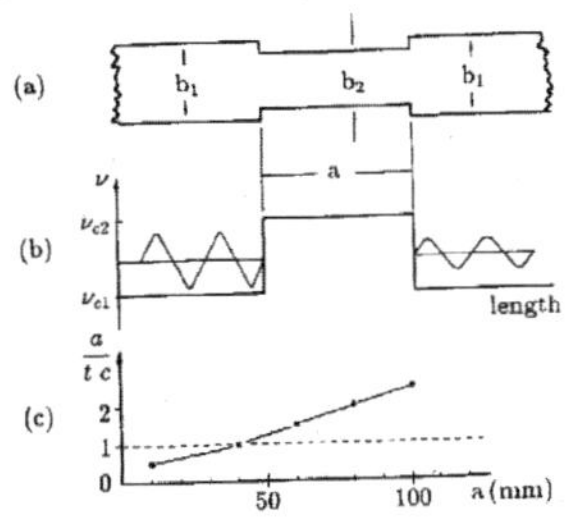

Figure 4. Simulation of quantum tunnelling by experiments with classical evanescent waves.

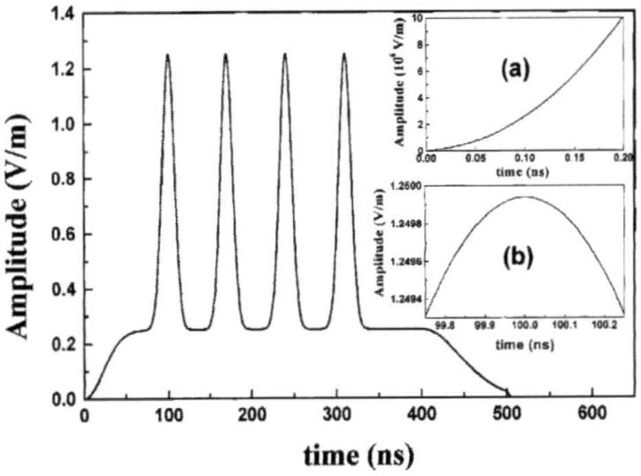

Figure 5. Experiment along a metallic waveguide with two barriers (undersized guide segments), i.e., with two evanescence regions

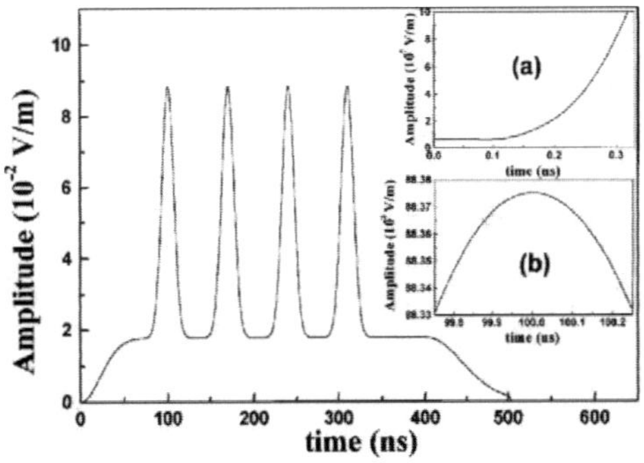

Figure 6. Appearances of a spherical (or pointlike, at the limit) object in vacuum at $v < c$ and $v > c$.

Actually, as we were saying, the ordinary formulation of SR has been restricted *too much*. For instance, even leaving Superluminal speeds aside, it can be easily so widened as to include antimatter[5]. Then, one finds space-time to be a priori populated by normal particles P (which travel forward in time carrying positive energy), *and* by dual particles Q "which travel backwards in time carrying negative energy". The latter shall appear to us as antiparticles, i.e., as particles – regularly travelling forward in time with positive energy, but – with all their "additive" charges (e.g., the electric charge) reversed in sign[5,1]: see Fig.2.

To clarify this point, we can here recall only what follows: We, as macroscopic observers, have to move in time along a single, well-defined direction, to such an extent that we cannot even see a motion backwards in time... and every object like Q, travelling backwards in time (with negative energy), will be *necessarily* reinterpreted by us as an anti-object, with opposite

Journal of Physics: Conference Series **196** (2009) 012020 doi:10.1088/1742-6596/196/1/012020

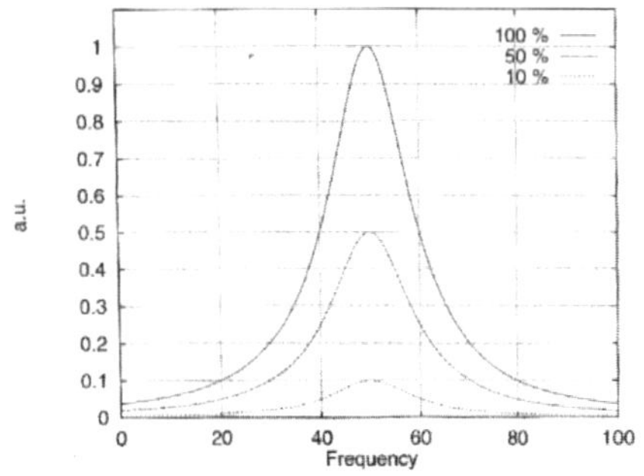

Figure 7. Intersections of an "X-shaped wave" with planes orthogonal to its motion line.

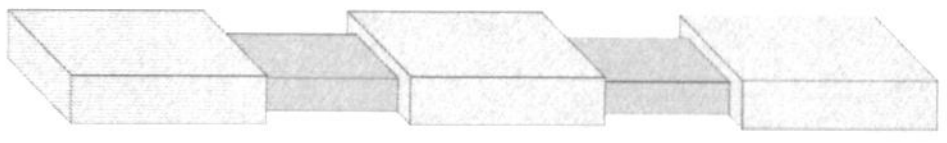

Figure 8. C

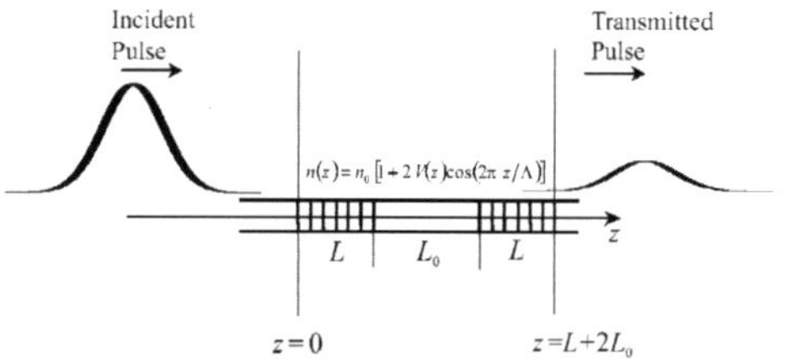

Figure 9. Theoretical prediction of Superluminal localized "X-shaped" waves for the electromagnetic case.

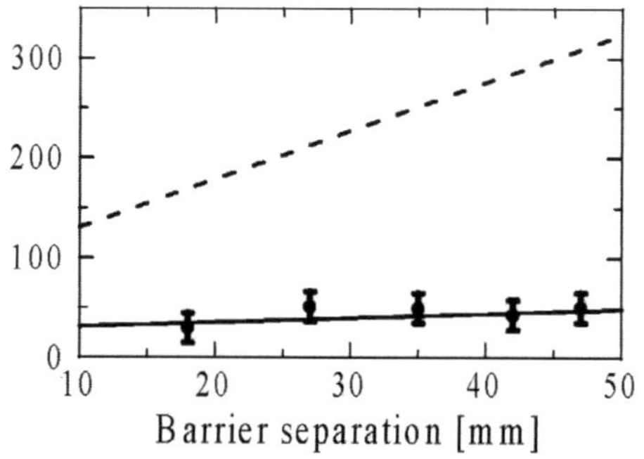

Figure 10. Scheme of the experiment by Saari *et al.*, who announced the production, in optics, of the waves depicted in Fig.8.

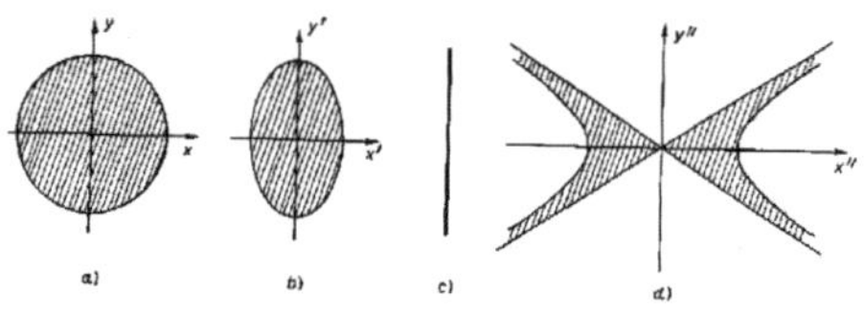

Figure 11. The square magnitude of the "classical", X-shaped Superluminal Localized Solution (SLS).

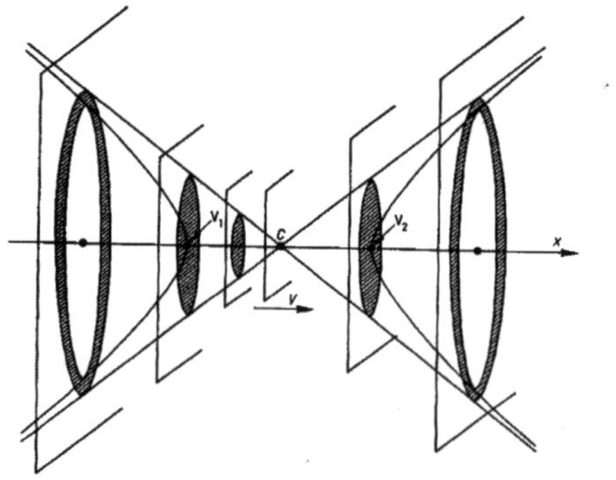

Figure 12. C

charges but travelling forward in time (with positive energy): cf. Fig.2 and refs.[3-5,1].

But let us forget about antimatter and go back to "tachyons". A strong objection against their existence is based on the opinion that by using tachyons it would be possible to send signals into the past, owing to the fact that a tachyon T which, say, appears to a first observer O as emitted by A and absorbed by B, can appear to a second observer O' as a tachyon T' which travels backwards in time with negative energy[1,3]. However, by applying (as it is obligatory to do) the same "reinterpretation rule" or switching procedure seen above, T' will appear to the new observer O' just as an antitachyon $\overline{\mathrm{T}}$ emitted by B and absorbed by A, and therefore travelling forward in time, even if in the contrary *space* direction. In such a way, every travel towards the past, and every negative energy, disappear[1,3-5]. The mentioned reinterpretation procedure[1,3-5] ought to be called the Sudarshan's principle, or the Stueckelberg-Feynman-Sudarshan principle: indeed, it was Sudarshan[1] who stated it clearly, by taking proper account of the interplay between the signs *both* of the motion direction in time *and* of the energy.

Starting from this observation, it is possible to solve[1,6] the so-called causal paradoxes associated with Superluminal motions: paradoxes which result to be the more instructive and amusing, the more sophisticated they are, but that cannot be re-examined here.

Let us mention here just the following. The reinterpretation principle, according to which, in simple words, signals are carried only by objects which appear to be endowed with positive energy, does eliminate any information transfer backwards in time; but this has a price: that of abandoning the ingrained conviction that the judgement about what is cause and what is effect is independent of the observer[1-6]. In fact, in the case examined above, the first observer O considers the event at A to be the cause of the event at B. By contrast, the second observer O' will consider the event at B as causing the event at A. All the observers will however see the cause to happen *before* its effect[1-6].

Taking new objects or entities into consideration always forces us to a criticism of our prejudices. If we require the phenomena to obey the *law* of (retarded) causality with respect to all the observers, then we cannot demand also the *description* "details" of the phenomena to be invariant: namely, we cannot demand in that case also the invariance of the "cause" and "effect" *labels*[6,2].

To illustrate the nature of our difficulties in accepting that, e.g., the labels of cause and effect depend on the observer, let us cite an analogous situation that does not imply present-day prejudices:

> For ancient Egyptians, who knew only the Nile and its tributaries, which all flow South to North, the meaning of the word "south" coincided with the one of "upstream", and the meaning of the word "north" coincided with the one of "downstream". When Egyptians discovered the Euphrates, which unfortunately happens to flow North to South, they passed through such a crisis that it is mentioned in the stele of Tuthmosis I, which tells us about *that inverted water that goes downstream (i.e. towards the North) in going upstream* [Csonka, 1970].

In the last century, theoretical physics led us in a natural way to suppose the existence of various types of objects: like magnetic monopoles, quarks, strings, tachyons, besides black-holes etcetera: and various sectors of physics could not go on without them, even if the existence of most of them is uncertain (perhaps, also because attention has not yet been paid to some links existing among them: e.g., a Superluminal electric charge is expected to behave as a magnetic monopole; and a black-hole a priori can be the source of tachyonic matter). According to Democritus of Abdera, everything that was thinkable without meeting contradictions had to exist somewhere in the unlimited universe. This point of view – which was given by M.Gell-Mann the name of "totalitarian principle" – was later on expressed (T.H.White) in the humorous form "Anything not forbidden is compulsory"...

3. A glance at the experimental status-of-the-art

Extended Relativity can allow a better understanding of many aspects also of *ordinary* physics; and this remains true independently of the circumstance that tachyons do or do not exist in our cosmos as asymptotically free objects (their existence as "intermediate states" is, of course, obvious[1,3]). As already said, we are dealing with Superluminal motions, however, since this topic has recently returned in fashion, especially because of the fact that at least three or four different experimental sectors of physics seem to suggest the possible existence of faster-than-light motions. Our first aim is putting forth in the following some information about the experimental results obtained in a couple of those different physics sectors, with a mere mention of the others.

3.1. Neutrinos

A long series of experiments, started in 1971, seems to show that the square ${m_0}^2$ of the mass m_0 of muon-neutrinos, and more recently of electron-neutrinos too, is negative; which, if confirmed, would mean that (when using a naïve language, commonly adopted) such neutrinos possess an "imaginary mass" and are therefore tachyonic, or mainly tachyonic[7,3]. Notice, incidentally, that in extended relativity the dispersion relation for a free Superluminal object becomes

$$\omega^2 - \boldsymbol{k}^2 = -\Omega^2, \text{ or } E^2 - \boldsymbol{p}^2 = -m_{\text{o}}^2,$$

and there is *no* need therefore of imaginary masses! The present author can testify that at least by 1971 —and probably some years before (as well as we ourselves, by the way)— George Sudarshan had got the idea that neutrinos could be tachyons: an idea proposed in print by Cawley[7] later on, in 1972.

3.2. Galactic Micro-quasars

As to the *apparent* Superluminal expansions observed in the core of quasars[8] and, recently, in the so-called galactic microquasars[9], we shall not deal here with that problem, because it is far from the other topics of this paper: not to mention that for those astronomical observations there exist orthodox interpretations, based on ref.[10], that —even if "statistically" weak— are accepted by the majority of the astrophysicists.

Here, let us mention only that simple geometrical considerations in Minkowski space show that a *single* Superluminal light source would appear[11,3]: (i) initially, as a source in the "optical boom" phase (analogous to the acoustic "boom" produced by an airplane travelling with constant supersonic speed): namely, as an intense source which suddenly comes into view; and that (ii) afterwards seems to split into TWO objects receding one from the other with speed $V > 2c$ [both phenomena being similar to those actually observed, according to refs.[9]].

3.3. Evanescent waves and "tunnelling photons"

Within quantum mechanics (and precisely in the *tunnelling* processes), it had been shown that the tunnelling time —firstly evaluated as a simple "phase time" and later on calculated through the analysis of the wavepacket behaviour— does not depend on the barrier width in the case of opaque barriers ("Hartman effect")[12]. This implies Superluminal and arbitrarily large (group) velocities V inside long enough barriers: see Fig.3.

Experiments that may verify this prediction by, say, electrons are difficult: And, in fact, only preliminary results for tunnelling neutrons exist[12,26]. Luckily enough, however, the Schroedinger equation in the presence of a potential barrier is mathematically identical to the Helmholtz equation for an electromagnetic wave propagating, for instance, down a metallic waveguide along the x-axis (as recalled, e.g., by R.Chiao *et al.*[13]); and a barrier height U greater than the electron energy E corresponds (for a given wave frequency) to a waveguide of transverse size lower than a cut-off value. A segment of "undersized" guide —to go on with our example— does therefore behave as a barrier for the wave (photonic barrier)[16,13], as well as any other photonic band-gap filters. The wave assumes therein —like a particle inside a quantum barrier— an imaginary momentum or wave-number and gets, as a consequence, exponentially damped along x. In other words, it becomes an *evanescent* wave (going back to normal propagation, even if with reduced amplitude, when the narrowing ends and the guide returns to its initial transverse size). Thus, a tunnelling experiment can be simulated[13,16] by having recourse to evanescent waves (for which the concept of group velocity can be properly extended[14]).

The fact that evanescent waves travel with Superluminal speeds (cf., e.g., Fig.4) has been actually *verified* in a series of famous experiments.

Namely, various experiments, performed since 1992 onwards by G.Nimtz *et al.* at Cologne[15], by R.Chiao, P.G.Kwiat and A.Steinberg's group at Berkeley[16], by A.Ranfagni and colleagues at Florence[17], and by others at Vienna, Orsay, Rennes[17] etc., verified that "tunnelling photons" travel with Superluminal group velocities. Let us add that also extended relativity had predicted[19] evanescent waves to be endowed with faster-than-c speeds; the whole matter appears to be therefore theoretically self-consistent. The debate in the current literature does not refer to the experimental results: which can be correctly reproduced even by numerical elaborations[20,21] based on Maxwell equations only (cf., e.g., the illuminating figures 11 and 13 in ref.[21], here reproduced as Figs.5* and 6* of this paper), but rather to the question whether they allow, or do not allow, sending signals or information with Superluminal speed[22,21,14]. Actually, a peaked wavepacket suffers a strong amplitude attenuation while travelling inside a quantum or classical barrier; its width, however, remains unaffected (cf. Fig.7*): Something that might have some relevance when thinking of attempting transmissions by Morse's alphabet). Moreover, many authors have emphasized that —at least in the case of quantum barriers— the tunnelling of particles is a statistical process, in the sense that one cannot know a priori which particle, or photon, will pass through the barrier. This is true, of course; but the weight of such a consideration becomes lower when the number of the particles at our disposal for attempting a "signal" transmission does increase. To remain within the Morse alphabet example, one can send out *dots* and *dashes* by emitting pulses of, say, one thousand and ten thousand particles each, respectively; the dot and dashes will then be recognized also after the tunnelling!... This becomes even more meaningful when one approaches the classical limit. The claim that superluminal tunnelling cannot be used to transmit any information is in need, therefore, of further discussion, especially at the light of what will follow below: More details can be found, anyway, in refs.[50,51].

Let us repeat that all the phenomena mentioned in this, as well as in the following, sub-sections can be accomodated into the standard frameworks of quantum physics or of classical relativistic physics: It is therefore obvious that such phenomena can receive explication or interpretation in terms of standard physics (e.g., sometimes but not always, in terms of suitable "reshapings"): but this does not eliminate the fact that Superluminal motions take place.

As we already said, in the above-mentioned experiments one meets a substantial attenuation of the considered pulses during tunnelling (or during propagation in an absorbing medium). However, by having recourse to suitable devices, as a "gain doublet", it has been recently reported the observation of undistorted pulses propagating with Superluminal group-velocity with a *small* change in amplitude[23].

Let us underline that some of the most interesting experiments of this series seem to be the ones with TWO "barriers" (e.g., with two gratings in an optical fiber, or with two segments of undersized waveguide separated by a piece of normal-sized waveguide: Fig.5=8*). For suitable frequency bands —i.e., for "tunnelling" far from resonances—, it was found that the total crossing time does *not* depend on the length of the intermediate (normal) guide: namely, that the wavepacket speed along it is infinite[24,25]... This agrees with what predicted by Quantum Mechanics for the non-resonant tunnelling through two successive opaque barriers (namely, the tunnelling phase time, which depends on the entering energy, has been shown by us to be *independent* of the distance between the two barriers[26]); something that has been theoretically confirmed, and generalized, by Y.Aharonov *et al.*[26]. Such a prediction has been experimentally verified a second time, with a cleaner experiment, taking advantage of the circumstance that quite interesting evanescence regions can be constructed in the most varied manners, like by means of different photonic band-gap materials or gratings (it being possible to use from multilayer dielectric mirrors, or semiconductors, to photonic crystals...). Indeed, a very recent confirmation came from an experiment having recourse to two gratings in an optical fiber[25]. On this respect, rather interesting are the figures 1 and 5 of ref.[25], here reported as Figs.9* and 10* of this paper (see especially the experimental results depicted in Fig.10*).

We cannot skip a further topic —which, being delicate, should not appear in a brief overview as this one— since some experimental contributions to it (like the one performed at Princeton by J.Wang *et al.*[23] and published in *Nature* on July 20, 2000) arose a general interest.

Even if in extended relativity all the ordinary causal paradoxes seem to be solvable[6,3,1] on the basis of the above-seen Stueckelberg-Feynman-Sudarshan reinterpretation rule, *nevertheless* —let us repeat— one has to remember that (whenever it has to be considered an object, $\mathcal{O}$, travelling with Superluminal speed) one can meet negative contributions to the *tunnelling times*[27,12]: and this should not be regarded as unphysical. In fact, whenever an "object" (electromagnetic pulse, particle,...) $\mathcal{O}$ *overcomes* the infinite speed[3,6] with respect to a certain observer, it will afterwards appear to the same observer as the "*anti*-object" $\overline{\mathcal{O}}$ travelling in the opposite *space* direction[1,3,6].

More precisely, when going on from the lab to a frame $\mathcal{F}$ moving in the *same* direction as the particles or waves entering the barrier region, the object $\mathcal{O}$ penetrating and travelling through the final part of the barrier (with almost infinite speed[12,21,26,27], like in Figs.3) will appear in the frame $\mathcal{F}$ as an anti-object $\overline{\mathcal{O}}$ crossing that portion of the barrier *in the opposite space-direction*[6,3,1]. In the new frame $\mathcal{F}$, therefore, such anti-object $\overline{\mathcal{O}}$ would yield a *negative* contribution to the tunnelling time: which could even result, in total, to be negative. For clarifications, see refs.[28]. What we want to stress here is that the appearance of such negative times is once more predicted by relativity itself, on the basis of the ordinary postulates[3,6,21,28]. (In the case of a non-polarized wave, the wave anti-packet coincides with the initial wave packet; if a photon is however endowed with helicity $\lambda = +1$, the anti-photon will bear the opposite helicity $\lambda = -1$).

From the theoretical point of view, besides refs.[3,6,12,21,27,28], see refs.[29]. On the (quite interesting!) experimental side, see papers [30].

Let us *add* here that, via quantum interference effects it is possible to obtain dielectrics with refraction indices very rapidly varying as a function of frequency, also in three-level atomic systems, with almost complete absence of light absorption (i.e., with quantum induced transparency)[31]. The group-velocity of a light pulse propagating in such a medium can decrease to very low values, either positive or negatives, with *no* pulse distortion. It is known that experiments have been performed both in atomic samples at room temperature, and in Bose-Einstein condensates, which showed the possibility of reducing the speed of light to a few meters per second. Similar, but negative group velocities, implying a propagation with Superluminal speeds thousands of times higher than the previously mentioned ones, have been recently predicted also in the presence of such an "electromagnetically induced transparency", for light moving in a rubidium condensate[32], while the corresponding experiments are being performed (for instance at the "LENS" laboratory in Florence).

Finally, let us recall that faster-than-c propagation of light pulses can be (and was, in same cases) observed also by taking advantage of anomalous dispersion near an absorbing line, or nonlinear and linear gain lines —as already seen—, or nondispersive dielectric media, or inverted two-level media, as well as of some parametric processes in nonlinear optics (cf., e.g., G.Kurizki *et al.*'s works).

3.4. Superluminal Localized Solutions (SLS) to the wave equations. The "X-shaped waves"

The fourth sector (to leave aside the others) is not less important. It came into fashion again, when some groups of scholars in engineering (for sociological reasons, most physicists had abandoned the field) rediscovered by a series of works that any wave equation —to fix the ideas, let us think of the electromagnetic case— admit also solutions as much sub-luminal as Super-luminal (besides the ordinary —plane, spherical,...— waves endowed with speed c/n).

Let us recall that, starting with the pioneering work by H.Bateman, it had slowly become known that all wave equations (in a general sense: scalar, electromagnetic, spinorial,...) admit wavelet-type solutions with sub-luminal group velocities[33]; namely, soliton-like solutions, even if they are linear equations. Subsequently, also Superluminal solutions started to be written down. A quite important feature of some new solutions of these (which attracted much attention for possible applications) is that they propagate as *localized*, non-diffracting pulses: namely, according to the Courant and Hilbert's terminology[33], as "undistorted progressing waves"; which possess the further property of "self-reconstructing" themselves after obstacles smaller than their *aperture* size (that is, smaller than the width of the antenna generating them, enourmously larger, in general, than their wavelenth). It is easy to realize the practical importance, for instance, of a radio transmission carried out by localized waves, independently of their being sub- or Super-luminal. But non-diffractive wave packets can be of use even in theoretical physics for a reasonable representation of elementary particles[37]; and so on.

Within extended relativity since 1980 it had been found that —whilst the simplest subluminal object conceivable is a small sphere, or a point as its limit— the simplest Superluminal objects turns out to be instead an "X-shaped" wave (see refs.[38], and Figs.6=11* and 7=12* of this paper), or a double cone as its limit, which moreover travels without deforming —i.e., rigidly— in a homogeneous medium[3]. Analogously, the equipotential surfaces of the electrostatic field, generated by a tiny charged sphere at rest, will assume[3,38] the shape represented in Fig.13* when the source is Superluminal. For clarifying the connection existing between what predicted by SR and the localized X-waves (mathematically and experimentally constructed in recent times, as we are going to see) let us refer ourselves to a paper recently appeared in Physical Review E, i.e., to the last one of refs.[38], where the issue of the (X-shaped) field created by a Superluminal electric charge has been tackled. Localized waves do exist, of course, with any group velocicty (the subluminal ones being ball-like, as expected); but it is not without meaning that the most interesting localized solutions happened to be just the Superluminal ones, and with a shape of that kind. Even more, since from Maxwell equations under simple hypotheses one goes on to the usual *scalar* wave equation for each electric or magnetic field component, one could expect the same solutions to exist also in the field of acoustic waves, and of seismic waves (and of gravitational waves too).

In other words, from the present point of view, it is rather interesting to note that, during the last fifteen years, "X-shaped" waves have been *actually* found as solutions to the Maxwell and to the wave equations [the form of any wave equations is intrinsically relativistic, by the way]. Actually, such waves (as suitable superpositions of Bessel beams[39], that is, of simple solutions to the wave equation already discovered[34] in 1941) were mathematically constructed for the first time by Lu *et al.*[40], *in acoustics*: and later on by Recami *et al.*[41] for electromagnetism; and were then called "X-waves" or rather X-shaped waves. In an elementary Appendix we briefly show how X-shaped solutions to the wave equation (in particular, the "classic" X-wave) can be constructed.

It is more important for us that the X-shaped waves have been indeed *produced in experiments* both with acoustic and with electromagnetic waves; indeed, X-waves were produced that, in their medium, travel undistorted with a speed larger than sound, in the first case, and than light, in the second case. In acoustics, the first experiment was performed by Lu *et al.* themselves[42] in 1992, at the Mayo Clinic (and their papers received the first IEEE 1992 award). In the electromagnetic case, certainly more intriguing, Superluminal localized X-shaped solutions were first mathematically constructed (cf., e.g., Fig.8=14*) in refs.[41], and later on experimentally produced by Saari *et al.*[43] in 1997 at Tartu by visible light (Fig.9=15*), as announced in their Physical Review Letters article, and more recently, as we already mentioned, by Mugnai, Ranfagni and Ruggeri at Florence by microwaves[44] (paper appeared in Physical Review letters too, three years later, in 2000). Further experimental activity is in progress; while in the theoretical sector the activity has been growing so intensely, that it is not possible to quote here the relevant recent literature; we might recall, e.g., the papers devoted to building up new analogous solutions with finite total energy or more suitable for high frequencies, on one hand, and localized solutions Superluminally propagating even along a normal waveguide, on the other hand[45,46]; or the attempts at focusing X-shaped waves, at a certain instant, in a small region[47]. But we cannot avoid mentioning that suitable superpositions of Bessel beams (which can originate also subluminal pulses) can produce even *stationary* intense wave-field: confined within a tiny region (a static envelope); while the field intensity outside that region is everywhere negligible[48]; such "frozen waves" can have (a patent is pending) very many important applications, for instance as a new kind of tweezers, and especially —of course— in medicine.

Before going on, let us eventually touch the problem of producing an X-shaped Superluminal wave like the one in Fig.7=12*, but truncated – of course – in space and in time (by the use of a finite, dynamic antenna, radiating for a finite time): in such a situation, the wave will keep its localization and Superluminality only along a certain "depth of field", decaying abruptly afterwards[39,41].

We can become convinced about the possibility of realizing it, by imagining the simple ideal case of a negligibly sized Superluminal source S endowed with speed $V > c$ in vacuum and emitting electromagnetic waves W (each one travelling with the invariant speed c). The electromagnetic waves will result to be internally tangent to an enveloping cone C having S as its vertex, and as its axis the propagation line x of the source[3].

This is analogous, as we know, to what happens for an airplane that moves in the air with constant supersonic speed. The waves W interfere mostly negatively inside the cone C, and constructively only on its surface. We can place a plane detector orthogonally to x, and record magnitude and direction of the W waves that hit on it, as (cylindrically symmetric) functions of position and of time. It will be enough, then, to replace the plane detector with a plane antenna which *emits* —instead of recording— exactly the same (axially symmetric) space-time pattern of waves W, for constructing a cone-shaped electromagnetic wave C that will propagate with the Superluminal speed V (of course, without a source any longer at its vertex): even if each wave W travels with the invariant speed c.

Here let us only remark that such localized Superluminal waves appear to keep their good properties only as long as they are fed by the waves arriving (with speed c) from the antenna: Taking account of the time needed for fostering such Superluminal pulses (i.e., for the arrival of the feeding speed-c waves coming from the aperture). one concludes that these localized Superluminal waves are probably unable to transmit *information* faster than c. However, they don't seem to have anything to do with the illusory "scissors effect", even if the energy feeding them appears to travel with the speed of light. In fact, the spot —endowed, as we know, with Superluminal group-velocity— is able to get, for instance, two (tiny) detectors at a distance L to click after a time *smaller* than L/c). A lot of discussion is still going on about the possible differences among group-velocity, signal-velocity and information speed.

As we mentioned above, the existence of all these X-shaped Superluminal (or "Super-sonic") waves seem to constitute at the moment, together, e.g., with the Superluminality of evanescent waves, some valuable confirmations of refs.[1], as well as of extended relativity: a theory[3], let us recall, based on the ordinary postulates of SR and that consequently does not appear to violate any of its fundamental principles. It is curious, moreover, that one of the first applications of such X-waves (that takes advantage of their propagation without deformation) is in advanced progress in the field of medicine, and precisely of ultrasound scanners[49].

Before ending, let us remark that a series of new SLSs to the Maxwell equations, suitable for arbitrary frequencies and arbitrary bandwidths have been recently constructed by us: many of them being extremely well localized in the surroundings of their vertex, and some of them being endowed with *finite* total energy. Among the others, we have set forth an infinite family of generalizations of the classical X-shaped wave; and shown how to deal with the case of a *dispersive* medium. Results of this kind may find application in other fields in which an essential role is played by a wave-equation.

Acknowledgements
The author is deeply indebted, besides to George Sudarshan, to all the Organizers of this Symposium for their very kind invitation and hospitality: in particular to Bhamathi Sudarshan, R.M.Walser, A.Valanju, P.Valanju; and acknowledges stimulating discussions along the years with E.C.G.Sudarshan M.Zamboni Rached, A.van der Merwe, D.Stauffer, A.Gigli Berzolari, as well as with G.Battistoni, J.D.Bekenstein, R.Bonifacio, R.Chiao, G.Degli Antoni, F.Fontana, A.Friberg, L.Horwitz, M.Ibison, G.Marmo, P.Milonni, D.Mugnai, G.Nimtz, V.S.Olkhovsky, P.Pizzochero, A.Ranfagni, P.Saari, A.Shaarawi and A.Steinberg.

APPENDIX

In this Appendix we want to show how localized superluminal solutions (SLS) to the wave equations, and in particular the X-shaped ones, can be mathematically constructed. Here we shall consider, for simplicity, only the case of a dispersionless medium like vacuum, and of free space (without boundaries).

It is known since more than a century that a particular axially symmetric solution to the wave equation in vacuum ($n = n_0$) is, in cylindrical coordinates, the function $\psi(\rho, z, t) = J_0(k_\rho \rho)\, e^{+ik_z z} e^{-i\omega t}$ with $k_\rho^2 = n_0^2 \frac{\omega^2}{c^2} - k_z^2$; $k_\rho^2 \geq 0$, where J_0 is the zeroth-order ordinary *Bessel function*, k_z and k_ρ are the axial and the transverse wavenumber respectively, ω is the angular frequency and c the velocity of light. Using the transformation

$$\begin{cases} k_\rho = \dfrac{\omega}{c}\, n_0 \sin\theta \\ \\ k_z = \dfrac{\omega}{c}\, n_0 \cos\theta \end{cases} \tag{1}$$

such particular solution $\psi(\rho, z, t)$ can be rewritten in the the well-known *Bessel beam* form:

$$\psi(\rho, \zeta) \;=\; J_0(n_0 \frac{\omega}{c} \rho \sin\theta)\; e^{+in_0 \frac{\omega}{c} \zeta \cos\theta} \tag{2}$$

where $\zeta \equiv z - Vt$ while $V = c/(n_0 \cos\theta)$ is the phase velocity, quantity θ ($0 < \theta < \pi/2$) being the cone angle of the Bessel beam.

More in general, SLSs (with axial symmetry) to the wave equation will be the following ones[40,50]:

$$\psi(\rho, \zeta) \;=\; \int_0^\infty S(\omega) J_0 \left(\frac{\omega}{V} \rho \sqrt{n_0^2 \frac{V^2}{c^2} - 1} \right) e^{+i \frac{\omega}{V} \zeta} d\omega, \tag{3}$$

where $S(\omega)$ is the adopted frequency spectrum.

Indeed, such solutions result to be pulses propagating in free space without distortion and with the Superluminal velocity $V = c/(n_0 \cos\theta)$. The most popular spectrum $S(\omega)$ is that one given by $S(\omega) = \mathrm{e}^{-a\omega}$, which provides the ordinary ("classic") X-shaped wave

$$X \equiv \psi(\rho,\zeta) \;=\; \frac{V}{\sqrt{(aV - i\zeta)^2 + \rho^2(n_0^2\frac{V^2}{c^2} - 1)}}. \tag{4}$$

Because of its non-diffractive properties and its low frequency spectrum, the X-wave is being particularly applied in fields like acoustics[42]. The "classic" X-wave is represented in Fig.10=16*. As we already said in the text, infinite series of SLSs can be constructed, more and more concentrated in the vicinity of their vertex, and corresponding to any desired frequency and bandwidth.

References

[1] Bilaniuk O M, Deshpande V K and Sudarshan E C G 1962 Am. J. Phys. vol. 30, 718; Sudarshan E C G 1963 J. Math. Phys. (NY) vol. 4, 1029; 1969 Ark. Fiz. 39, 585; 1969 Proc. Indian Acad. Sci., vo.69, 133; 1970 in *Symposia on Theoretical Physics and Mathematics* vol. 10, 129 (New York, NY); Phys. Rev. vol. D1, 2473; 1970 in *Proceedings of the VIII Nobel Symposium*, ed. by N.Swartholm (New York, NY), p.385; 1969 Reports NYO-3399-191/SU-1206-191 (Syracuse University), 1970 ORO-3992-5 (UT, Austin), 1970 CPT-81/AEC-30 (UT, Austin) and 1972 CPT-166 (UT, Austin); Arons M E and Sudarshan E C G 1968 Phys. Rev. vol. 173, 1622; Dhar J and Sudarshan E C G 1968 Phys. Rev. vol. 174, 1808; Bilaniuk O M and Sudarshan E C G 1969 Nature vol. 223, 386; Phys. Today vol. 22, 43; Sudarshan E C G and Mukunda N 1970 Phys. Rev. vol. D1, 571; Gleeson A M and Sudarshan E C G 1970 Phys. Rev. D1, 474; Gleeson A M, Gundzig M G, Sudarshan E C G and Pagnamenta A 1972 Phys. Rev., vol. A6, 807; 1972 Fields Quanta vol. 2, 175; Sudarshan E C G and Narlikar J V 1976 Montly Notices of Royal Astron. Soc., vol. 175, 105; Sudarshan E C G 1978 in Tachyons, Monopoles, and Related Topics, ed. by Recami E (North-Holland; Amsterdam) pp.43-46

[2] For an early review, see Recami E and Mignani R 1974 Rivista N. Cim. vol. 4, 209-290, E398, and refs. therein, Recami E (editor), 1978 *Tachyons, Monopoles, and Related Topics* (North-Holland; Amsterdam); Alväger T and Kreisler M N 1968 Phys. Rev. vol. 171, 1357 and refs. therein

[3] Recami E 1986 Rivista N. Cim. vol. 9(6), 1÷178, issue no.6 [178 printed pages], and refs. therein

[4] Recami E 1973 in *Annuario '73, Enciclopedia EST*, ed. by Macorini E (Mondadori; Milano), pp.85-94; and 1986 Nuovo Saggiatore, vol. 2(3), 20-29, issue no.3

[5] Recami E 1993 in I Concetti della Fisica, ed. by Pollini F and Tarozzi G (Acc. Naz. Sc. Lett. Arti; Modena), pp.125-138; Recami E and Rodrigues W A 1983 *Antiparticles from Special Relativity*, Found. Physics vol. 12, 709-718; vol. 13, E533

[6] Recami E 1987 Found. Physics vol. 17, 239-296. See also 1985 Lett. Nuovo Cimento vol. 44, 587-593; Caldirola P and Recami E 1980 in *Italian Studies in the Philosophy of Science*, ed. by Dalla Chiara M (Reidel; Boston), pp.249-298; Shaarawi A M and Besieris I M 2000 J. Phys. A vol. 33, 7255-7263

[7] Baldo Ceolin M 1993 *Review of neutrino physics*, invited talk at the*XXIII Int. Symp. on Multiparticle Dynamics (Aspen, CO)*; Otten E W 1995 Nucl. Phys. News vol. 5, 11. For the theoretical point of view, see Giannetto E, Maccarrone G D, Mignani R and Recami E 1986 Phys. Lett. B vol. 178, 115-120 and refs. therein; Giani S 1999 *Experimental evidence of superluminal velocities in astrophysics and proposed experiments*, CP458, in *Space Technology and Applications International Forum*, ed. by El-Genk M S (A.I.P.; Melville, 1999), pp.881-888; Cawley R G, Lett. Nuovo Cim. vol. 3, 523

[8] Zensus J A and Pearson T J (editors) 1987 *Superluminal Radio Sources* (Cambridge Univ. Press; Cambridge, UK)

[9] Mirabel I F and Rodriguez L F 1994 *A superluminal source in the Galaxy*, Nature vol. 371, 46 [with an editorial comment, *A galactic speed record*, by Gisler G at page 18 of the same issue]; Tingay S J *et. al.* 1995 *Relativistic motion in a nearby bright X-ray source*, Nature vol. 374, 141

[10] Rees M J 1966 Nature vol. 211, 46; Cavaliere A, Morrison P, and Sartori L 1971 Science vol. 173, 525.

[11] Recami E, Castellino A, Maccarrone G D, and Rodonò M 1986 *Considerations about the apparent Superluminal expansions observed in astrophysics*, Nuovo Cimento B vol. 93, 119; Mignani R and Recami E 1974 Gen. Relat. Grav. vol. 5, 615.

[12] Olkhovsky V S, Recami E, and Jakiel J 2004 *Unified time analysis of photon and particle tunnelling*, Phys. Reports vol. 398, pp.133-178; Olkhovsky V S and Recami E 1992 Phys. Reports vol. 214, 339, and refs. therein, in particular Hartman T E 1962 J. Appl. Phys. vol. 33, 3427; MacColl L A 1932 Phys. Rev. vol. 40,

621 (1932); Olkhovsky V S, Recami E, Raciti F and Zaichenko A K 1995 J. de Phys.-I vol. 5, 1351-1365; Privitera G, Recami E, Salesi G, and Olkhovsky V S 2003 *Tunnelling Times: An Elementary Introduction*, Rivista Nuovo Cim. vol. 26, monographic issue no.4.

[13] Chiao R Y, Kwiat P G, and Steinberg A M 1991 Physica B vol. 175, 257; Ranfagni A, Mugnai D, Fabeni P and Pazzi G P 1991 Appl. Phys. Lett. vol. 58, 774; Martin Th and Landauer R 1992 Phys. Rev. A vol. 45, 2611; Japha Y and Kurizki G 1996 Phys. Rev. A ,vol. 53, 586; Kurizki G, Kozhekin A E and Kofman A G 1998 Europhys. Lett. vol. 42, 499; Kurizki G, Kozhekin A E, Kofman A G and Blaauboer M 1998 paper delivered at the VII Seminar on Quantum Optics, Raubichi, BELARUS (May, 1998)

[14] Recami E, Fontana F, and Garavaglia R 2000 Int. J. Mod. Phys. A vol. 15, 2793, and refs. therein

[15] Nimtz G and Enders A 1992 J. de Physique-I vol. 2, 1693; 1993 vol. 3, 1089, 1994 vol. 4, 1379, 1993 Phys. Rev. E , vol. 48, 632; Brodowsky H M, Heitmann W, and Nimtz G 1994 J. de Physique-I vol. 4, 565, 1996 Phys. Lett. A vol. 222, 125, 1994 vol. 196, 154; For a review, see Nimtz G and Heitmann W 1997 Prog. Quant. Electr. vol. 21, 81.

[16] Steinberg A M, Kwiat P G, and Chiao R Y 1993 Phys. Rev. Lett. , vol. 71, 708, and refs. therein; 1993 Scientific American vol. 269(2), 38, issue no.2; for a review, see Chiao R Y and Steinberg A M 1997 in Progress in Optics, ed. by Wolf E (Elsevier; Amsterdam), p.345, Japha Y and Kurizki G 1996 Phys. Rev. A vol. 53, 586.

[17] Ranfagni A, Fabeni P, Pazzi G P, and Mugnai D 1993 Phys. Rev. E , vol. 48, 1453; Spielmann Ch, Szipocs R, Stingl A, and Krausz F 1994 Phys. Rev. Lett. vol. 73, 2308, Balcou Ph and Dutriaux L1997 Phys. Rev. Lett. vol. 78, 851; Laude V and Tournois P 1999 J. Opt. Soc. Am. B vol. 16, 194

[18] Aug. 1993 Scientific American ; Oct. 21, 1993 Nature ; Apr. 1995 New Scientist; 19 June 1995 Newsweek

[19] also Mugnai D, Ranfagni A, Ruggeri R, Agresti A, and Recami E 1995 Phys. Lett. A vol. 209, 227 and Ref.[3], p.158 and pp.116-117,

[20] Brodowsky H M, Heitmann W and Nimtz G 1996 Phys. Lett. A , vol. 222, 125.

[21] Barbero A P L, Figueroa H E H, and Recami E 2000 *On the propagation speed of evanescent modes*, Phys. Rev. E vol. 62, 8628, and refs. therein.

[22] Nimtz G and Haibel A 2002 Ann. der Phys. vol. 11 163-171; Ziolkowski R W 2001 Phys. Rev. E vol. 63 no.046604; Shaarawi A M and Besieris I M 2000 J. Phys. A vol. 33 7227-7254; 7255-7263; Recami E, Fontana F and Garavaglia R 2000 Int. J. Mod. Phys. A, vol. 15, 2793, and refs. therein.

[23] Wang L J, Kuzmich A, and Dogariu A 2000 Nature vol. 406 277.

[24] Nimtz G, Enders A and Spieker H 1993 in *Wave and Particle in Light and Matter*, ed. by van der Merwe A and Garuccio A (Plenum; New York); 1994 J. de Physique-I vol. 4, 565, Enders A and Nimtz G 1993 Phys. Rev. B vol. 47, 9605

[25] Longhi S, Laporta P, Belmonte M, and Recami E 2002 Phys. Rev. E vol. 65 no.046610.

[26] Olkhovsky V S, Recami E and Salesi G 2002 Europhysics Letters 57 879-884, *Tunneling through two successive barriers and the Hartman (Superluminal) effect*, e-print quant-ph/0002022; Aharonov Y, Erez N, and Reznik B 2002 Phys. Rev. A vol. 65 no.052124; Recami E 2004 *Superluminal tunneling through successive barriers: Does QM predict infinite group-velocities?*, J. Modern Optics, vol. 51, pp.913-923; Olkhovsky V S, Recami E, and Zaichenko A K 2005 *Resonant and non-resonant tunneling through a double barrier*, Europhysics Letters vol. 70, 712-718

[27] Olkhovsky V S, Recami E, and Salesi G 2003 refs.[26]; Esposito S, Phys. Rev. E vol. 67 no.016609.

[28] Olkhovsky V S, Recami E, Raciti F, and Zaichenko A K, ref.[12], page 1361 and refs. therein; refs.[3,6]; and Recami E, Fontana F and Garavaglia R, ref.[14], page 2807 and refs. therein.

[29] Chiao R Y, Kozhekin A E, and Kurizki G 1996 Phys. Rev. Lett. vol. 77, 1254; Bolda E L *et. al.* 1993 Phys. Rev. A vol. 48, 3890; Garret C G B and McCumber D E 1970 Phys. Rev. A vol. 1, 305.

[30] Wang L J, Kuzmich A and Dogariu A 2000 Nature vol. 406 277; Mitchell M W and Chiao R Y 1997 Phys. Lett. A vol. 230 133-138; Chu S and Wong W 1982 Phys. Rev. Lett. vol. 48 738; Segard B and Macke B 1985 Phys. Lett. A vol. 109 213-216; Macke B*et. al.* 1987 J. Physique vol. 48 797-808; Nimtz G Eur. Phys. Journal B (to appear as a Rapid Note)

[31] Alzetta G, Gozzini A, Moi L, and Orriols G 1976 Nuovo Cimento B , vol. 36, 5

[32] Artoni M, La Rocca G C, Cataliotti F S and Bassani F 2006 Phys. Rev. A (in press).

[33] Bateman H 1915 *Electrical and Optical Wave Motion* (Cambridge Univ. Press; Cambridge); Courant R and Hilbert D 1966 *Methods of Mathematical Physics* (Wiley; New York, 1966) vol. 2, p.760; Brittingham J N 1983 J. Appl. Phys. vol. 54, 1179; Ziolkowski R W 1985 J. Math. Phys. vol. 26, 861; Durnin J, Miceli J J and Eberly J H 1987 Phys. Rev. Lett. vol. 58, 1499; Barut A O *et. al.* 1990 Phys. Lett. A vol. 143, 349; 1990 Found. Phys. Lett. vol. 3, 303; 1992 Found. Phys. vol. 22, 1267; Hillion P 1993 Acta Applicandae Matematicae, vol. 30, pp 35

[34] Stratton J A 1941 *Electromagnetic Theory* (McGraw-Hill; New York), p.356; Barut A O A O *et. al.* 1993 Phys. Lett. A vol. 180, 5, 1994 vol. 189, 277. For a review-article about Localized Superluminal Waves, see

Recami E, Rached M Z, Nóbrega K Z, Dartora C A, and Figueroa H E H 2003 *On the localized superluminal solutions to the Maxwell equations*, IEEE Journal of Selected Topics in Quantum Electronics vol. 9(1) 59-73

[35] Donnelly R and Ziolkowski R W 1993 Proc. Roy. Soc. London A vol. 440, 541; Besieris I M, Shaarawi A M and Ziolkowski R W 1989 J. Math. Phys., vol. 30, 1254; Esposito S 1997 Phys. Lett. A vol. 225, 203; Vaz J and Rodrigues W A 1997 Adv. Appl. Cliff. Alg. vol. S-7, 457.

[36] See also Recami E and Rodrigues W A 1985 *A model theory for tachyons in two dimensions*, in *Gravitational Radiation and Relativity*, ed. by Weber J and Karade (M World Scient.; Singapore), pp.151-203, and refs. therein.

[37] Shaarawi A M, Besieris I M and Ziolkowski R W 1990 J. Math. Phys. vol. 31, 2511, Sect.VI, 1989 Nucl Phys. (Proc.Suppl.) B vol. 6, 255, 1994 Phys. Lett. A vol. 188, 218; Ignatovich V K 1978 Found. Phys. vol. 8, 565; Barut A O 1992 Phys. Lett. A vol. 171, 1, 1994 vol. 189, 277, 1994 Ann. Foundation de Broglie L, and 1990 *Quantum theory of single events, Localized de Broglie–wavelets, Schroedinger waves and classical trajectories*, preprint IC/90/99 (ICTP; Trieste)

[38] Barut A O, Maccarrone G D and Recami E 1982 Nuovo Cimento A vol. 71, 509; Caldirola P, Maccarrone G D, and Recami E 1980 Lett. Nuovo Cim. vol. 29, 241; Recami E and Maccarrone G D 1980 Lett. Nuovo Cim. vol. 28, 151; Recami E, refs.[3,4,41]; Recami E, Rached M Z and Dartora C A 2004 *The X-shaped, localized field generated by a Superluminal electric charge*, Phys. Rev. E vol. 69 no.027602

[39] Durnin J, Miceli J J, and Eberly J H 1987 Phys. Rev. Lett. vol. 58 1499; 1988 Opt. Lett. vol. 13, 79

[40] Lu J Y and Greenleaf J F 1992 IEEE Trans. Ultrason. Ferroelectr. Freq. Control vol. 39, 19

[41] Recami E 1998 Physica A vol. 252, 586; Lu J Y, Greenleaf J F, and Recami E 1996 *Limited diffraction solutions to Maxwell (and Schroedinger) equations* [Lanl Archives # physics/9610012], Report INFN/FM–96/01 (I.N.F.N.; Frascati); Ziolkowski R W, Besieris I M and Shaarawi A M 1993 J. Opt. Soc. Am., A vol. 10, 75, 2000 J. Phys. A vol. 33, 7227-7254; Friberg A T, Vasara A and Turunen J 1991 Phys. Rev. A vol. 43, 7079

[42] Lu J Y and Greenleaf J F 1992 IEEE Trans. Ultrason. Ferroelectr. Freq. Control vol. 39, 441. In this case the wave speed is larger than the *sound* speed in the considered medium

[43] Saari P and Reivelt K 1997 *Evidence of X-shaped propagation-invariant localized light waves*, Phys. Rev. Lett. vol. 79, 4135-4138

[44] Mugnai D, Ranfagni A and Ruggeri R 2000 Phys. Rev. Lett. , vol. 84, 4830

[45] Rached M Z, Recami E, and Figueroa H E H 2002 *New localized Superluminal solutions to the wave equations with finite total energies and arbitrary frequencies*, Eur. Phys. Journal D vol. 21, pp.217-228; Rached M Z, Nobrega K Z, Figueroa H E H, and Recami E 2003 *Localized Superluminal solutions to the wave equation in (vacuum or) dispersive media, for arbitrary frequencies and with adjustable bandwidth*, Optics Communications vol. 226 15-23; Rached M Z, Recami E and Fontana F 2003 *Superluminal localized solutions to Maxwell equations propagating along a waveguide: The finite-energy case*, Physical Review E vol. 67 no.036620

[46] Rached M Z, Recami E and Fontana F 2001 *Localized Superluminal solutions to Maxwell equations propagating along a normal-sized waveguide*, Phys. Rev. E vol. 64 no.066603; Rached M Z, Nobrega K Z, Recami E and Figueroa H E H 2002 *Superluminal X-shaped beams propagating without distortion along a co-axial guide*, Physical Review E vol. 66 no.046617.; Besieris I M, Abdel-Rahman M, Shaarawi A M, and Chatzipetros A 1998 Progress in Electromagnetic Research (PIER) vol. 19, 1-48

[47] Rached M Z, Shaarawi A M, and Recami E 2004 *Focused X-shaped pulses*, Journal of the Optical Society of America A vol. 21, pp.1564-1574, and refs. therein

[48] Rached M Z 2004 *Stationary optical wave-fields with arbitrary longitudinal shape*, Optics Express vol. 12, pp.4001-4006

[49] Lu J Y, Zou H, and Greenleaf J F 1994 *Ultrasound in Medicine and Biology* vol. 20, 403, 1993 Ultrasonic Imaging vol. 15, 134

[50] Sõnajalg H and Saari P 1996 *Suppression of temporal spread of ultrashort pulses in dispersive media by Bessel beam generators*, Opt. Letters vol. 21, pp.1162-1164

Journal of Physics: Conference Series **196** (2009) 012021 doi:10.1088/1742-6596/196/1/012021

Tachyons

Oleksa-Myron Bilaniuk

Swarthmore College, Pennsylvania

Email: obilani1@swarthmore.edu

Let me begin by expressing my wholehearted admiration for the extraordinary genius and the wonderful personality of George Sudarshn. It was not only a great privilege but also a genuine pleasure working with him on the tachyon problem during our Rochester days.

Before the publication by Albert Einstein of his paper "Zur Electrodynamik bewegter Korper" ("On the Electrodynamics of Moving Bodies") in June 1905 [1], physicists had no reason to restrict their consideration of motions to velocities limited by that of light in vacuum. Some of such publications are listed in references [2-8]. One of the preeminent physicists who wrote extensively on electrons "moving in any given way" was Arnold Sommerfeld [5].

All such speculation came to an abrupt halt in 1905, after Einstein's publication of June 1905, where he explicitly stated: "...speeds in excess of light have no possibility of existence...". The main reasons for the $v > c$ restriction are:

- Even with infinite amount of energy v cannot exceed c : $v < c$.
- With $m = m_o(1 - v^2/c^2)^{-1/2}$ and a real, positive m, $v > c$ implies an "unphysical" imaginary proper mass m_o.
- Causality violation : travel into the past would become possible.

The concept of an imaginary proper mass was dismissed offhand as "unphysical". The need for an infinite amount of energy to accelerate a mass close to the velocity of light, the requirement for the mass traveling at $v > c$ to have an "unphysical" imaginary value, and the apparent possibility of traveling into the past at $v > c$ -- these three conundrums had a chilling effect on anybody who was impudent enough to take seriously the possibility of a particle moving faster than light in vacuum.

The causality violation by anything purportedly traveling backward in time was evocatively described in 1930 by Reginald Buller in his limerick:

There was a young lady named Bright
Who traveled far faster than light.
She went out one day
In a relative way
And returned the previous night.

Worldwide, until 1962, trying to publish anything that seemingly contradicted Einstein automatically ran the risk of being considered a crank. An interesting 1963 Spanish version of the above limerick is attributed to "Traductor Anonimo". In 1974, when George Sudarshan and I published "*Beyond the Light Barrier* " in Russian in "Einsteinovskyi Sbornik 1973" ("Einstein Collection"), Acad.Sci.USSR, Moscow, 1974, we found an imaginative Russian version by a well-known poet S. Ya. Marshak.

To get the first paper on tachyons by George Sudarshen, his student V.K. Deshpande, and myself published – we had to resort to some diplomatic maneuvering. At a meeting of the American Physical Society in 1961, I spotted Walter Michels, the editor of the American Journal of Physics, chatting with the chairman of the Rochester Physics Department, Robert Marshak. I managed to join their conversation and found an opportunity to ask Bob Marshak to assure professor Michels that George Sudarshan and Oleksa Bilaniuk were respectable, *bona fide* physicists. Moreover, Bob Marshak asked Michels to examine our prospective submission on "Metarelativity" thoroughly and seriously, and not throw it automatically into the waste basket, which on first sight might appear the normal thing to do. As a result, our publication appeared in record time.

Our paper rapidly became widely quoted and a number of intrepid experimentalists started looking for faster-than-light particles, particularly after Gerald Feinberg published, in 1967, an attempt at quantization of our "metaparticles", which he aptly renamed tachyons (from Greek tacis=swift).

But the acceptance was not universal. A challenge came, of all places, from the "dean" of science-fiction writers, Isaac Asimov, who published an article [12] on the subject, entitled "*Impossible, that's all.*" Improbably, who came to our defense but another renowned science fiction writer, Arthur C. Clarke, with the article [13] "*Possible, that's all.*"

In the end, Isaac Asimov wrote a contrite article [14] "*The Luxon Wall* ", in which he admitted that he had been "left flat-footed by the advance of physics." He even called me and as a result he was invited to give a talk at Swarthmore. He was a sincere and deep-thinking person and we parted as best of friends.

On a more serious note, over 600 publication on tachyons, experimental and theoretical, appeared in the first 18 years following our 1962 "Metarelativity" paper. Please see Ref. 15 for a full listing. A number of these dealt with the resolution of the causality problem, which remained vexing for a while. In the end, it became clear that tachyons do not lead to causality violation either in the framework of special or of general relativity [16,17].

As soon as it became evident that no laws of physics prohibit the existence of faster-than-light objects, experimentalists, as if mindful of Gell-Mann's witty "Totalitarian Principle" [18], envisioned searches for the mystifying tachyons.

Another jocular aphorism, the "Alvarez Criterion", may be quoted to envision the considerations which the potential tachyon searchers must have before embarking on their experiments. The Alvarez criterion is stated in terms of : $\mu = P \times \sigma$ where μ = merit of an experiment, P = probability of its success, and σ = significance of the result. For tachyons, P=0 and σ = infinity, so that μ = 0 x infinity, which is indeterminate. The outcome of tachyon experiments being indeterminate, the decision rested with the gumption of the experimenters. Considering the number of tachyon search experiments (see a partial list in Refs. 19-31), the gumption was indeed high.

The searches included: 1. search for tachyonic magnetic monopoles, 2. tachyon Michelson experiment, 3. missing mass searches in bubble chamber and emulsion records. There were over 300 papers during 1967-1980 on the quantum mechanics of tachyons. For a full bibliography, please see Ref. 15. Indications that electron neutrinos may be tachyonic was published in 1985, by Alan Chodos, Avi I. Hauser and V. Alan Kostelecky in 1985 [22], and other papers exploring this hypothesis, have re-energized

both the experimental and theoretical interest in tachyons.

Preliminary results of measurements of the endpoint of the electron energy spectrum of tritium decay yield negative values of $(\text{neutrino mass})^2 = -27 \pm 20$ eV . This would indicate that the proper mass of the neutrino is imaginary and that the neutrino is in fact a tachyon. Experimenters now face the challenge to obtain a result which would prove beyond any doubt that the mass of the electron neutrino is indeed imaginary. A summary of experimental results for (neutrino mass)^2, obtained in 1991 – 2000 is given in the following table:

EXPERIMENT	MEASURED M^2	FORMAL LIMIT	C.L. %	YEAR
Mainz	$-1.6 \pm 2.5 \pm 2.1$ eV^2	2.2	95	2000
Troitsk	$-1.0 \pm 3.0 \pm 2.1$ eV^2	2.5	95	2000
Zurich	$-24 \pm 48 \pm 61$ eV^2	11.7	95	1992
Tokyo INS	$-65 \pm 85 \pm 65$ eV^2	13.1	95	1991
Los Alamos	$-147 \pm 68 \pm 41$ eV^2	9.3	95	1991
Livermore	$-130 \pm 20 \pm 15$ eV^2	7.0	95	1995
China	$-31 \pm 75 \pm 48$ eV^2	12.4	95	1995
Avegare of PDG (98)	-27 ± 20 eV^2	15.0	95	1998

Another approach to ascertain if electron neutrinos are tachyons involves verifying if the "knee" in the cosmic ray spectrum is due to high-energy proton decay

$p \to n + e^+ + \nu$, with concomitant emergence of a neutron spike where ν would have negative energy in proton's rest frame, hence would be tachyonic. The possibility of the "knee" being due to such "tachyonic" proton decay has been discussed in a number of papers, including Ref. [31].

The examination of the relativistic velocity addition $v = (u + v)(1 + uw/c^2)^{-1/2}$ involving superluminal velocities leads to a curious conclusion. If there is a parallel tachyonic universe, all velocities within that universe would be seen by us as greater than c, but would be seen as less than c by the inhabitants of that universe. Conversely, to the inhabitants of that universe, all velocities within our world would appear to be tachyonic.

Tachyons have so far eluded physicists, but not enterprising vendors. There is a number of "tachyonized" products on sale and books that provide guidance how to stay healthy by using tachyons [32-33]. Astrologers too seem to have a keen interest in tachyons. In their Australian journal *Alternate Universe* they have published an article "*Tachyon Universe Discovered*". But in contrast to the tachyon vendors, they give a clue that this discovery should not be taken too seriously: The date of the publication is April 1, 2006.

References

[1] Einstein A 1905 *Zur Electrodynamik bewegter Korper* (*On the Electrodynamics of Moving Bodies*), Ann. Phys. 17, 891
[2] Thomson J J 1889 Phil. Mag., 23, 13
[3] Heaviside O1892 Electr. Papers, VII, Chap.47, 497, MacMillan & Co., London
[4] Des Courdes TH., Arch. Neerland. Sci., II S, 652, 1900.
[5] Sommerfeld A 1904 *Simplified deduction of the field and the forces of an electron moving in any given way*, Knkl. Acad. Wetensch, 7, 345-367
[6] Sommerfeld A 1904 K. Akad. Wet., Amsterdam, Proc. 8, 346
[7] Sommerfeld A 1904 Nachr. Ges. Wiss. Gottingen, 99, 363
[8] Sommerfeld A 1904 Proc. Amst. Acad. 26, 357
[9] Sommerfeld A 1905 Nachr. Ges. Wiss. Gottingen, 201-235
[10] Sudarshan E C G and Bilaniuk O M 1974 *Beyond the Light Barrier* in *Einsteinovskyi Sbornik 1973* (*Einstein Collection*), Acad. Sci. USSR, Moscow
[11] Bilaniuk O M, Deshpande V K and Sudarshan E C G 1962 *Metarelativity*, Am. J. Phys. 30, 718-723

[12] Isaac Asimov: *Impossible, that's all* in *Magazine of Fantasy and Science Fiction*, December 1967, p.113.
[13] Arthur C Clarke: *Possible, that's all* in Magazine of Fantasy and Science Fiction, October 1968
[14] Isaac Asimov: *The luxon wall* in *Magazine of Fantasy and Science Fiction*, December 1969, p.96. "...suddenly I sounded like a fuddy-duddy left flat-footed by the advance of physics..."
[15] Perepelitsa V P 1980 *Full bibliography on faster-than-light particles*, Institute of Theoretical and Experimental Physics, ITEP-100 & 165,Moscow
[16] Bilaniuk O M and Sudarshan E C G 1969 *Causality and spacelike signals*. Nature, 223, 386-387, 1969 and Physics Today 22/12, 50-52
[17] Perepelitsa V F 1981 *Principle of causality, relativity theory and superluminal signals*, Institute of Theoretical and Experimental Physics, Moscow, ITEP 104
[18] Gell-Mann's *Totalitarian Principle*: *Anything which is not prohibited is compulsory*. (After White T H)
[19] Alvager T, Blomquist J, Erman P 1963 Ann. Rep. of Nobel Res. Inst., Stockholm, p.95-97
[20] Alvager T, Erman P, Kerek A 1965 Ann. Rep. Of Nobel Res. Inst., Stockholm
[21] Alvager T, Kreisler M N 1968 *Quest for faster-than-light particles*, Phys. Rev. 171, p. 1357-1361
[22] Davis M B, Kreisler M N, and Alvager T 1969 *Search for faster-than-light particles* Phys. Rev. 183, p. 1132-1133
[23] Bartlett D F, Lahana M D 1972 *Search for tachyon monolpoles*, Phys. Rev. D6, p.1817-1823
[24] Baltay C, Feinberg G 1970 N. Yeh, R. Linsker, *Search for uncharged faster-than-light particles*, Phys. Rev. D1, p.759-770
[25] Danburg J S, Kalbfleish G R, Borenstein S R, Strand R C, Vanderburg V, and Chapman J W 1971 J. Lys, *Search for ionizing tachyon pairs from 2.2 GeV/c K-p interactions*, Phys. Rev. D4, p.53-65
[26] Clay R W and Crouch P C 1974 *Possible observation of tachyons associated with extensive air showers*, Nature 248, p. 28-30
[27] Bartlett D F, Soo D, and White M G 1978 *Search for tachyon monopoles in cosmic rays*, Phys. Rev. D18, p. 2253-2261
[28] Bhat P N, Gopalkrishnan N V, Gupta S K, and Tonwar S C 1979 *Search for tachyons in EAS*, J. Phys. G5, p. L13-L18
[29] Chodos A, Hauser A I, and Kostelecky V A 1985 *Neutrino as a Tachyon*, Phys. Lett., 150B/6
[30] Ciborowski J and Rembielinski J 1999 *Tritium decay and the hypothesis of tachyonic neutrinos*, Eur. Phys. J C8, 157-161
[31] Robert Ehrlich 2000 *Neutrino mass*^{2} *inferred from the cosmic ray spectrum and tritium beta decay*, Phys. Rev. D 60, 229-232
[32] Martina Bochnik & Tommy Thomsen 2005 *Holistic Wellness with Tachyons: A practical guide to the use of Tachyons* 224 pages
[33] David Wagner and Gabriel Cousens 2000 *Tachyon Energy: A New Paradigm in Holistic Healing*, MD, North Atlantic Books

Aspects of Tachyon theory

S K Bose

Department of Physics, University of Notre Dame, Notre Dame, IN 46556

E-mail: bose.1@nd.edu

Abstract. Does the Special Theory of Relativity (STR) forbid the existence of particles traveling with speed greater than the speed of light in vacuo (Tachyons)? Prof. Sudarshan and collaborators O.M. Bilaniuk and V.K. Despande examined this question in 1962 and concluded that STR does not rule out such objects. Now, the momentum 4-vector of a Tachyon is necessarily space-like and consequently, the sign of energy no longer Lorentz-invariant. Thus a Tachyon will be found to have negative energy in certain inertial frames. The authors noted that in the latter, the sign of time-intervals will also be reversed. A negative energy Tachyon traveling backward in time could now be reinterpreted as a positive energy particle traveling forward in time. This reinterpretation could be done consistently as was shown by several examples. A quantum field theory of free spinless Tachyons was suggested by Feinberg (1967). M. Arons and Sudarshan reexamined this model and showed that the model did not possess invariance under the Poincaré group. An alternative version was constructed by them that possessed the desired invariance property but no local commutativity. Subsequently, Dhar and Sudarshan constructed a model of a neutral scalar Tachyon with Yukawa coupling to Fermions. The model had unusual features such as the emission of a Tachyon by a Fermion as a real process and the need for an additional non-local interaction. Prof. Sudarshan, in association with J. Narlikar, studied Tachyons in the context of cosmology in 1976. One of their conclusions is that any primordial Tachyons that might have been created at the beginning are unlikely to have survived to the present era.

The question of the possible existence of particles that travel with speed in excess of c, the speed of light in vacuo (hereafter we take c=1), was considered by Sommerfeld,[1] who uncovered many unusual properties of these hypothetical objects. For instance, application of a force will decelerate such a particle even as it gains energy. With the advent of the special theory of relativity (STR) an impression gained ground, supported by Einstein himself,[2] that the existence of such objects was forbidden by the STR. The question was reexamined by Sudarshan and collaborators.[3] A result of this exercise was the conclusion: STR permits particles with speed in excess of light-speed, provided only that they are created that way. The speed of light barrier may not be breached either from below or from above.

For an ordinary particle – a tardyon – the energy-momentum 4-vector is time-like with

$$P_i p^i = E^2 - p^2 = m_0^2 \quad , \quad m_0 > 0 \tag{1}$$

and the usual expressions

$$E = \tilde{a} m_0 \quad , \quad \vec{p} = \tilde{a} m_0 \vec{v} \quad ; \quad \tilde{a} = (1-v^2)^{-1/2} \tag{2}$$

in terms of particle velocity $\vec{v}$ may be derived via a suitable Lorentz transformation from the rest frame $(E = m_0 \ , \ \vec{p} = 0)$. For a particle with speed above the speed of light in a vacuum – a tachyon – the momentum 4-vector is space-like

$$P_i p^i = E^2 - p^2 = -m_*^2 \tag{1a}$$

and expressions for energy and momentum are

$$E = \Gamma m_* \quad , \quad \vec{p} = \Gamma m_* \vec{v} \quad , \quad \Gamma = (v^2 - 1)^{-1/2} \tag{2a}$$

The above (2a) may be obtained from (2) by formally letting $v > 1$ and simultaneously $m_0 = im_*$. This 'unitary trick' was devised in Ref. 3, the authors named m_* the meta mass. A more direct interpretation is that m_* is the minimum value of p, i.e. value in a frame in which $E = 0 \ , \ p = m_*$. Indeed eq.(2a) may be derived via a Lorentz transformation from the latter frame. For a transformation along the direction of $\vec{p}$, the desired Lorentz transformation has parameter $\tan h^{-1}\beta$ given by $v\beta = 1$. Here β is the relative speed between the two frames and v the particle velocity.

Quite unlike the case with a time-like 4-momentum; for a space-like momentum 4-vector, the sign of energy is not Lorentz-invariant. Thus for a tachyon there will exist inertial frames in which it will be found to have negative energy. Thus, let a tachyon with positive energy E have velocity $\vec{v}$ in a frame Σ. In the frame Σ', moving with velocity $\vec{\beta}$ with respect to Σ, the energy E' will be $E' = \gamma E(1 - \vec{v}\cdot\vec{\beta})$ where $\gamma = (1-\beta^2)^{-1/2}$. Thus E' will be negative if $\vec{v}\cdot\vec{\beta} > 1$. What is the meaning of a negative energy particle? The authors of ref. 3 noticed that whenever energy changes sign so will a time interval. That is, let $\Delta t = t_2 - t_1$, be positive, $t_2 > t_1$. Then, in frame Σ', $\Delta t' < 0$, $t'_2 < t'_1$ when $\vec{v}\cdot\vec{\beta} > 1$. Thus a negative energy particle will necessarily move backward in time. An observer in Σ' could simply then interpret the tachyon as having a positive energy and moving forward in time (according to his sense of time-ordering). Specifically, consider the sequence of events as seen from Σ: a tachyon with $E>0$ is emitted at source S_1 at time t_1, and subsequently absorbed at sinк S_2 at time t_2, with $t_2 > t_1$. In Σ' a negative energy tachyon will be found to be absorbed first at the instant t'_2 at S_2 and emitted *later* at t'_1 at S_1. Since the absorption of a negative energy is the same as the emission of a positive energy, in so far as the effect on the absorber is concerned, it is possible for an observer to simply describe the sequence as the emission of a positive energy tachyon at S_2 followed by the absorption of the same at S_1. This is the *reinterpretation principle*, illustrated in ref. 3 by a number of examples. This principle is the magic wand that makes possible a consistent physical explanation of phenomena associated with a tachyon. Without it, one could not seriously entertain the hypothesis of the existence of tachyons. It should be emphasized that the principle operates at a purely classical level. Furthermore, it is called upon to explain phenomena only in certain (but not all!) inertial frames. These features distinguish it from, say, the Feynman picture of a positron. In subsequent publications , Bilaniuk and Sudarshan[4] showed that the reinterpretation principle also resolved the apparent paradoxes and a few persistent misconceptions regarding the alleged violation of the causality principle by tachyons.

A deeper connection with STR will require that the states of a tachyon furnish a projective unitary irreducible representation of the Poincaré group. According to the general theory,[5] these are to be 'induced' from those of the corresponding stability group (a subgroup that leaves fixed a chosen 4-momentum). For a space-like 4-momentum, the stability group is S0(2,1), whose unitary representations are all infinite dimensional, except for the one-dimensional singlet representation.[6] Thus we have the following: a tachyon will come either as a single component object or as an infinite supermultiplet whose members are distinguished by a quantum number, which can be interpreted as the helicity. The supermultiplets themselves (there are four classes of these) are distinguished by the value of the Casimir invariant, built out of operators from the enveloping algebra of the Lie algebra of the stability group SO(2,1). These remarks should be kept in mind while thinking of a possible quantum field theory of a tachyon.

Field theories have been proposed for the singlet tachyon, represented by a scalar field. The first to attempt this was Tanaka.[7] As Tanaka noted, his model had difficulties with Lorentz-invariance: a Lorentz-invariant unique vacuum state did not exist, nor did an unitary operator implementing a Lorentz transformation. Next attempt is due to Feinberg.[8] The most unusual feature of this model is that the free, scalar field had to be quantized in conformity with Fermi (rather than Bose!) statistics. This comes about as follows: the free scalar field satisfied a Klein-Gordon-like equation (with m_0^2 replaced by $-m_*^2$) that follows from eq. (1a). A set of plane wave solutions of this is indexed by the pair $(\omega_\kappa, \vec{\kappa})$, where $\omega_\kappa = \sqrt{\kappa^2 - m_*^2}$ and the condition $\kappa \equiv |\vec{\kappa}| \geq m_*$ is imposed to ensure that ω_κ stays real. The hermitian field $\phi(x,t)$ is then expanded in terms of these solutions, in the usual way, and the coefficients of positive-(negative-) frequency waves are identified with annihilation (creation) operators. However, the separation of the field into positive- and negative-frequency parts is not Lorentz-invariant, for much the same reason as the sign of energy is not. Concomitantly, the distinction between creation (a^+) and destruction (a) operators also is not Lorentz-invariant. The latter can mix them up; in particular, there exists Lorentz-transformations under which $a \rightarrow a^+$ and $a^+ \rightarrow a$. This situation makes it impossible to impose the canonical commutation relations (CCR)

$$[a(\kappa), a^+(\kappa')] = \delta^3(\kappa - \kappa') \tag{3}$$

Feinberg, accordingly, replaced the above commutator by an anti-commutator. Although, one difficulty was thus resolved, the others remained. The fact that a(κ) could be transformed into $a^+(\kappa)$ means that the vacuum state of one observer was non-vacuum for certain other observers. A careful reexamination of the Feinberg model was next undertaken by Arons and Sudarshan.[9] These authors chose to quantize in an infinite (spatial) volume, to keep the translational invariance intact and also used normalisable wave packet states instead of plane-waves. What these authors then found is that the vacuum state may be transformed by a Lorentz-transformation into a state with an *infinite* number of particles. The corresponding Fock spaces must be inequivalent representations of the canonical anti-commutation relations and hence, the Lorentz-transformation was not unitarily implementable. Arons and Sudarshan also gave a more direct proof that an unitary operator U(Λ,a) representing the element (Λ,a) of the Poincaré group and acting as an automorphism

$$U(\Lambda, a)\phi(x)U^{-1}(\Lambda, a) = \phi(\Lambda x + a) \tag{4}$$

does not exist in the Feinberg model. An alternative quantization procedure was next proposed by these authors.[9] Since the root of all troubles was in the separation of the field into positive- and negative-frequency parts, and interpreting the expansion coefficients as annihilation and creation operators, it was proposed not to do it at all. The complex scalar field $\phi(x,t)$ was now expanded into annihilation operators alone, for both positive and negative energies, and its hermitian adjoint $\phi^+(x,t)$ in terms of creation operators of positive and negative energies. This procedure, while leading to a Lorentz-invariant formulation, was at the expense of the explicit appearance of states of negative energy. To interpret these the following *physical postulate* was proposed: In a transition amplitude any negative energy tachyon in the initial state is to be interpreted as an outgoing positive energy anti-tachyon with the sign of all additive quantum numbers reversed and similarly for negative energy particles in the final state. This *physical postulate* was the quantum mechanical analogue of the classical *reinterpretation principle*. The scalar field was quantized via commutators, in conformity with the spin-statistics theorem. The resulting field theory did not have the property of local commutativity. This last feature is not unexpected, it is, indeed, necessary for the inner consistency of theory. In a field theory with local commutativity, the field could not annihilate the vacuum state without itself vanishing identically.[10] The field theory of tachyons, proposed in ref. 9, was fully developed in a subsequent publication by Dhar and Sudarshan.[11] Quantization procedure was carried out for a neutral as well as for a complex scalar field. An Yukawa interaction connecting fermions to the scalar tachyon was introduced. The assumed coupling permits emission of a tachyon by a fermion, of course. This could occur as a *real* process (permitted by energy-momentum conservation) provided that the energy of the initial fermion is suitably high. This has the implication that what appears to be a single fermion in its rest frame will be seen as a fermion plus an emitted tachyon in suitable other Lorentz frames. This is another instance of a general feature involving tachyons; the frame dependence of phenomena observed. As another example, we may consider tachyon-fermion elastic scattering as seen in one frame. In a different Lorentz-frame, where the initial tachyon energy might be negative, the process will be observed as the decay of the initial fermion into the final fermion plus a tachyon-antitachyon pair. To date no field theory based on infinite component fields that provide projective unitary representations of SO(2,1) has been constructed.

Let us now consider the gravitational interaction of a tachyon. We may extend the Einstein principle of equivalence to a tachyon, whose motion will then obey the geodesic equation (for a free particle). Raychaudhuri[12] obtained the following two results: 1) In a weak gravitational field a tachyon will experience a repulsive inverse square force and 2) for a tachyon falling radially into a Schwarzschild black-hole there exists a minimum value of the Schwarzschild radial coordinate at which the tachyon will be forced to turn around and move in the opposite direction. This happens at a point inside the event horizon. The first result is immediately understandable. As a tachyon falls in a gravitational field, its energy will increase and consequently, its speed must decrease (according to eq. (2a)). The effective repulsive force must, moreover, be inverse square in the weak field limit, since the gravitational potential in that limit is just the Newtonian potential. Consider now the motion of a tachyon in the Schwarzschild geometry. For a purely radial motion, the expression for the *proper* velocity is

$$\left(\frac{d\ell}{d\tau}\right)^2 = 1 - \frac{\eta}{\kappa^2} B(r) \quad ; \quad B(r) = g_{00}(r) = 1 - \frac{2M}{r} \tag{5}$$

where $d\tau = B^{1/2}dt$, $d\ell = B^{-1/2}dr$ are elements of proper time and length and r and t are the Schwarzschild radial and time coordinates. In the above we use units G = 1, c = 1. In eq. (5). κ is a constant of motion (the energy-at-infinity per unit mass/meta mass) and $\eta = +1$ for time-like, $\eta = -1$ for space-like and $\eta = 0$ for light-like geodesics. The above eq. (5) shows that there exists minimum value of r given by

$$r_{\min} = \frac{2M}{1+\kappa^2} \tag{6}$$

at which a tachyon ($\eta = -1$) will be forced to turn back. The bounce occurs inside the event horizon. Additional results are obtained by Dhurandhar and Narlikar[13] including the one that for a tachyon the event horizon (r = 2M) is no longer a one-way membrane. These authors also studied tachyon geodesics in the Kerr metric.[14] One of their conclusions is that the capture of a tachyon by a Kerr black hole may lead to a violation of the second law[15] of black hole physics, i.e. the black hole surface area may decrease.

In the context of standard cosmology, the problem of tachyons was considered by Narlikar and Sudarshan[16] and by others.[12,17] An integration of the equations of motion yielded the result that a tachyon obeys the same relaxation rule as ordinary tardyons or massless particles; namely

$$P\,R(t) = \text{const} \tag{7}$$

where P is the magnitude of particle momentum, in the co-moving frame, and R(t) the cosmic scale factor. The result is not altogether unexpected since it arises, essentially, from the very high degree of symmetry of standard cosmology. However, for tachyons there is a problem with eqn. (7). Here P falls as the universe expands. But the momentum of a tachyon cannot fall below m_*, the value of its meta-mass. What happens when $P = m_*$ is reached (say at time $t = t_m$)? At that instant of 'Nirvana' ($E = 0$, $v \to \infty$) a tachyon will simply disappear from view. Alternatively,[16] one may notice from the equation of motion, that t_m is actually a maximum on the trajectory and hence the tachyon will be forced to travel backward in time starting $t = t_m$. This could then be interpreted as tachyon-antitachyon annihilation with zero energy release. Narlikar and Sudarshan go on to show that these conclusions remain valid when the tachyon is treated via a quantum field; a scalar field coupled conformally to the background metric. These authors further considered the question if any tachyons that might have been produced in the early universe (say, at the era of helium synthesis or earlier) could have survived to the present cosmic era. The answer depends on the assumed (meta)-mass[16](M) of the tachyon. For survival till present it was estimated $M/m_e \sim 10^{-11}$, where m_e is the electron mass. Thus it appears highly unlikely that any possible primordial tachyon survives today. An immediate consequence of this is that tachyons have no role in the resolution of the 'missing-mass' problem.

Let us finally consider the situation where a tachyon field may exist without an attendant tachyon. In fact, Tanaka's original publication was an attempt in that direction; to arrange to have tachyons only as virtual particles around more conventional particles (tardyons). A tachyon-cloud around a slower-than-light particle is also treated in ref. 18. Another interesting phenomenon involves the concept of tachyon-condensation. The point is this: a Lagrangian may contain a term with the wrong sign of the mass-square. Instead of accepting it to signify a tachyon, one interprets it to herald the onset of an instability, caused by the use of a 'wrong' vacuum state which is a maximum of the given potential. Any sort of disturbance will make the field roll over to the 'true vacuum' state, which is a minimum of the potential. When the field theory is reformulated around the true vacuum, tachyon states disappear from the spectrum. An example of such behavior is the usual Higg's mechanism involving a quartic potential of a scalar field. Actually, excitations with wrong sign of mass-square term appear copiously in the string theories. These, according to Sen [19] are to be interpreted via tachyon condensation. This suggestion has interesting cosmological consequences, particularly relating to the question of dark energy. For a discussion of these speculations, see refs. 20 and 21.

My initiation into the community of practitioners of the art of doing physics was at the hands of Prof. E.C.G. Sudarshan. It gives me genuine pleasure to dedicate this modest contribution to him.

References

[1] Sommerfeld A 1904 *K. Akad. Wet. Amsterdam, Proc.* **8** 346
[2] Einstein A 1905 *Ann. Physik* **17**
[3] Bilaniuk O M, Despande V K and Sudarshan E C G 1962 *Am. J. Physics* **30** 718
[4] Bilaniuk O M and Sudarshan E C G 1969 *Nature* **223** 386; 1969 *Physics Today* **22** 43
[5] Mackey G 1968 *Induced Representation of Groups and Quantum Mechanics* (New York: W.A. Benjamin)
[6] Bergmann V 1947 *Ann. Math.* **48** 568
[7] Tanaka S 1960 *Prog. Theo. Physics (Kyoto)* **24** 171
[8] Feinberg G 1967 *Phys. Rev.* **159** 1089
[9] Arons M E and Sudarshan E C G 1968 *Phys. Rev.* **173** 1622
[10] Streater R F and Wightman A S 1964 *TCP Spin, Statistics and All That*, (New York: W.A. Benjamin, Inc.)
[11] Dhar J and Sudarshan E C G 1968 *Phys. Rev.* **174** 1808
[12] Raychaudhuri A 1974 *J. Math. Phys.* **15** 856
[13] Dhurandhar S V and Narlikar J V 1976 *Pramana* **6** 388
[14] Dhurandhar S V and Narlikar J V 1978 *Gen. Rel. and Gravit.* **9** 1089; Dhurandhar S V 1979 *Phys. Rev. D* **19** 2310
[15] Hawking S W 1972 *Comm. Math. Phys.* **25** 152
[16] Narlikar J V and Sudarshan E C G 1976 *Mon. Not. R. Astr. Soc.* **175** 105
[17] Davies P C W 1975 *Nuovo Cimento* **258** 571; 2004 *Int. J. Theo. Phys.* **43** 141
[18] Boulware D S 1970 *Phys. Rev. D* **1** 2426; Sudarshan E C G 1970 *Phys. Rev. D* **1** 2428
[19] Sen A 2002 *J. High Energy Phys.* **04** 048; **07** 065; 2002 *Mod. Phys. Lett. A* **17** 1797
[20] Gibbons G W 2003 *Class. Quan. Gravity* **20** 5321
[21] Bagla J S, Jassal H K and Padmanabhan T 2003 *Phys. Rev. D* **67** 063504

Journal of Physics: Conference Series **196** (2009) 011012 doi:10.1088/1742-6596/196/1/011012

Introduction to "Open Systems" : the seventh quest

Gianfausto DellAntonio

Email: dellanto@mat.uniroma1.it

Abstract: Opening remarks at the Open Systems Session (the seventh quest).

This session will explore Open Systems: the seventh among Sudarshan's continuing quests. The first talk will be by Dr. Asorey.

$$\rho' = \sum_{\alpha} C(\alpha)\rho C^{\dagger}(\alpha)$$

Dinner hosted by the Physics Department in honor of E C G Sudarshan

Journal of Physics: Conference Series **196** (2009) 011012 doi:10.1088/1742-6596/196/1/011012

Speakers at open systems: Sudarshan's seventh quest

DellAntonio

Asorey

Hegerfeldt

Chisolm

$$\rho' = \sum_{\alpha} C(\alpha)\rho C^{\dagger}(\alpha)$$

Journal of Physics: Conference Series **196** (2009) 012023 doi:10.1088/1742-6596/196/1/012023

Dynamical Maps and Density Matrices

M Asorey[1], A Kossakowski[2], G Marmo[3], E C G Sudarshan[4]

[1] Departamento de Física Teórica. Facultad de Ciencias. Universidad de Zaragoza, 50009 Zaragoza. Spain

[2] Institute of Physics, Nicolaus Copernicus University Torún 87-100, Poland

[3] Dipartimento di Scienze Fisiche, Universitá Federico II di Napoli and INFN, Sezione di Napoli Complesso Univ. di Monte Sant' Angelo, Via Cintia, 80125 Napoli, Italy

[4] Department of Physics. University of Texas at Austin Austin, Texas 78712-1081

Abstract.

The relations between dynamical maps and quantum states of bipartite systems are analyzed from the perspective of quantum conditional probability. In particular, we explore new interesting relations between completely positive maps, which correspond to quantum channels, and states of bipartite systems which correspond to correlations between the initial and final states. The new connection emerges in a natural way from the generalisation of the classical concept of conditional probability.

We develop applications of these relations which prove to be very useful in both directions, either for the classification of positive maps which are not completely positive, the classification of non-decomposable dynamical maps or for the classification of positive partial transpose and entangled states.

1. Introduction

The classic dichotomy between being and becoming has been translated into the description of physical systems as the dichotomy between bipartite states and dynamical maps. This dichotomy between statics and dynamics has been weakly questioned in both schemes, but persists as a key element of fundamental theories. In this respect quantum physics only differs from classical physics in the change of the nature of quantum states and quantum maps.

A surprising genuine feature of quantum physics is that it reinforces the existence of relations between these two different facets of quantum systems for the case of composite bipartite systems. The analysis of such relations has some additional interest because it permits to relate entanglement properties of quantum states of composite systems and the positivity properties of quantum maps. The aim of the present paper is to further analyse the relations between quantum states and dynamical maps from this perspective.

2. Quantum states and Unital Dynamical Maps

The basic elements for the description of quantum systems are the quantum states. Pure quantum states can be identified with rays of vectors $|\psi>$ in some Hilbert space $\mathcal{H}$. In the probabilistic interpretation, pure states are associated with the corresponding projectors $\rho_\psi = |\psi\rangle\langle\psi|/\langle\psi|\psi\rangle$.

However, the knowledge of the state of a quantum system is not in general maximal and various pure states can be compatible with the experimental results. In this situation one has to describe quantum states as statistical mixtures or density states [1].

If we assume that the results of our measurements are compatible with the possibility that our system may be found in the pure state ψ_1 with probability p_1, or in the state ψ_2 with probability p_2, probability p_n for the state ψ_n and so on, the state of our system is represented by the convex combination, usually called density matrix,

$$\rho = \sum_j p_j |\psi_j\rangle\langle\psi_j| = \sum_j p_j \rho_j$$

with

$$0 \leq p_j \leq 1 \quad \sum_j p_j = 1\,.$$

The family $(p_1, p_2, \ldots, p_n, \ldots)$ is usually said to represent a probability vector. Genuine quantum states involve therefore mixed states as well as pure states.

In general, for quantum system with only n level states $\mathcal{H} = \mathbb{C}^n$ any quantum state is described by an $n \times n$ density matrix ρ, i.e. a positive, selfadjoint matrix with unit trace,

$$i) \qquad \rho^+ = \rho \tag{1}$$
$$ii) \qquad \rho \geq 0 \tag{2}$$
$$iii) \qquad \mathrm{tr}\,\rho = 1. \tag{3}$$

We shall denote by $\mathcal{D}(\mathcal{H})$ the space of linear maps ρ associated to density matrices, i.e. matrices satisfying the properties (1)-(3). $\mathcal{D}(\mathcal{H})$ is convex, i.e. for any $\rho_1, \rho_2 \in \mathcal{D}(\mathcal{H})$ and $0 \leq s \leq 1$

$$\rho = s\rho_1 + (1-s)\rho_2 \in \mathcal{D}(\mathcal{H}). \tag{4}$$

Pure states are extremal states in $\mathcal{D}(\mathcal{H})$, i.e. states that cannot be decomposed as a convex sum of two different states (4).

Any evolution process or dynamical map has to transform quantum states into quantum states and preserve the superposition principle. Therefore a quantum dynamical map between two systems associated to the Hilbert spaces $\mathcal{H}_1$, $\mathcal{H}_2$ can be identified with a linear homomorphism φ mapping $\mathcal{D}(\mathcal{H}_1)$ into $\mathcal{D}(\mathcal{H}_2)$, which satisfies very intricate positivity conditions which complicate considerably its classification. If pure states are transformed into pure states these requirements imply that φ is a projective map $\varphi : \mathcal{H}_1 \to \mathcal{H}_2$. However, in general a dynamical map can be any map $\varphi : \mathcal{L}(\mathcal{H}_1) \to \mathcal{L}(\mathcal{H}_2)$ which preserves the basic properties of the quantum states [2]

$$i) \qquad \text{Selfadjointness} \qquad \varphi(\rho^\dagger)^\dagger = \varphi(\rho) \tag{5}$$
$$ii) \qquad \text{Positivity} \qquad \varphi(\rho) > 0 \tag{6}$$
$$iii) \qquad \text{Trace Condition} \qquad \mathrm{Tr}\,\varphi(\rho) = 1 \tag{7}$$
$$iv) \qquad \text{Linearity} \qquad \varphi(s\rho_1 + (1-s)\rho_2) = s\varphi(\rho_1) + (1-s)\varphi(\rho_2) \tag{8}$$

A simple example is the unitary map defined by any unitary matrix U as

$$\varphi(\rho) = U\rho U^\dagger.$$

The set of quantum maps $Maps_Q(\mathcal{H}_1, \mathcal{H}_2)$ is convex since for any pair of quantum maps $\varphi_1, \varphi_2 \in Maps_Q(\mathcal{H}_1, \mathcal{H}_2)$ and $0 \leq s \leq 1$

$$\varphi = (s\varphi_1 + (1-s)\varphi_2) \in Maps_Q(\mathcal{H}_1, \mathcal{H}_2) \tag{9}$$

is a quantum map.

Sometimes it is convenient to consider a larger class of maps. Those which satisfy only the conditions (5) (6) and (8), but not the normalisation condition (7). In particular, maps with $\varphi(\mathbb{I}) = \mathbb{I}$ are called unital maps and they differ by a normalisation factor from trace preserving maps $\mathrm{Tr}\,\varphi(a) = \mathrm{Tr}\,a$ which satisfy the property (7).

3. Positive and Completely Positive Maps

A map $\varphi : \mathcal{L}(\mathcal{H}_1) \to \mathcal{L}(\mathcal{H}_2)$ is said to be positive if $\varphi(a) > 0$ for any positive operator $a \in \mathcal{L}(\mathcal{H}_1)$. Any dynamical map is positive, and because of the isomorphism $\mathcal{L}(\mathcal{H}) = \mathcal{H} \otimes \mathcal{H} = \mathcal{H}^{\otimes 2}$ can be identified with a homomorphism $\widetilde{\varphi}$ from $\mathcal{H}_1^{\otimes 2} = \mathcal{H}_1 \otimes \mathcal{H}_1$ into $\mathcal{H}_2^{\otimes 2} = \mathcal{H}_2 \otimes \mathcal{H}_2$. Moreover, by using the canonical Hilbert products of $\mathcal{H}_1$ and $\mathcal{H}_2$, it can also be uniquely identified, with an endomorphism $\widehat{\varphi}$ of $\mathcal{H}_1 \otimes \mathcal{H}_2$. Thus, any dynamical map φ which is an homomorphism from $\mathcal{L}(\mathcal{H}_1)$ into $\mathcal{L}(\mathcal{H}_2)$ can also be also identified with an endomorphism $\widehat{\varphi}$ of $\mathcal{H}_1 \otimes \mathcal{H}_2$. The endomorphisms $\widehat{\varphi}$ corresponding to dynamical maps are selfadjoint endomophisms of $\mathcal{H}_1 \otimes \mathcal{H}_2$. A dynamical map φ is said to be completely positive (CP) if the associated map $\widehat{\varphi}$ is a positive endomorphism of $\mathcal{H}_1 \otimes \mathcal{H}_2$.

It is obvious that the class of completely positive maps is a subclass in the space of all positive maps. Indeed, any completely positive map is positive, but not any positive map needs to be completely positive. A simple counterexample is the transposition map

$$\phi_T(a) = a^T$$

which is positive but not completely positive.

Thus, CP maps satisfy stronger conditions than simple positive maps. Simply positivity only requires that $\varphi(a) > 0$ for any linear positive operator $a > 0$, but completely positivity requires a more stringent condition $(b, \widehat{\varphi}(b)) > 0$ for any $b \in \mathcal{H}_1 \otimes \mathcal{H}_2$, which not any positive map satisfies.

Another structural difference between the two classes of maps is the behaviour under tensor product operations. Indeed, simple positivity is not preserved under tensorial products because in general $\varphi = \varphi_1 \otimes \varphi_2$ is not positive even if $\varphi_1$and φ_2 are positive maps, whereas complete positivity is preserved under tensor products. For any completely positive maps φ_1 and φ_2 its tensor product $\varphi = \varphi_1 \otimes \varphi_2$ is always completely positive. In particular, it can be shown that a dynamical map φ is completely positive if and only if $\varphi \otimes \mathbb{I}_n$ is a positive map for any identity matrix of arbitrary dimensions $n \times n$ [3, 4].

In order to characterise the difference between these two classes of dynamical maps it is convenient to introduce some structures in the space of density matrices.

Let us restrict ourselves for simplicity from now on to finite-dimensional Hilbert spaces. For finite dimensions the space of linear maps $\mathcal{L}(\mathbb{C}^n)$, which can be identified with the algebra of $n \times n$ complex matrices M_n, has an hermitian product defined by

$$(a, b) = \mathrm{Tr}(a^\dagger\, b) \qquad \text{for any } a, b \in M_n.$$

This product induces another product $\langle\ \cdot,\ \cdot\rangle$ in the space of dynamical maps $Maps_Q(\mathbb{C}^n, \mathbb{C}^n)$ by the following mechanism. Let $\{f_\alpha; \alpha = 1, 2 \cdots n^2\}$ be an orthonormal basis in M_n

$$(\, f_\alpha, f_\beta\,) = \delta_{\alpha\beta}.$$

Then, we define

$$\langle \varphi_1, \varphi_2 \rangle = \sum_{\alpha=1}^{n^2} \mathrm{Tr}\, \varphi_1(\mathrm{f}_\alpha)^\dagger \varphi_2(\mathrm{f}_\alpha).$$

Using this correspondence it is possible to introduce two different bases associated to the basis f_α in the space of linear maps:

1. A basis: defined by

$$e_{\alpha\beta}(a) = f_\alpha(f_\beta^\dagger, a)$$

and,
2. B basis:

$$\varphi_{\alpha\beta}(a) = f_\alpha \, a \, f_\beta^\dagger.$$

Any map can be expressed as

$$\varphi(a) = \sum_{\alpha,\beta} A_{\alpha\beta} \, e_{\alpha\beta}(a) \qquad \text{A matrix form}$$

or

$$\varphi(a) = \sum_{\alpha,\beta} B_{\alpha\beta} \, \varphi_{\alpha\beta}(a) \qquad \text{B matrix form}$$

By definition B matrix is hermitian: $B^*_{\beta\alpha} = B_{\alpha\beta}$ [2]
Any orthonormal basis $|i>; i = 1, 2, \cdots n$ in $\mathbb{C}^n$ induces a basis in M_n by

$$f_\alpha = e_{ij} = |i><j|.$$

In this basis the matrices A and B satisfy the following interesting properties,
1. *A Matrix* :

$$A_{\alpha\alpha'} = A_{ij,i'j'} \qquad \begin{cases} A^*_{ij,i'j'} = A_{ji,j'i'} \\ A_{ij,i'j'} \; x_i x_j^* \; y_{i'} y_{j'}^* > 0 \\ A_{ii,i'j'} = \delta_{i'j'} \end{cases} \tag{10}$$

2. *B Matrix* :

$$B_{\alpha\alpha'} = B_{ii',jj'} \qquad \begin{cases} B^*_{ii',jj'} = B_{jj',ii'} \\ B_{ii',jj'} \; x_i^* \, y_{i'} \, x_j \, y_{j'}^* > 0 \\ B_{ii',ij'} = \delta_{i'j'}. \end{cases} \tag{11}$$

In terms of the hermitian matrix B it is very easy to characterize a map as completely positive or simply positive.
For any selfadjoint endomorphism φ of $\mathcal{L}(\mathbb{C}^n) = M_n = \mathbb{C}^{n^2}$ it follows that

a) φ is completely positive iff the matrix $B_{\alpha\beta}$ is positive
b) φ is completely positive iff

$$\widehat{\varphi \otimes \mathbb{I}_k} : \mathbb{C}^n \otimes \mathbb{C}^k \to \mathbb{C}^n \otimes \mathbb{C}^k$$

is positive for any k, where I_k denotes the identity matrix in $\mathbb{C}^k$.
c) φ is completely positive if there exist a family of operators $D_i; i = 1, 2 \cdots r$ in M_n such that φ can be decomposed as [5, 6]

$$\varphi(a) = \sum_{i=1}^{N} D_i \, a \, D_i^\dagger$$

for any $a \in M_n$. If $N = 1, \varphi$ is extremal.

Journal of Physics: Conference Series **196** (2009) 012023 doi:10.1088/1742-6596/196/1/012023

The most relevant property of completely positive maps arises from the following result. If $\rho = \rho_1 \otimes \rho_2 \in \mathcal{D}(\mathbb{C}^n \otimes \mathbb{C}^n)$ is a separable state the unitary evolution map

$$\mathrm{Tr}_2 \mathrm{U}\, \rho\, \mathrm{U}^\dagger = \varphi_{\mathrm{U}}(\rho_1)$$

is completely positive. This means, that the evolution of a system due to the coupling to any kind of environment is given always by a positive map. In particular, this implies that if the dynamical map has to be embedded into a map for composition of our system with any kind of environment the map has to be completely positive. But this is not a necessary physical requirement, therefore, dynamical maps which are not completely positive can have a physical meaning although their classification is more involved than that of completely positive maps (CP).

For positive maps in $\mathcal{L}(\mathbb{C}^n) = M_n$ we have the following properties

i) φ is not completely positive if the matrix $B_{\alpha\beta}$ has at least one negative eigenvalue

ii) φ is not completely positive if there is a $k \in \mathbb{N}$ such that

$$\widehat{\varphi \otimes \mathbb{I}_k} : \mathbb{C}^n \otimes \mathbb{C}^k \to \mathbb{C}^n \otimes \mathbb{C}^k$$

is not positive.

iii) φ is not completely positive if there are two families of operators $D_i, ; i = 1, 2 \cdots N$ and $C_j; j = 1, 2 \cdots M$ in M_n such that φ can be decomposed as

$$\varphi(a) = \sum_{i=1}^{N} D_i\, a\, D_i^\dagger - \sum_{j=1}^{M} C_j\, a\, C_j^\dagger$$

The possibility that the transposition operator might be useful to distinguish between positive and completely positive suggests the utility of consider a new class of maps: a map is said to be copositive if under composition with transposition becomes positive. More generally,

A map $\varphi : M_n \to M_n$ is k-positive if

$$\varphi \otimes \mathbb{I}_k(a) \geq 0 \quad \text{for any} \quad a \in M_n,$$

φ is k-copositive if under transposition

$$\varphi^T(a) = \varphi(a^T)$$

becomes k-positive, i.e. $\varphi^T \otimes \mathbb{I}_k(a) > 0$ for any $a > 0$. φ is k-decomposable if both

$$\varphi \otimes \mathbb{I}_k \quad \text{and} \quad \varphi^T \otimes \mathbb{I}_k$$

are positive, and completely positive if it is k-positive for any k. Similarly, φ is decomposable if it is k-decomposable for any k.

A positive map φ is decomposable if can be expressed as the sum of positive and copositive maps [7]

$$\varphi(a) = \sum_{i=1}^{N} D_i\, a\, D_i^\dagger + \sum_{i=1}^{M} C_i\, a^T\, C_i^\dagger$$

In n=2 dimensions every positive map is decomposable [8, 3, 9, 10]. In higher dimensions positive maps which are not decomposable are called atomic [11]. A first example is provided by Choi map [9]

$$\varphi_{\text{Choi}}\begin{pmatrix} a_{11} & a_{12} & a_{13} \\ a_{21} & a_{22} & a_{23} \\ a_{31} & a_{32} & a_{33} \end{pmatrix} = \begin{pmatrix} a_{11}+a_{22} & -a_{12} & -a_{13} \\ -a_{21} & a_{22}+a_{33} & -a_{23} \\ -a_{31} & -a_{32} & a_{33}+a_{11}. \end{pmatrix} \quad (12)$$

The physical relevance of positive maps which are not completely positive was though to be meaningless for a long time because they could not appear from unitary evolution of separable states of system and environment. However, this property turned out to be crucial for pointing out the relevant role of those maps in the detection of entanglement. Non CP positive maps can arise in the unitary evolution of bipartite systems only from entangled states. In particular, they can appear from the interaction with the environment only if the system and the environment are entangled.

The relevance of positive maps for the entanglement detection is mainly due to the following result. A mixed state ρ is separable if and only if

$$(\varphi\otimes\mathbb{I})\rho\geq 0$$

for any positive map φ [12]

If there are negative modes in the associated map $\varphi\otimes\mathbb{I}_m$ they correspond to entangled states. Thus, positive but not CP maps can be used as detectors of entanglement. In particular, they can be used as good discriminators in the classification of entangled states.

This relation extends to a more general parallelism in the approaches to the classification the degree of entanglement of bipartite systems and the classification of positive but not CP maps, which might be useful in both directions [13, 15, 14].

Indeed in the space of physical states of bipartite systems we have three main classes of states: Separable (non-entangled states), partial positive transpose (PPT) states and entangled states. The class of PPT states overlaps with the class of separable states but in higher dimensions also with that of entangled states. The entagled states can be further subdivided into those of zero discord or non-zero discord[1]. On the space of dynamical maps we have: CP maps, decomposable maps and indecomposable maps.

4. Unital Quantum Maps and States of Bipartite Systems

Let us now explore further the properties of the afore mentioned one-to-one correspondence [7] between elements of $\mathbb{C}^{n^2}\otimes\mathbb{C}^{m^2}$ and linear maps from $\mathbb{C}^{n^2}$ into $\mathbb{C}^{m^2}$ which permits to establish a correspondence between quantum maps φ from $\mathbb{C}^{n^2}$ into $\mathbb{C}^{m^2}$ and quantum states ω of the composite system $\mathbb{C}^{n^2}\otimes\mathbb{C}^{m^2}$ [4, 20]

$$\widehat{\varphi}:\mathbb{C}^{n^2}\otimes\mathbb{C}^{m^2}\Longleftrightarrow\varphi\in\mathcal{L}(\mathbb{C}^{n^2},\mathbb{C}^{m^2})$$

4.1. Bipartite Mixed states

If the quantum system under consideration is a bipartite compound, density states will be described by selfadjoint positive matrices R in $\mathcal{H}_1\otimes\mathcal{H}_2$ which have unit trace, i.e

$$R^+=R,\quad R\geq 0\quad\text{and}\quad\operatorname{tr}R=1. \quad (13)$$

[1] The concept of quantum discord was introduced by W. Zurek [16]-[18]

[7] Similar results hold for infinite dimensional systems thanks to nuclear theorems [19].

As the Hilbert space is a tensor product we may introduce partial traces with respect to both factor spaces $\mathcal{H}_1$ and $\mathcal{H}_2$,

$$\mathrm{tr}_1\, R = \rho^{(2)}, \quad \mathrm{tr}_2\, R = \rho^{(1)} \tag{14}$$

giving rise to density states for the subsystems.

We can concatenate any number of subsystems to construct a multipartite density state [2] whose partial traces (all the traces except one) will give density states for the remaining subsystem. Here we shall restrict our interest to bipartite systems where we have

$$\mathrm{tr}_1\, \mathrm{tr}_2\, R = \mathrm{tr}_1\, \rho^{(2)} = \mathrm{tr}_2\, \rho^{(1)} = 1. \tag{15}$$

The fact that R is a density matrix guarantees that all the partial traces are also density matrices. In particular,

$$\begin{array}{lll} R^+ = R & \text{implies} & \rho^{(1)+} = \rho^{(1)}\,, \rho^{(2)+} = \rho^{(2)} \\ R \geq 0 & \text{implies} & \rho^{(1)} \geq 0\,,\, \rho^{(2)} \geq 0 \\ \mathrm{tr}\, R = 1 & \text{implies} & \mathrm{tr}\, \rho^{(1)} = \mathrm{tr}\, \rho^{(2)} = \mathbb{I}. \end{array} \tag{16}$$

If we introduce suitable bases e_a in $\mathcal{H}_1$ and e_r in $\mathcal{H}_2$, we could write the matrix for R in component form $R_{ar,sb} = (e_s \otimes e_b, R\, e_a \otimes e_r)$. This matrix, satisfies the following identity

$$\begin{array}{l} R_{ar,sb} = R^*_{bs,ra} \\ z_{ar}\, R_{ar,sb}\, z^*_{bs} \geq 0 \\ R_{ar,ra} = 1, \end{array} \tag{17}$$

for any matrix $(z_{ab}) \in M_{mn}$

The canonical connection between bipartite systems and positive maps between the two subsystems follows from a very general principle. Indeed, in any Hilbert space $\mathcal{H}$ there is an one-to-one correspondence between the set $\mathcal{P}^p_q$ of p–contravariant q–covariant tensors and the set $\mathcal{P}_{p+q}$ of $p+q$ covariant tensors. The equivalence is due to the identification of $\mathcal{H}$ and its dual space $\mathcal{H}^*$ by means of the hermitian product. Consequently, endomorphisms of $\mathcal{H}$ which are in $\mathcal{P}^1_1$ are in one-to-one correspondence with 2–covariant tensors of $\mathcal{P}_2$.

In the theory of quantum information, completely positive maps correspond to quantum channels, and its relation with states of bipartite systems correspond to the existence of correlations between the initial and final states. In the classical information theory the relation is based on the concept of conditional probability. A generalisation of that idea for quantum systems is possible. The analogy requires the introduction of quantum conditional probability operators.

Let us consider two systems with physical states in $\mathcal{D}(\mathbb{C}^n)$ and $\mathcal{D}(\mathbb{C}^m)$, respectively. The first one describes an initial (*input*) states and the second one a final (*output*) states. Let us consider a map $\varphi : \mathcal{D}(\mathbb{C}^n) \to \mathcal{D}(\mathbb{C}^m)$, such that its dual map $\varphi^\dagger : \mathcal{D}(\mathbb{C}^m) \to \mathcal{D}(\mathbb{C}^n)$ is completely positive and unital, i.e. $\varphi(I_m) = I_n$.

For any pair of initial $\rho \in \mathcal{D}(\mathbb{C}^n)$ and final states $\varphi(\rho) \in \mathcal{D}(\mathbb{C}^m)$, the interesting composite states $\omega^\rho_\varphi \in \mathcal{D}(\mathbb{C}^n \otimes \mathbb{C}^m)$ encoding the correlation between the input and output of φ should satisfy the following two conditions:

$$i) \qquad \mathrm{Tr}\, \omega^{\rho\dagger}_\varphi\, a \otimes \mathbb{I}_m = \mathrm{Tr}\, \rho^\dagger\, a, \qquad \text{for all } a \in \mathcal{D}(\mathbb{C}^n) \tag{18}$$

$$ii) \qquad \mathrm{Tr}\, \omega^{\rho\dagger}_\varphi\, \mathbb{I}_n \otimes b = \mathrm{Tr}\, \varphi(\rho)^\dagger\, b, \qquad \text{for all } b \in \mathcal{D}(\mathbb{C}^m). \tag{19}$$

It is well known that conditional probability does not generally exist for quantum systems, therefore it is difficult to define a compound state ω^ρ_φ satisfying the above conditions.

The first construction of a composite state ω_φ^ρ satisfying the above two conditions is due to Ohya [21, 22] and proceeds as follows. Let $\rho \in \mathcal{D}(\mathbb{C}^m)$ be density matrix with the following spectral decomposition

$$\rho = \sum_k \lambda_k \, m_k \rho_k \tag{20}$$

with

$$\rho_k = \frac{1}{m_k} \mathbb{P}_k, m_k = \mathrm{Tr}\, \mathbb{P}_k \,,$$

where λ_k are the eigenvalues of ρ, and $\mathbb{P}_k$ are eigenprojectors of ρ, respectively.

Then, for any $\varphi : \mathcal{D}(\mathbb{C}^n) \to \mathcal{D}(\mathbb{C}^m)$ the Ohya composite state $\omega_\varphi \in \mathcal{D}(\mathbb{C}^n \otimes \mathbb{C}^m)$ is defined by

$$\omega_\varphi^\rho = \sum_k \lambda_k \, m_k \, \rho_k \otimes \varphi(\rho_k). \tag{21}$$

Let us observe that by construction the Ohya state ω_φ^ρ is non–linear with respect to ρ, but since φ is a positive unital map, ω_φ^ρ is a separable state in $\mathcal{D}(\mathbb{C}^n \otimes \mathbb{C}^m)$.

The construction of the Ohya map is very similar to that of the Jamiolkowski map but it satisfies the properties (18) (19) which are not fulfilled in the later case. We recall that Jamiolkowski map [4, 3] is nothing but the image under $\varphi^\dagger \otimes \mathbb{I}_m$ of the maximally entangled element

$$\sum_{\alpha,\beta} e_{\alpha\beta} \otimes e_{\alpha\beta}$$

of the tensor product $M_n \otimes M_m$ defined by any orthonormal basis $e_{\alpha\beta}$ of M_m, i.e.

$$\omega_\varphi = \sum_{\alpha,\beta=1}^{n} \varphi^\dagger(e_{\alpha\beta}) \otimes e_{\alpha\beta} \tag{22}$$

4.2. Quantum Conditional Probability

There are other ways of generating composite states from dynamical maps satisfying all the conditions (18) (19). One interesting family is provided by quantum conditional probability operators [20].

A map

$$\pi \, : \, \mathcal{L}(\mathbb{C}^n \otimes \mathbb{C}^m) \to \mathcal{L}(\mathbb{C}^n \otimes \mathbb{C}^m)$$

is said to be a quantum conditional probability operator (QCPO) if for any $\sigma \in \mathcal{D}(\mathbb{C}^n \otimes \mathbb{C}^m)$ satisfies that

$$\pi(\sigma) > 0 \tag{23}$$
$$\mathrm{Tr}_2 \; \pi(\sigma) = \mathbb{I}_n. \tag{24}$$

Such an operator π is the quantum analogue of the classical conditional probability. Another definition of quantum conditional probability has been given in Ref. [24].

One way of building QCPO is to consider the map $\pi : \mathcal{L}(\mathbb{C}^n \otimes \mathbb{C}^m) \to \mathcal{L}(\mathbb{C}^n \otimes \mathbb{C}^m)$ which transforms any $\sigma \in \mathcal{D}(\mathbb{C}^n \otimes \mathbb{C}^m)$ into the operator $\pi(\sigma) : \mathbb{C}^n \otimes \mathbb{C}^m \to \mathbb{C}^n \otimes \mathbb{C}^m$, defined by

$$\pi(\sigma) = (\sigma_1^{-1/2} \otimes I_m) \, \sigma \, (\sigma_1^{-1/2} \otimes I_m),$$

where

$$\mathrm{tr}_2\, \sigma = \sigma_1. \tag{25}$$

Since $\sigma \in \mathcal{D}(\mathbb{C}^n \otimes \mathbb{C}^m)$ is a state of the bipartite system $\mathbb{C}^n \otimes \mathbb{C}^m$, $\sigma_1 > 0$ and π satisfies the properties (23) (24) required for QCPO.

Now, given a QCPO π one can define a composite state $\omega^\rho \in \mathcal{D}(\mathbb{C}^n \otimes \mathbb{C}^m)$ for any $\rho \in \mathcal{D}(\mathbb{C}^n)$ by

$$\omega^\rho = (\rho^{1/2} \otimes I_m)\, \pi\, (\rho^{1/2} \otimes I_m)$$

which verifies that

$$\mathrm{tr}_2\, \omega^\rho = \rho \tag{26}$$

and defines by tracing out the $\mathbb{C}^n$ component

$$\mathrm{tr}_1\, \omega^\rho = \mathrm{tr}_1\, \pi(\rho \otimes I_m) = \varphi(\rho) \tag{27}$$

a dynamical φ

$$\varphi\,:\, D(\mathbb{C}^n) \to D(\mathbb{C}^m)\,.$$

This correspondence defines in an explicit way a relation between QCPO and dynamical maps

The study of this relation between φ and π is based on duality between quantum maps and composite states which has been investigated Refs. [23, 24, 25, 26] (see also references therein). In the case of CP maps the correspondence is invertible. Let $n = m$ and φ a CP map, by using the image of the Jamiolkowski map associated to the basis e_{ij} we can define

$$\sigma_\varphi = \sum_{ij=1}^{n} \varphi(e_{ij}) \otimes e_{ij} = (\varphi \otimes \mathbb{I}_n) \sum_{ij=1}^{n} e_{ij} \otimes e_{ij} \in M_n \otimes M_n$$

and, since φ is CP, $\sigma_1 = \mathrm{Tr}_2\, \sigma_\varphi > 0$, by

$$\pi_{\sigma_\varphi} = (\sigma_1^{-1/2} \otimes \mathbb{I}_n)\, \sigma_\varphi\, (\sigma_1^{-1/2} \otimes \mathbb{I}_n)$$

a quantum conditional probability operator. Finally, we define the bipartite state

$$\omega_\varphi^\rho = (\rho^{1/2} \otimes \mathbb{I}_n)\, \pi_{\sigma_\varphi}\, (\rho^{1/2} \otimes \mathbb{I}_n)$$

which closes the relation.

We can apply this construction to obtain relations between of entanglement properties of ω_φ^ρ and properties of the positive map φ. Indeed, one can show that [20]

1. The bipartite state ω_φ^ρ has a positive partial transpose (PPT) iff φ is k-completely positive and k'-completely copositive
2. The bipartite state ω_φ^ρ has negative partial transpose (NPT) iff φ is k-completely positive and k'-copositive ($k' < n$) provided that $\mathrm{rank}\, \rho = n$

Let us consider some applications to point out the utility of QPCO.
Example 1:
Indeed, the previous properties of QPCO can be used to show that

$$\varphi(a) = \sum_{ij} c_{ij}\, e_{ij}\, a\, e_{ij}^T + \mu\, a \tag{28}$$

is completely positive iff

$$c_{ij} \geq 0, \quad i \neq j \qquad \text{and} \qquad |c_{ii}\delta_{ij} + \mu| \geq 0$$

and completely copositive if

$$c_{ii} + \mu \geq 0 : \quad c_{ij} + c_{ji} > 2|\mu|, \quad i \neq j$$

$$c_{ij}c_{ji} \geq \mu^2, \quad i \neq j$$

The same techniques can be applied to the following two examples are based on SU(n) symmetry. Let $\lambda_1, \lambda_2, \cdots, \lambda_{n^2-1}$ be the Gell-Mann-Tilma basis of M_n

Example 2. The map [27]

$$\varphi(a) = \sum_{\alpha=1}^{n^2-1} x_\alpha \, \lambda_\alpha \operatorname{Tr} a \, \lambda_\alpha + \frac{1}{n} \mathbb{I} \operatorname{Tr} a \tag{29}$$

is unital $\varphi(\mathbb{I}) = \mathbb{I}$, invariant under transposition

$$\varphi^T(a) = \varphi(a)$$

and positive, provided that

$$|x_\alpha| \leq 1 \quad \alpha = 1, 2, \ldots n^2 - 1,$$

because

$$\operatorname{Tr}|x >< x|\varphi(|y >< y|) = \sum_{\alpha=1}^{n^2-1} x_\alpha \, (x, \lambda_\alpha x)(y, \lambda_\alpha y) + \frac{1}{n} \geq 0 \tag{30}$$

Moreover, φ is also completely positive because in that case

$$\sum \varphi(e_{ij}) \otimes e_{ij} = \sum_{\alpha=1}^{n^2-1} x_\alpha \, \lambda_\alpha \, \otimes \lambda_\alpha^T + \frac{1}{n} \mathbb{I}_n \otimes \mathbb{I}_n \geq 0. \tag{31}$$

Example 3. Given a a mixed state ω in $\mathbb{C}^n$: $\omega \geq 0$ with $\operatorname{Tr} \omega = 1$ we can consider two bases of M_n defined

$$f_\alpha = \begin{cases} \lambda_\alpha & \text{for} \quad \alpha = 1, 2, \cdots, n^2 - 1 \\ \omega & \text{for} \quad \alpha = n^2. \end{cases} \tag{32}$$

and

$$g_\alpha = \begin{cases} \lambda_\alpha - \mathbb{I} \operatorname{Tr}(\omega^\dagger \lambda_\alpha); & \text{for} \quad \alpha = 1, 2, \cdots, n^2 - 1 \\ \mathbb{I}; & \text{for} \quad \alpha = n^2 \end{cases} \tag{33}$$

with $x_{n^2} = 1$ and x_α satisfying (30) for $\alpha = 1, 2, \cdots, n^2 - 1$. Notice that the bases (32)(33) are dual one to each other

$$\operatorname{Tr} f_\alpha g_\beta = \delta_{\alpha\beta}.$$

Then, the unital maps φ and φ^T defined by

$$\varphi(a) = \sum_{\alpha=1}^{n^2} x_\alpha\, f_\alpha\, \mathrm{Tr}\,(a\, g_\alpha) \tag{34}$$

and

$$\varphi^T(a) = \sum_{\alpha=1}^{n^2} x_\alpha\, g_\alpha\, \mathrm{Tr}\,(a\, f_\alpha) \tag{35}$$

are both positive

$$\sum_{\alpha=1}^{n^2} x_\alpha\, (x, g_\alpha x)(y, f_\alpha y) \geq 0$$

and completely positive

$$\sum_{\alpha=1}^{n^2 1} x_\alpha\, g_\alpha \otimes f_\alpha^T \geq 0$$

provided that

$$|x_\alpha| \leq 1 \quad \alpha = 1, 2, \ldots n^2 - 1.$$

Acknowledgements

Three of us dedicate this work to George Sudarshan on the ocasion of his 75th birthday for his friendship, collaboration, inspiration and many enlightening discussions. The work of M.A. and G. M. has been partially supported by a cooperation grant INFN-CICYT. M.A. has also been partially supported by the Spanish MCyT grant FPA2000-1252 and DGIID-DGA (grant 2006-E24/2). A. K. has been supported by the Grant PBZ-M

References

[1] von Neumann J 1932 Mathematische Grund lagen der Quantenmechank, Berlin: Springer
[2] Sudarshan E C G, Mathews P M, and Rau J 1961 Phys. Rev. 121 920
[3] Choi M D 1975 Linear Alg. and Appl. 10 285
[4] Jamiolkowski A 1972 Rep. Math. Phys. 3, 275
[5] Stinespring W F 1955 Proc. Amer. Math. Soc. 6 211
[6] Kraus K 1971 Ann. Phys. 64 311
[7] Størmer E 1963 Acta Math. 110 233
[8] Størmer E 1982 Proc. Amer. Math. Soc. 86 402
[9] Choi M D 1975 Linear Alg. and Appl. 12 95
[10] Woronovicz S L 1976 Rep. Math. Phys. 10, 165
[11] Ha K C 1998 Publ. RIMS Univ. 34, 591
[12] Horodecki M, Horodecki P, and Horodecki R 1996 Phys. Lett. A 223, 1
[13] Ha K C, Kye S H, and Park V S 2003 Phys. Lett. A 313, 163
[14] Arrighi P and Patricot C 2004 On quantum operations as quantum states, Ann. Phys. 311, 26
[15] Zyczkowski K and Bengtsson I 2004 Open Syst. Inf. Dyn. 11, 3
[16] Zurek W H 2000 Ann. der Physik 9 (5) 853-862
[17] Ollivier H and Zurek W H 2002 Phys. Rev. Lett. 88 17901
[18] Zurek W H 2003 Phys. Rev. A 67 012320
[19] Gelfand I M and Vilenkin N 1968 Generalized Functions vol. IV, Academic Press, NY

[20] Asorey M, Kossakowski A, Marmo G, and Sudarshan E C G 2005 Open Syst. Inf. Dyn. 12, 319
[21] Ohya M 1983 IEEE T. Inform. Theory 29, 770
[22] Ohya M 1983 Lett. Nuovo Cimento 38, 402
[23] Takasaki T 1983, Tomiyama J, Math. Z. 184, 101
[24] Cerf N J and Adami C 1997 Phys. Rev. Lett. 79, 5194
[25] Cerf N J and Adami C 1998 Physica D 120, 62
[26] Cerf N J and Adami C 1999 Phys. Rev. A 60, 893
[27] Kossakowski A 2003 Open Syst. Inf. Dyn. 10, 213

The quantum jump approach — applications to quantum optics and to spin-boson detector models

Gerhard C. Hegerfeldt

Institut für Theoretische Physik, Universität Göttingen, Friedrich-Hund-Platz 1, 37077 Göttingen, Germany

E-mail: hegerf@theorie.physik.uni-goettingen.de

Abstract. The quantum jump approach (QJA), a discretized version of continuous observations of systems coupled to a bath, such as photons, is outlined and it is shown how it avoids the quantum Zeno effect by temporal coarse-graining. Applications to quantum counting processes and spin-boson models for detecting a particle are given.

1. Introduction

What happens when an excited atom is waiting to decay and to emit a photon? *When* does this happen? How is one sure it is happening at a particular time, t say, and not before? Clearly, to be sure of this one must monitor the atom and watch for the first appearance of a photon. This monitoring should, in principle, be done continuously because there is a difference between the observation of a photon *at* time t and the *first* observation of a photon at t, i.e. *no* photon observation before t and then eventually an observation at t. The quantum mechanical Hamiltonian does not contain explicit emission times, and the state of the complete system — atom plus radiation field (photons) — develops into a complicated superposition which also does not contain an explicit time for the emission of a photon. Thus, instead of speaking of the "emission of a photon"as an objective event one should rather speak of the "(first) detection of a photon".

The difficulty with this, though, is that the quantum mechanical description of continuous measurements is notoriously complicated in practical applications. This difficulty was highlighted by the widely publicized so-called quantum Zeno effect (QZE) which was derived in a general form by Misra and Sudarshan [1]. The quantum Zeno effect states that, under the assumption of the von Neumann-Lüders projection rule[1], of a semi-bounded Hamiltonian and of time-reversal invariance, rapidly repeated measurements will slow down the time development of a system, with an eventual freezing of the state in the limit. It has been pointed out by Lüders [5] that the von Neumann-Lüders rule can only yield an approximate description of a measurement and that taking it at face value is an over-idealization. In consequence, the "freezing of the state"

[1] The projection rule as currently used has been formulated by Lüders [2]. For observables with degenerate eigenvalues his formulation differs from that of von Neumann [3]. It is noteworthy that in the first edition of his book Dirac [4] defines observations which cause minimal disturbance and which seem to be related to Lüders' prescription; curiously, though, in later editions this passage has been omitted.

alluded to in the formulation of the QZE is an unreachable limit. However, the slow-down of the time development has been experimentally verified [6]. The von Neumann-Lüders projection rule successfully tries to describe the actions of measurements which interfere as little as possible with the system, and since it is simple and easy to apply it is widely used in modern developments of quantum mechanics like quantum teleportation and quantum information theory.

In the first part of this article we will outline an approach how to describe the watching for the first photon, then the next and so on. This is the so-called quantum jump approach (QJA)[7, 8, 9, 10]. It is based on the simple idea of replacing continuous measurements by discrete measurements separated by a small time interval Δt. Ideally one would like to take the limit $\Delta t \to 0$, but in the calculation one then immediately runs into the QZE and finds an early example of freezing to a subspace, not a freezing of the state (see Section 2.3). Keeping Δt small, but finite, one arrives at a description which, on a *coarse-grained* time scale, can be regarded as continuous for all practical purposes. In the approach, everything can be calculated explicitly, and the results are independent of the particular choice of Δt. The use of a coarse-grained time scale by-passes the QZE. This approach is similar in spirit to the Monte-Carlo wavefunction approach [11] and to quantum trajectories [12]. The QJA can be carried over to other systems which are coupled to a continuous number of degrees of freedom. Moreover, as will be seen in an example, it can also be applied to systems coupled to a finite or discrete number of degrees of freedom where it can yield considerable simplifications in the otherwise very involved numerical treatment, simplifications which are valid until revivals occur. As will be explained later, mathematically the underlying reason why the QJA works is the validity or, in the discrete case, the approximate validity of the Markov property.

There are numerous applications of the QJA. These include applications to electron shelving [13], to quantum beats [14], to quantum counting processes, to efficient solutions of Bloch equations by simulation[11], to an analysis of the widely discussed experiment in [6] relating to the quantum Zeno effect, to the problem of arrival times of particles, to double jumps of dipole-dipole interacting atoms and to spin-boson detector models for quantum mechanical particles [15, 16].

In the application to the QZE experiment of Itano et al. [6] the QJA helped to resolve a controversy in connection with the interpretation of this experiment. It had been argued that everything could be described by a Hamiltonian dynamics [17] or by Bloch equations [18], without invoking measurements. In my opinion this is indeed true. However, in [19] the experiment of Itano et al. was analyzed using the QJA and it was shown that, at least in an approximate way, it could also be described in terms of measurements and by the von Neumann-Lüders rule, with the ensuing applicability of the QZE, and thus explaining the slow-down of the time development. Moreover, this latter viewpoint was shown to be more fruitful in that it yielded a quick intuitive understanding of the dynamics involved, without recourse to lengthy numerical calculations involving Bloch equations.

After the overview of the QJA we will indicate in the remainder applications to quantum counting processes and describe recent results for spin-boson detector models. Quantum counting processes were introduced axiomatically by Davies and Srinivas [20]. It will be shown here that these axioms are not needed if one goes to a coarse-grained time-scale and uses the QJA. Moreover, in this case the counting operators do not have to be arrived at heuristically but can be calculated explicitly. The spin-boson detector models were originally introduced in [21]. The application of the QJA approach yields considerable simplifications in the continuum limit of these models and leads to a connection with fluorescence detector models, i.e. models based on interactions with lasers.

Journal of Physics: Conference Series **196** (2009) 012024 doi:10.1088/1742-6596/196/1/012024

2. Overview of quantum jump approach

2.1. The conditional Hamiltonian $H_{\rm cond}$

The Hamiltonian for an N-level atom plus quantized electromagnetic field $\mathbf{E} = \mathbf{E}^{(+)} + \mathbf{E}^{(-)}$ is given in the Schrödinger picture, in the limit of long wave-lengths and in the rotating-wave approximation, by

$$\begin{aligned} H &= \sum \hbar\omega_i \left|i\right\rangle \left\langle i\right| + H_F^0 + \left[e\mathbf{D}^{(-)}\cdot\mathbf{E}_e^{(+)}(t) + \text{h.c.}\right] + \left[e\mathbf{D}^{(-)}\cdot\mathbf{E}^{(+)} + \text{h.c.}\right] \\ &\equiv H_A^0 + H_F^0 + H_{AL}(t) + H_{AF} \end{aligned} \tag{1}$$

where $\mathbf{E}_e(t)$ is an external field and

$$\mathbf{D}^{(-)} = \sum_{i>j} \mathbf{D}_{ij}|i\rangle\langle j|, \qquad \mathbf{D}_{ij} = \langle i|\mathbf{X}|j\rangle, \qquad \mathbf{E}^{(+)} = \sum_{\mathbf{k}\lambda} \mathrm{i}e\left\{\frac{\hbar\omega_{\mathbf{k}}}{2\varepsilon_0 V}\right\}^{1/2} \varepsilon_{\mathbf{k}\lambda}a_{\mathbf{k}\lambda}$$

and a frequency cutoff is included. V is the quantization volume, later taken to infinity, and $i > j$ means $\omega_i > \omega_j$.

Let photon measurements be performed at times $t_1 = \Delta t, t_2 = 2\Delta t, \cdots$ on an ensemble of systems which, at time $t_0 = 0$, is supposed to be in the initial state $|0_{ph}\rangle|\psi\rangle$. By the von Neumann-Lüders projection rule the *subensemble* for which no photons are detected until time t_n is described, with $P_0 \equiv |0_{ph}\rangle\mathbf{1}_A\langle 0_{ph}|$, by

$$P_0 U(t_n, t_n - \Delta t)P_0 \cdots P_0 U(t_1, 0)|0_{ph}\rangle|\psi\rangle \equiv |0_{ph}\rangle U_{\rm cond}(t_n, 0)|\psi\rangle \ . \tag{2}$$

The norm squared of this is the probability of finding no photons for the measurements between 0 and t_n. $U_{\rm cond}$ gives the time development of an atom under the condition that no photon is observed until time t, and it will now be determined explicitly by simple second order perturbation theory. Going over to the interaction picture with respect to $H_A^0 + H_F^0$ one has

$$H_I(t) = H_{AL}^I(t) + H_{AF}^I(t) \tag{3}$$

which is obtained by replacing $|i\rangle\langle j|$ and $a_{\mathbf{k}\lambda}$ in the original interaction Hamiltonian by $|i\rangle\langle j|e^{i\omega_{ij}t}$ and $a_{\mathbf{k}\lambda}e^{-i\omega_{\mathbf{k}}t}$, respectively, with

$$\omega_{ij} \equiv \omega_i - \omega_j \ .$$

We now calculate, for $t_i \le t' < t_{i+1}$,

$$\langle 0_{ph}|\frac{d}{dt'}U_I(t', t_i)|0_{ph}\rangle \ . \tag{4}$$

In the first-order contribution only $H_{AL}^I(t)$ remains since

$$\langle 0_{ph}|H_{AF}^I(t)|0_{ph}\rangle = 0 \ . \tag{5}$$

The second order is, by Eq. (5),

$$-\hbar^{-2}\int_{t_i}^{t'} dt''\left\{\langle 0_{ph}|H_{AF}^I(t')H_{AF}^I(t'')|0_{ph}\rangle + H_{AL}^I(t')H_{AL}^I(t'')\right\} \ . \tag{6}$$

Now, if the external field $\mathbf{E}_e(t)$ is smooth in time, then the second part in Eq. (6) contributes a term of higher order in Δt and can therefore be omitted[2]. Thus the second-order contribution

[2] For thermal or a chaotic external field, however, this part may give rise to a contribution linear in Δt and then has to be retained. In this case one can no longer work with state vectors or wavefunctions but has to use (conditional) density matrices. A particular example of this is treated in Ref. [14].

becomes

$$-\hbar^{-2}\int_{t_i}^{t'} dt'' \sum_{\substack{ij\ell m\\ i>j,\ \ell>m}} |i\rangle\langle j|m\rangle\langle\ell| \sum_{\mathbf{k}\lambda}\frac{e^2\hbar\omega_k}{2\varepsilon_0 V}(\mathbf{D}_{ij}\cdot\varepsilon_{\mathbf{k}\lambda})(\varepsilon_{\mathbf{k}\lambda}\cdot\mathbf{D}_{m\ell})e^{-\mathrm{i}(\omega_{\mathbf{k}}-\omega_{ij})t'+\mathrm{i}(\omega_{\mathbf{k}}-\omega_{\ell m})t''}$$

$$= -\hbar^{-2}\sum_{\substack{ij\ell\\ i,\ell>j}} e^{\mathrm{i}(\omega_{ij}-\omega_{\ell j})t'}|i\rangle\langle\ell|\int_0^{t'-t_i} d\tau \sum_{\mathbf{k}\lambda}\frac{e^2\hbar\omega_k}{2\varepsilon_0 V}(\mathbf{D}_{ij}\cdot\varepsilon_{\mathbf{k}\lambda})(\varepsilon_{\mathbf{k}\lambda}\cdot\mathbf{D}_{j\ell})e^{-\mathrm{i}(\omega_{\mathbf{k}}-\omega_{\ell j})\tau} \quad (7)$$

One can now use properties of the correlation function

$$\kappa_{ji\ell m}(\tau) \equiv \sum_{\mathbf{k}\lambda}\frac{e^2\omega_{\mathbf{k}}}{2\varepsilon_0\hbar V}(\mathbf{D}_{ji}\cdot\varepsilon_{\mathbf{k}\lambda})(\varepsilon_{\mathbf{k}\lambda}\cdot\mathbf{D}_{\ell m})\ e^{-\mathrm{i}(\omega_{\mathbf{k}}-\omega_{\ell m})\tau}\ . \quad (8)$$

With $V^{-1} = \Delta^3 k/(2\pi)^3$ one can perform the limit $V \to \infty$, and the sum over $\mathbf{k}$ becomes an integral over ω, with a suitable frequency cutoff, and an integral over the unit sphere. The correlation function has an effective width of the order of $\omega_{\ell m}^{-1}$ around $\tau = 0$, and for $t' - t_i \gg \omega_{\ell m}^{-1}$ one can therefore extend the τ integration in Eq. (7) to infinity. This is equivalent to the approximation

$$\int_0^{t'-t_i} d\tau e^{\mathrm{i}(\omega_{\ell m}-\omega_{\mathbf{k}})\tau} \cong \pi\delta(\omega_{\mathbf{k}}-\omega_{\ell m}) + \mathrm{i}\mathcal{P}\frac{1}{\omega_{\mathbf{k}}-\omega_{\ell m}} \quad (9)$$

and corresponds to the usual Markov approximation in the derivation of the Bloch equations. For the second-order contribution one then obtains

$$-\sum_{\substack{ij\ell\\ i,\ell>j}} |i\rangle\langle\ell|e^{i\omega_{i\ell}t'}\int d^3k\frac{e^2\omega_{\mathbf{k}}}{(2\pi)^3\hbar 2\epsilon_0}\mathbf{D}_{ij}\cdot\sum_{\lambda=1}^{2}\varepsilon_{\mathbf{k}\lambda}\varepsilon_{\mathbf{k}\lambda}\cdot\mathbf{D}_{j\ell}\left(\pi\delta(\omega_{\mathbf{k}}-\omega_{\ell m})\ +\mathrm{i}\mathcal{P}\frac{1}{\omega_{\mathbf{k}}-\omega_{\ell m}}\right)\ .$$

The principal-value term is analogous to a level shift and is, as usual, omitted. The last integral equals $\Gamma_{ijj\ell}$ where

$$\Gamma_{ijk\ell} \equiv \frac{e^2}{6\pi\varepsilon_0\hbar c^3}\mathbf{D}_{ij}\cdot\mathbf{D}_{k\ell}|\omega_{k\ell}|^3 \quad + \quad \text{principal value term.} \quad (10)$$

Hence, integrating over t' from t_i to t_{i+1} and using $1+\delta \approx e^\delta$ for small δ, we obtain

$$\langle 0_{ph}|U_I(t_{i+1},t_i)|0_{ph}\rangle = \exp\left\{-\frac{\mathrm{i}}{\hbar}\int_{t_i}^{t_{i+1}} dt'\left\{H_{AL}^I(t') - \mathrm{i}\hbar\sum_{\substack{ij\ell\\ i,\ell>j}}\Gamma_{ijj\ell}e^{\mathrm{i}\omega_{i\ell}t'}|i\rangle\langle\ell|\right\}\right\}\ . \quad (11)$$

For small Δt this can be replaced by a time-ordered exponential and thus, with $t \equiv t_n$,

$$\prod_1^n\langle 0_{ph}|(U_I(t_i,t_{i-1})|0_{ph}\rangle \cong \mathcal{T}\exp\left\{-\frac{\mathrm{i}}{\hbar}\int_0^t dt'\left\{H_{AL}^I(t') - \mathrm{i}\hbar\sum_{\substack{ij\ell\\ i,\ell>j}}\Gamma_{ijj\ell}e^{\mathrm{i}\omega_{i\ell}t'}|i\rangle\langle\ell|\right\}\right\} \quad (12)$$

where the product sign on the l.h.s. includes an ordering in an obvious way. Since

$$\langle 0_{ph}|U(t_i,t_{i-1})|0_{ph}\rangle = e^{-\mathrm{i}H_A^0 t_i/\hbar}\langle 0_{ph}|U_I(t_i,t_{i-1})|0_{ph}\rangle e^{\mathrm{i}H_A^0 t_{i-1}/\hbar} \quad (13)$$

and since, for $t \equiv t_n = n\Delta t$,

$$U_{\rm cond}(t,0) = \prod_1^n \langle 0_{ph}|U(t_i,t_{i-1})|0_{ph}\rangle,$$

we obtain, on a coarse-grained time scale, from Eqs. (12) and (13)

$$U_{\rm cond}(t,0) = \mathcal{T}\exp\left\{-\frac{\rm i}{\hbar}\int_0^t dt'\left\{H_A^0 + H_{AL}(t') - {\rm i}\hbar\sum_{\substack{i\ell j\\ i,\ell>j}}\Gamma_{ijj\ell}|i\rangle\langle\ell|\right\}\right\} \tag{14}$$

which is the transformation of Eq. (12) back to the Schrödinger picture.

Thus, with the atomic operator Γ defined as

$$\Gamma \equiv \sum_{\substack{i\ell j\\ i,\ell>j}}\Gamma_{ijj\ell}|i\rangle\langle\ell| \tag{15}$$

the conditional Hamiltonian for an N-level atom with no photon emission until time t is, on the coarse-grained time scale, given by

$$H_{\rm cond}(t) = H_A^0 + H_{AL}(t) - {\rm i}\hbar\Gamma\ . \tag{16}$$

For initial atomic state $|\psi\rangle$ the probability to find no photon until time t is thus given by

$$\|U_{\rm cond}(t,0)|\psi\rangle\|^2\ , \tag{17}$$

and the probability to find the first photon in $(t, t+\Delta t)$ is the difference of this expression for t and $t+\Delta t$. Thus, on the coarse grained time scale, the probability density $w(t)$ for the first photon is the negative derivative of Eq. (16),

$$\begin{aligned} w(\tau) &= -\frac{d}{dt}||U_{\rm cond}(t,0)|\psi\rangle||^2 \\ &= \langle\psi|\Gamma + \Gamma^\dagger|\psi\rangle\ . \end{aligned} \tag{18}$$

2.2. Connection with the quantum Zeno effect

When one lets Δt become smaller and smaller the above derivation shows very nicely how and where the quantum Zeno effect turns up in a very natural way. If Δt is chosen much smaller than the inverse optical frequencies, the last exponential in Eq. (7) can be replaced by 1, and the integral becomes proportional to $t' - t_i$. Eq. (11) is then replaced by

$$\langle 0_{ph}|U_I(t_{i+1},t_i)|0_{ph}\rangle = \exp\left\{-\frac{\rm i}{\hbar}\left\{H_{AL}(t_i)\Delta t - {\rm i}\hbar\ {\rm const}\sum_{\substack{ij\ell\\ i,\ell>j}}\Gamma_{ijj\ell}e^{{\rm i}\omega_{i\ell}t_i}|i\rangle\langle\ell|(\Delta t)^2\right\}\right\}.$$

The product of these operators then becomes, for $\Delta t \to 0$,

$$\mathcal{T}\exp\left\{-\frac{\rm i}{\hbar}\int_0^t dt'\left\{H_{AL}(t')\right\}\right\}\ . \tag{19}$$

This is a purely atomic operator, and hence for $\Delta t \to 0$ the time development takes place in the atomic subspace only. The atomic state does not freeze, only the field becomes frozen. This is an example of freezing to a subspace the possibility of which was already pointed out in [1]. For this reason one cannot choose Δt arbitrarily small in the quantum jump approach.

2.3. The reset operation $\mathcal{R}$

We consider an ensemble where no photons are present at time t_0 and assume that a short time Δt later a photon has been found. In this section we going to determine the state (or density matrix) of an atom after a such a broadband detection of a photon. The initial ensemble is thus described by $\rho = |0_{ph}\rangle\rho_A\langle 0_{ph}|$ where ρ_A is the atomic density matrix. The state of the subensemble for which photons are detected by a non-absorptive measurement at time $t_1 = t_0 + \Delta t$ is given, in view of the von Neumann-Lüders projection rule, by

$$P_1\rho(t_1)P_1/\mathrm{tr}(\cdot) \tag{20}$$

where

$$P_1 \equiv \mathbf{1} - |0_{ph}\rangle\mathbf{1}_A\langle 0_{ph}| \ . \tag{21}$$

Note that Eq. (20) still contains the photons.

After a photon measurement by absorption no photons are present any longer and it was argued in Ref. [8] that the resulting reset state is obtained from Eq. (20) by a partial trace over the photons, i.e. by

$$|0_{ph}\rangle \left(\mathrm{tr}_{ph}P_1\rho(t_1)P_1\right) \langle 0_{ph}|/\mathrm{tr}(\cdot) \ . \tag{22}$$

The physical reason for this is that for the atomic description alone it should make no difference in infinite space whether or not the photons are absorbed, as long as they are sufficiently far away from the atom and no longer interacting with it. Eq. (22) can be calculated by perturbation theory for $U_I(\Delta t, 0)$, as in Section 2.1. Going up to second order in the interaction one obtains in a straightforward way for the reset state

$$\begin{aligned}
\hat{\mathcal{R}}\rho_A \ \Delta t &\equiv \mathrm{tr}_{ph}P_1U(\Delta t,0)|0_{ph}\rangle\rho_A\langle 0_{ph}|U(\Delta t,0)^\dagger P_1 \\
&= e^{-iH_A^0\Delta t/\hbar}\sum_{\substack{ij\ell m\\ i>j,\ell>m}} |j\rangle\langle i|\rho_A|\ell\rangle\langle m|e^{iH_A^0\Delta t/\hbar} \\
&\times \int_0^{\Delta t} dt' \int_0^{\Delta t} dt'' e^{i\omega_{\mathbf{k}}(t'-t'')-i\omega_{ij}t'+i\omega_{\ell m}t''} \sum_{\mathbf{k}\lambda} \frac{e^2\omega_{\mathbf{k}}}{2\varepsilon_0\hbar V}(\mathbf{D}_{ji}\cdot\varepsilon_{\mathbf{k}\lambda})(\varepsilon_{\mathbf{k}\lambda}\cdot\mathbf{D}_{\ell m}) \ .
\end{aligned} \tag{23}$$

Note that the atomic trace $\mathrm{tr}_A\hat{\mathcal{R}}\rho_A \ \Delta t$ gives the probability for a photon to be found at time $t_0 + \Delta t$. The external field drops out since its action is of second order in Δt. We decompose the rectangular integration domain over t' and t'' in Eq. (23) into two triangles, and using the Markov property as in Eqs. (7) and (9) one obtains in the limit $V \to \infty$

$$\hat{\mathcal{R}}\rho_A \ \Delta t = e^{-iH_A^0\Delta t/\hbar}\sum_{\substack{ij\ell m\\ i>j,\ell>m}} \{\Gamma_{ji\ell m} + \Gamma_{\ell mji}\}\,|j\rangle\langle i|\rho_A(t_n)|\ell\rangle\langle m|e^{iH_A^0\Delta t/\hbar}\int_0^{\Delta t} dt' e^{-\mathrm{i}(\omega_{ij}-\omega_{\ell m})t'},$$

with $\Gamma_{ijk\ell}$ given by Eq. (10). For simplicity we now assume that the atomic transition frequencies are well separated. Then the last integral contributes only for $ij = \ell m$ and one obtains

$$\hat{\mathcal{R}}\rho_A = \sum_{\substack{ij\ell m\\ i>j,\ell>m}} \{\Gamma_{ji\ell m} + \Gamma_{\ell mji}\}\,|j\rangle\langle i|\rho_A|\ell\rangle\langle m| \ . \tag{24}$$

For the general case one can use the same expression, as shown in [9]. Up to normalization this is the state of an atom after a photon detection at time $t_0 + \Delta t$, under the condition that no photon was present at time t_0. Its trace, multiplied by Δt, gives the probability for this event.

3. Application to quantum counting processes

In [20] Davies-Srinivas proposed axioms of for 'homogeneous quantum counting processes'. These axioms imply the *existence* of two superoperators $\hat{J}$ and $\hat{S}_t$ which map trace class operators to trace class operators and give the probability density for finding exactly n photons at times $t_1 < t_2 \cdots < t_n$ in $[t_0, t]$ in the form

$$w(t_1, \cdots t_n; [t_0, t]) \;=\; \mathrm{tr}\left(\hat{S}_{t-t_n}\; \hat{J}\; \hat{S}_{t_n - t_{n-1}} \;\cdots\; \hat{J}\; \hat{S}_{t_1 - t_0}\; \rho_{\mathrm{A}}(t_0)\;\right).$$

The explicit form of the operators is not known but have to be obtained by an educated guess. In the QJA one can obtain the same structure, but all operators can be calculated explicitly.

In the QJA we define the superoperator in atomic space

$$\mathcal{S}_{t,t'}\; \rho_{\mathrm{A}} \;:=\; U_{\mathrm{cond}}(t,t')\; \rho_{\mathrm{A}}\; U_{\mathrm{cond}}(t,t')^{\dagger}$$

Then the probability density for finding exactly n photons at times $t_1 < t_2 \cdots < t_n$ in $[t_0, t]$ can be calculated to be given by

$$w(t_1, \cdots t_n; [t_0, t]) \;=\; \mathrm{tr}\left(\mathcal{S}_{t,t_n}\; \mathcal{R}\; \mathcal{S}_{t_n,t_{n-1}} \;\cdots\; \mathcal{R}\; \mathcal{S}_{t_1,t_0}\; \rho_{\mathrm{A}}(t_0)\;\right).$$

The superoperators $\mathcal{R}$ and $\mathcal{S}_{t,t'}$ of the QJA not only automatically satisfy the properties of the operators $\hat{J}$ and $\hat{S}_{t-t'}$ of Davies-Srinivas, but are also explicitly known. The only difference is that in the QJA time is course-grained while the axioms of Davies-Srinivas assume continuous time from the start.

4. Application to spin-boson detector models

In this model [21] there are spins (in a meta-stable state) which are weakly coupled to a bath of bosons (small $\gamma_l^{(j)}$ in (25)). In addition there is a moving particle, and where its wavefunction overlaps with a spin there is a high spin flip probability (large $g_l^{(j)}$ in (25)). First there will be a single spin flip due to the passing particle (accompanied by a boson emission). This leads to a large energetic gain for flipping neighbors i.e. a large flip probability. This then leads to a sudden flip of all spins from metastability, a sort of domino effect (together with a burst of bosons). The first boson detection can then be regarded as a detection of the particle. Note that in this model there is no direct measurement on the particle. The Hamiltonian is of the form

$$\begin{aligned} H \;=\; & \frac{\hat{\mathbf{p}}^2}{2m} + \sum_j \frac{\varepsilon^{(j)}}{2}\hat{\sigma}_z^{(j)} - \sum_{j<j'} \frac{J^{(jj')}}{2}\hat{\sigma}_z^{(j)} \otimes \hat{\sigma}_z^{(jj')} + \hbar \sum_\ell \omega_l \hat{a}_\ell^\dagger \hat{a}_\ell \\ & + \hbar \sum_j \sum_l \left(\gamma_l^{(j)} e^{i f_l^{(j)}}\, \hat{a}_\ell^\dagger \hat{\sigma}_-^{(j)} + \text{h.c.}\right) \;+\; \hbar \sum_j \chi^{(j)}(\hat{\mathbf{x}}) \sum_\ell \left(g_l^{(j)} e^{i f_l^{(j)}}\, \hat{a}_\ell^\dagger \hat{\sigma}_-^{(j)} + \text{h.c.}\right) \end{aligned} \quad (25)$$

where $\chi^{(j)}(\mathbf{x})$ is a position dependent sensitivity function for the increased coupling due to the particle. It should be noted that the excitation number (i.e. $\pm$ for spins plus number of photons) is conserved. Therefore, the first detection of the particle, taken as the first spin flip, is equivalent to the detection of the first boson.

In the corresponding continuum limit ($N \to \infty$) and with the simplifying assumption that all sensitivity functions $\chi^{(j)}$ are equal, i.e. $\chi^{(j)}(\mathbf{x}) \equiv \chi(\mathbf{x})$, an analysis similar to that in Section 2.1 yields that the time development under the condition of no spin flip (no boson) is given by the conditional Hamiltonian

$$H_{\mathrm{cond}} \equiv \hat{\mathbf{p}}^2/2m + \hbar/2\; (\delta_{\mathrm{shift}} - \mathrm{i}\, A)\chi(\hat{\mathbf{x}})^2$$

where the constants A and δ_{shift}, corresponding to a decay (Einstein coefficient) and level shift, respectively, are obtained from the correlation function

$$\kappa(\tau) \equiv \sum_{\ell} |g_\ell|^2 \, e^{-i(\omega_\ell - \omega_0)\tau}.$$

Here the Markov property is needed, i.e. $\kappa(\tau) \approx 0$ if τ is larger than some small correlation time τ_c. In H_{cond} the spin-boson detector enters mainly through an imaginary term which decreases the norm (since the non-detection probability decreases) and which mathematically mimics a position dependent absorption. This is similar to the fluorescence model [22].

For illustration we consider a 1d example with a single spin and N discrete boson modes $\omega_\ell = \omega_{\text{M}} n/N$, $n = 1, \cdots, N$, coupling constants $\gamma_l^{(j)} = 0$ and $g_\ell = -iG\sqrt{\omega_\ell/N}$ where ω_{M} is a maximal boson frequency. A particle wave packet is coming from the left. The probability of finding the detector spin in state $|\downarrow\rangle$ at time t is given by

$$P_1^{\text{discrete}}(t) = \sum_{\ell} \int_{-\infty}^{\infty} dx \, |\langle x \downarrow 1_\ell | \Psi_t \rangle|^2 \equiv 1 - P_0^{\text{discrete}}(t),$$

As long as no recurrences occur due to the discrete nature of the bath (i.e. no transitions $|\downarrow 1_\ell\rangle \mapsto |\uparrow 0\rangle$) one can regard

$$w_1^{\text{discrete}}(t) = \frac{d}{dt} P_1^{\text{discrete}}(t) = -\frac{d}{dt} P_0^{\text{discrete}}(t).$$

as the probability density for a spin flip (i.e. for a detection) at time t. The numerical evaluation is extremely time consuming. An example is given in Fig. 1 (dots) for $N = 40$, $\chi(x) = \theta(x)$ (Heaviside function) and an incoming Gaussian wave packet. The solid line is the result of the QJA for the continuum limit $N \to \infty$ which is much easier to evaluate.

Reset operation: Let the complete system (bath, detector, particle) be described at a particular time by a density matrix of the form

$$\varrho \equiv |0\rangle \, |\uparrow_1 \ldots \uparrow_D\rangle \, \varrho_{\text{p}} \, \langle \uparrow_1 \ldots \uparrow_D| \, \langle 0| \, , \tag{26}$$

(with ϱ_{p} the particle density matrix), i.e., no boson and all spins up. If a boson is found in a broadband boson measurement a time Δt later, the density matrix for the corresponding subensemble is obtained by sandwiching the above expression with

$$\mathbb{P}_1 \equiv \sum_{\boldsymbol{\ell}} |1_{\boldsymbol{\ell}}\rangle\langle 1_{\boldsymbol{\ell}}| \, , \tag{27}$$

by the von Neumann-Lüders projection rule, and the trace gives the probability. The subsystem consisting solely of the particle is then described after the detection of a boson by a partial trace,

$$\text{tr}_{\text{det}} \, \text{tr}_{\text{bath}} \, \mathbb{P}_1 \, U(\Delta t, 0) \varrho \, U^\dagger(\Delta t, 0) \, \mathbb{P}_1 \equiv \mathcal{R}\varrho_{\text{p}} \, \Delta t \, , \tag{28}$$

which defines the operation $\mathcal{R}$. This can be calculated by second-order perturbation theory. Note that even if ϱ_{p} is a pure state the reset state will in general not be pure.

In the continuum limit, $N \to \infty$, one can proceed similarly to Section 2.3. With the Markov property and $\chi^{(j)}(\mathbf{x}) \equiv \chi(\mathbf{x})$, one obtains to first order in Δt

$$\mathcal{R}\varrho_{\text{p}} = A \, \chi(\hat{\mathbf{x}}) \varrho_{\text{p}} \chi(\hat{\mathbf{x}}) \tag{29}$$

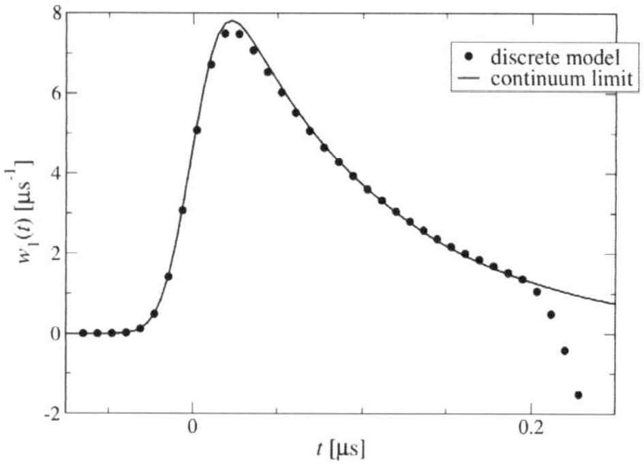

Figure 1. Dots: spin-flip probability density $w_1^{\mathrm{discrete}}(t)$ for $N = 40$, $\omega_0 = 2.39 \times 10^8 \mathrm{s}^{-1}$, $\omega_{\mathrm{M}} = 4.6\,\omega_0$, $G = 2.782 \times 10^3 \mathrm{s}^{-1/2}$, incoming Gaussian wave packet with velocity $v_0 = 1.79\mathrm{m/s}$ and momentum width $\Delta p = 20\mu\mathrm{m}^{-1} \times \hbar$. Solid line: $w_1(t)$ for the corresponding continuum limit. Up to the time of recurrences, $|\downarrow\ 1_\ell\rangle \mapsto |\uparrow\ 0\rangle$ (due to the discrete nature of the bath), the discrete and continuum probability densities are in good agreement.

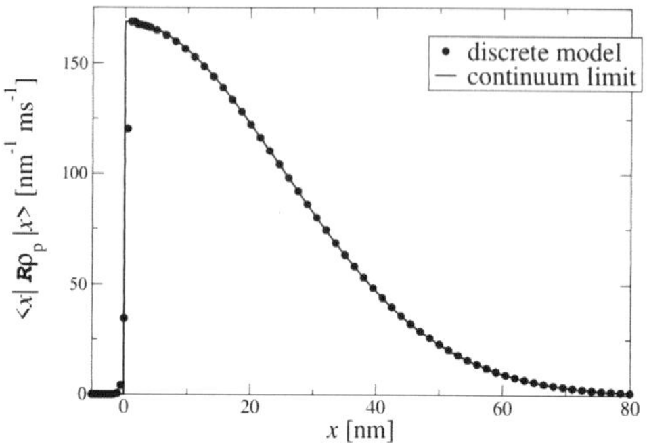

Figure 2. Dots: reset state $\langle x\,|\mathcal{R}\varrho_{\mathrm{p}}|\,x\rangle$ for $N = 15$, $\omega_0 = 2.38 \cdot 10^{12}\ \mathrm{s}^{-1}$, $\omega_{\mathrm{M}} = 4.6\,\omega_0$, $G = 2.782 \times 10^3\ \mathrm{s}^{-1/2}$, $\omega_0^{-1} = 4.185 \cdot 10^{-11}$ s, initial pure state $\varrho_{\mathrm{p}} = |\psi\rangle\langle\psi|$ with $|\psi\rangle$ a Gaussian wave packet ($v_0 = 1.79$ m/s and $\Delta x = 50$ nm) and $\Delta t = 100\ \omega_0^{-1}$. Solid line: reset state $A\ |\langle x\,|\Theta(\hat{x})|\,\psi\rangle|^2$ from Eq. (30) for the corresponding continuum limit. Up to small deviations around $x = 0$, the reset states from the discrete and the continuum model are in very good agreement.

for the state immediately after a detection. If ϱ_{p} is a pure state, then the reset state is also a pure state. In particular, if $\varrho_{\mathrm{p}} = |\psi^t_{\mathrm{cond}}\rangle\langle\psi^t_{\mathrm{cond}}|$, then the reset state is given by the wave function

$$|\psi^t_{\mathrm{reset}}\rangle \equiv A^{1/2}\chi(\hat{\mathbf{x}})\,|\psi^t_{\mathrm{cond}}\rangle\,. \tag{30}$$

This result is physically very reasonable since it means that right after a detection of the particle by a detector located in a specific region the particle is localized there. In particular, if χ is the characteristic function of the region the state is just projected onto this region by the detection.

We again consider the above 1d example with N (discrete) boson modes and $\chi(x) = \theta(x)$, but now $N = 15$. In Fig. 2 (dots) we have plotted $\langle x\,|\mathcal{R}\varrho_{\mathrm{p}}|\,x\rangle$ from Eq. (28) as a function of x for a Gaussian wave packet. For the corresponding continuum limit the solid line of Fig. 2 depicts

$$\langle x\,|\mathcal{R}\varrho_{\mathrm{p}}|\,x\rangle = A\ |\langle x\,|\Theta(\hat{x})|\,\psi\rangle|^2 \tag{31}$$

from Eq. (30). Up to small deviations around $x = 0$, the reset states from the discrete and the continuum model are in very good agreement.

In the present model, with $f_{\boldsymbol{\ell}}^{(j)}$ independent of $\mathbf{x}$, there is no explicit recoil on the particle from the created boson. This is in line with the original idea of a minimally invasive measurement. The absence of an explicit recoil distinguishes the present detector model from other models which are based on the direct interaction with the particle's internal degrees of freedom, such as interaction with laser light. It appears reasonable that in the present model no such recoil on the particle occurs: After all, the boson is emitted not by the particle but by the spin lattice. Hence, the recoil should be experienced by this lattice, rather than by the particle, similar to what occurs in the Mössbauer effect. Of course, the projection of the wave packet onto the detector region by means of the reset operation also changes the momentum distribution of the wave packet.

5. Discussion and summary

The quantum jump approach is applicable also to other systems which are coupled to a bath. As the analysis has shown one needs that the second order perturbation theory is proportional to Δt, which is essentially the Markov property. The QJA is not applicable for times in which revivals occur, such as for discrete bath modes. The temporal coarse graining in the approach is due to the use of the measurement theory of von Neumann-Lüders and their projection rule. Without coarse-graining this would lead to the quantum Zeno effect. The angular distribution of photons or other bosons can be included included in the approach [10]. In general there are no quantum mechanical predictions for a single measurement on a single system, but statements may be possible for repeated measurements on a single system.

In summary, we have shown the following.

- Repeated, semi-continuous, selective null (photon or boson) measurements lead to a *nonhermitean conditional* time-development operator between photon or boson detections.
- With this conditional time development the probability density of first photon or boson is given by the decrease of the norm-square of the state.
- After a photon or boson detection one has to reset the state ('jump'), then there is again the conditional time development until the next detection, and so on. This leads to the so-called quantum trajectories.
- Applications of this approach include light and dark periods, photon statistics, quantum Zeno effect, quantum beats, quantum counting processes, and quantum arrival times.

Acknowledgments

I would like to dedicate this paper to George Sudarshan on the occasion of his 75th birthday.

References

[1] Misra B and Sudarshan E C G 1977 *J. Math. Phys.* **18** 756
[2] Lüders G 1951 *Ann. Phys. (Leipzig)* **443** 322
[3] von Neumann J 1932 *Mathematische Grundlagen der Quantenmechanik* (Berlin: Springer) Chapter V.1. (English edition: 1955 *Mathematical Foundations of Quantum Mechanics* (Princeton: University Press))
[4] Dirac P A M 1930 *The Principles of Quantum Mechanics* 1st Ed. (Oxford: Clarendon Press) p 49
[5] Lüders G , private communication
[6] Itano W M, Heinzen D J, Bollinger J J, and Wineland D J 1990 *Phys. Rev.* A **41** 2295
[7] Hegerfeldt G C and Wilser T S 1992 *Classical and Quantum Systems. Proc. of the II. Intern. Wigner Symposium, Goslar 1991* ed Doebner H D, Scherer W, and Schroeck F (Singapur: World Scientific) p 104
[8] Hegerfeldt G C 1993 *Phys. Rev.* A **47** 449.
[9] Hegerfeldt G C and Sondermann D 1996 *Quantum Semicl. Opt.* **8** 121
[10] Hegerfeldt G C 2003 *Lecture Notes in Physics* vol 622 *Irreversible Quantum Dynamics* ed F Benatti and R Floreanini (Berlin Heidelberg: Springer) p 233
[11] Dalibard J, Castin Y, and Mølmer K 1992 *Phys. Rev. Lett.* **68** 580
[12] Carmichael H 1993 *Lecture Notes in Physics* vol m 18 *An Open Systems Approach to Quantum Optics* (Berlin Heidelberg: Springer)
[13] Dehmelt H G 1975 *Bull. Am. Phys. Soc.* **20** 60; Bergquist J C, Hulet R G, Itano W M and Wineland D J 1986 *Phys. Rev. Lett.* **57** 1699; Sauter T, Blatt R, Neuhauser W and Toschek P E 1986 *Opt. Comm.* **60** 287; Nagourney W, Sandberg J, and Dehmelt H 1986 *Phys. Rev. Lett.* **56** 2797
[14] Hegerfeldt G C and Plenio M B 1993 *Phys. Rev.* A **47** 2186
[15] Hegerfeldt G C, Neumann J T, and Schulman L S 2006 *J. Phys. A: Math. Gen.* **39** 14447
[16] Hegerfeldt G C, Neumann J T, and Schulman L S 2006 *quant-ph*/0610041 (to be published in *Phys. Rev.* A)
[17] Petrosky T, Tasaki S, and Prigogine I 1990 *Phys. Lett.* A **151** 109
[18] Frerichs V and Schenzle A 1991 *Phys. Rev.* A **44** 1962
[19] Beige A and Hegerfeldt G C 1996 *Phys. Rev.* A **53** 53
[20] Srinivas M D and Davies E B 1981 *Opt. Acta* **28** 981
[21] Gaveau B and Schulman L S 1990 *J. Stat. Phys.* **58** 1209
[22] Ruschhaupt A, Damborenea J A, Navarro B, Muga J G and Hegerfeldt G C 2004 *Europhys. Lett.* **67** 1

Journal of Physics: Conference Series **196** (2009) 012025 doi:10.1088/1742-6596/196/1/012025

Decohering histories and open quantum systems

Eric D. Chisolm

Theoretical Division, Los Alamos National Laboratory, Los Alamos, NM 87545

E-mail: echisolm@lanl.gov

Abstract. I briefly review the "decohering histories" or "consistent histories" formulation of quantum theory, due to Griffiths, Omnès, and Gell-Mann and Hartle (and the subject of my graduate work with George Sudarshan). I also sift through the many meanings that have been attached to decohering histories, with an emphasis on the most basic one: Decoherence of appropriate histories is needed to establish that quantum mechanics has the correct classical limit. Then I will describe efforts to find physical mechanisms that do this. Since most work has focused on density matrix versions of decoherence, I'll consider the relation between the two formulations, which historically has not been straightforward. Finally, I'll suggest a line of research that would use recent results by Sudarshan to illuminate this aspect of the classical limit of quantum theory.

1. Introduction

Many of the most interesting developments in quantum theory since Bell's Theorem have centered on the notion of "decoherence," understood most generally as the absence of interference effects. Decoherence is to be avoided at all costs when constructing mesoscopic or macroscopic quantum devices, such as SQUIDs or quantum computers, while its unavoidability in certain domains is said to be an important part of why large parts of the world appear classical. One area where decoherence has played an important role is the so-called "decohering histories" or "consistent histories" formulation of quantum theory, due to Griffiths [1], Omnès [2], and Gell-Mann and Hartle [3]. While it has not been adopted universally as an interpretive framework for quantum mechanics, which is what it was originally proposed for, it is certainly a useful tool for clarifying certain aspects of the classical limit of quantum theory, which remains a challenging problem despite the enormous progress made in the last few decades. In this article I will review the basic notions underlying decohering histories, and I will also spend a little time discussing various interpretations of this formalism. Having settled for application to the classical limit, leaving questions of interpretation aside, I will then compare the decohering histories approach to decoherence as seen in the time evolution of reduced density matrices. Since I argue that demonstrating decoherence of appropriate histories is needed to properly establish the classical limit of quantum theory, connecting these two approaches is of some importance. I will run across a problem which Sudarshan and I discussed at some length near the end of my time with him, without our coming to any good solution. Finally, I will apply recent results due to Sudarshan and coworkers to suggest a potentially very useful resolution of the issue, one that could illuminate not just this area but other aspects of the classical limit of quantum mechanics.

2. Histories and decoherence

The formalism of quantum mechanics is related to empirical results by the *Born rule*, which states the following: If a system is in state $|\Psi\rangle$ at time 0, then the probability at time t that the system will be found in state $|\alpha\rangle$ is given by

$$p(\alpha) = |\langle\alpha|U(t)|\Psi\rangle|^2 \tag{1}$$

where $U(t)$ is the time development operator which evolves the state from time 0 to time t. This expression can be rewritten

$$\begin{aligned} p(\alpha) &= \mathrm{Tr}[P_\alpha(t)\rho] \\ &= \mathrm{Tr}[P_\alpha(t)\rho P_\alpha(t)], \end{aligned} \tag{2}$$

where $\rho = |\Psi\rangle\langle\Psi|$ is the initial density operator of the system, P_α is the projection operator onto state $|\alpha\rangle$, and I have switched to the Heisenberg picture, in which observables evolve in time but states do not. The second line follows from the first because projection operators are idempotent and the trace is cyclic in its arguments. This expression is actually more general than Eq. (1) because it remains valid when ρ is a general density operator, representing an incoherent statistical mixture of pure states given by projectors.

An obvious generalization of this rule gives the probability not for a single measurement result, but for a series of measurement results at a succession of times:

$$p(\alpha_1, \alpha_2, \ldots, \alpha_n) = \mathrm{Tr}[P_{\alpha_n}(t_n)\cdots P_{\alpha_2}(t_2)P_{\alpha_1}(t_1)\rho P_{\alpha_1}(t_1)P_{\alpha_2}(t_2)\cdots P_{\alpha_n}(t_n)]. \tag{3}$$

This rule was given by Wigner [4], and Aharanov, Bergmann, and Lebowitz discussed both this rule and its time-symmetric generalization [5], but it is best known now because of its use in describing decohering histories. Before I get to decoherence, though, let me introduce some simplifying notation. Let α denote the sequence $\alpha_1, \alpha_2, \ldots, \alpha_n$, and given a sequence of measurements as above I define the *history* C_α by

$$C_\alpha = P_{\alpha_1}(t_1)P_{\alpha_2}(t_2)\cdots P_{\alpha_n}(t_n), \tag{4}$$

in terms of which Eq. (3) becomes the much tidier result

$$p(\alpha) = \mathrm{Tr}(C_\alpha^\dagger \rho C_\alpha). \tag{5}$$

This expression allows for much more conceptual contact between quantum and classical theory, as a history is the quantum mechanical analog of a trajectory of a classical particle. We routinely speak of such a particle occupying a succession of positions at a series of times, perhaps as it orbits another body, and in standard expositions of quantum theory it is not obvious what the corresponding quantum object is. This immediately suggests that histories may be of use in elucidating the classical limit of quantum mechanics, and that hope will be borne out below.

Since histories are products of projection operators, they inherit some (but not all) of the properties of their factors. Projection operators come in sets that satisfy

$$\begin{aligned} P_\alpha P_\beta &= P_\alpha \delta_{\alpha\beta} \\ \sum P_\alpha &= \mathbb{I} \end{aligned} \tag{6}$$

where $\mathbb{I}$ is the identity operator. The first condition is that the set $\{P_\alpha\}$ is *exclusive*; measurement results corresponding to two different P_α cannot occur simultaneously. [This restriction is lifted when one replaces projection operators with positive operator valued measures (POVM), but as I'll show in a moment exclusivity is important for defining decoherence, so how one would construct decohering histories using POVM is not clear.] The second condition guarantees that $\{P_\alpha\}$ is *exhaustive*, so every possible measurement result is included. Given a set of projection operators $\{P_{\alpha_i}\}$ at each of a predetermined set of times $t_1, t_2, \ldots, t_n$, one can construct a set of histories $\{C_\alpha\}$ that satisfy

$$\sum C_\alpha = \mathbb{I}, \tag{7}$$

so the histories are also exhaustive. Since a given set of histories defines a complete statistical sample space for a system, it makes sense to ask whether the probabilities associated with the histories by Eq. (5) satisfy the classical Kolmogorov axioms of probability:

(1) $p(\alpha) \geq 0$ for all α.

(2) $\sum_\alpha p(\alpha) = 1$.

(3) $p(\alpha \text{ or } \beta) = p(\alpha) + p(\beta)$ for all $\alpha \neq \beta$.

Axiom 1 is clearly satisfied by Eq. (5), and Axiom 2 is satisfied thanks to Eqs. (6) and (7), but we're in trouble with Axiom 3. This is easy to see if we recall that composite measurements in quantum theory are defined by taking the sum of the corresponding projection operators; e.g. if P_1 represents the spin of a spin-1 system being found up in the z direction, and P_{-1} represents the spin being found down, then

$$P = P_1 + P_{-1} \tag{8}$$

represents the spin being found to be nonzero. If we apply this same logic to histories then the operator representing either history α or history β is given by

$$C_{\alpha \text{ or } \beta} = C_\alpha + C_\beta. \tag{9}$$

(Notice that this is obviously correct if the two histories have the same events at all times but one, by the remarks that led to Eq. (8).) Therefore

$$\begin{aligned} p(\alpha \text{ or } \beta) &= \mathrm{Tr}(C^\dagger_{\alpha \text{ or } \beta}\,\rho\, C_{\alpha \text{ or } \beta}) \\ &= \mathrm{Tr}(C^\dagger_\alpha \rho C_\alpha) + \mathrm{Tr}(C^\dagger_\beta \rho C_\beta) + \mathrm{Tr}(C^\dagger_\alpha \rho C_\beta) + \mathrm{Tr}(C^\dagger_\beta \rho C_\alpha) \\ &= p(\alpha) + p(\beta) + 2\mathrm{Re}\,\mathrm{Tr}(C^\dagger_\alpha \rho C_\beta), \end{aligned} \tag{10}$$

and Axiom 3 holds for all histories if and only if the final term vanishes for any $\alpha \neq \beta$. Now we recognize this term; it is nothing more than a generalization of the *interference* terms we see ubiquitously in quantum theory. In fact, if we were to take a two-slit interferometer as our system and consider two-event histories that consisted of a particle passing through one of the two slits and then striking the screen, this term would be exactly the usual interference term. Thus we have generalized not only the Born rule but the expression for interference, and we have seen that in general it is the presence of interference that prevents quantum mechanical probabilities from satisfying the classical axioms of probability.

All of this inspires some new definitions. Given a complete set of histories, I define the *decoherence functional* between any two of them to be

$$D(\alpha, \beta) = \mathrm{Tr}(C^\dagger_\alpha \rho C_\beta). \tag{11}$$

Now $D(\alpha, \alpha)$ is the probability that history α occurs (cf. Eq. (5)), and if $\alpha \neq \beta$ then the real part of $D(\alpha, \beta)$ is the interference term between histories α and β. Thus a set of histories obeys the Kolmogorov rules of probability if and only if

$$\mathrm{Re}D(\alpha, \beta) = p(\alpha)\delta_{\alpha\beta}. \tag{12}$$

This condition is called *weak decoherence.* In many practical calculations the stronger condition

$$D(\alpha, \beta) = p(\alpha)\delta_{\alpha\beta}. \tag{13}$$

is actually achieved; this is called *medium decoherence.* (These names suggest that there ought to be some notion of strong decoherence, and several definitions have been proposed, but none have caught on, so I won't bother describing them.) Exactly what decoherence means physically is the subject of the next section.

3. Interpretation

When decoherence of histories was introduced by Griffiths [1], who called it *consistency* of histories, his goal was to develop a language for speaking meaningfully about the dynamics of a *closed* quantum system, for which the Copenhagen interpretation has nothing to say because a closed system by definition is not being measured. His idea was to allow the mathematical formalism of the theory tell him how to speak meaningfully about its dynamics; he derived Eq. (5) for the probability of a history, and he concluded that the formalism allows only those sets of histories which obey the Kolmogorov axioms to be assigned probabilities in a meaningful way. Thus he was led to Eq. (12), and he argued that if a set of histories of a closed quantum system exhibits weak decoherence, then it is meaningful to say that one of the histories in question actually happens in the system, with probabilities given by Eq. (5). Soon after, Omnès [2] arrived at similar expressions but with more of a focus on the logic of quantum theory, and he later integrated them into his own interpretation of quantum mechanics, which he describes as "a consistent and complete reformulation of the Copenhagen interpretation" which lacks Copenhagen's flaws [6]. In a slightly different vein, Gell-Mann and Hartle [3] had their sights set on quantum cosmology, where it had been realized that the Copenhagen interpretation was singularly inappropriate, the universe as a whole being the very model of a system with no outside observers. They have developed a quantum theory of closed systems, in which decoherence (they first applied this term to histories) is the sieve that determines à la Griffiths which histories are candidates for reality and which are not. Various objections have been raised to the use of decohering histories in this fashion, most famously by Dowker and Kent [7], and less famously by Sudarshan, Jordan, and me at about the same time [8], which elicited thorough responses from Griffiths [9] that were further elaborated in his extensive development of his program in [10].

In contrast, in this paper I would like to focus less on interpretation and more on the relevance of decohering histories to the classical limit. Roughly speaking, the classical limit of quantum mechanics consists of two components: Classical statistics and classical dynamics. By classical statistics, I mean that the probabilities associated with histories in classical physics obey the Kolmogorov rules of probability given above; Axiom 3 in particular is an essential part of our classical intuition, from coin tossing to weather prediction. By classical dynamics, I mean that the probabilities assigned by theory are strongly peaked around histories that correspond to solutions of the classical equations of motion; it is this aspect of the classical limit that received the lion's share of attention until about 30 years ago, when decoherence in various forms made its appearance. Both components, however, are equally necessary, which means that decoherence of an appropriately chosen set of histories is a precondition for a quantum system to have a sensible classical limit, and it is in precisely those histories that classical behavior will be seen. Thus even if all we care about is establishing classical behavior for all practical purposes (precisely John Bell's famous FAPP [11]), decoherence of appropriate histories must be demonstrated. Now I'll describe how this has been attempted.

4. Mechanisms of decoherence

By far the most popular option in the literature for achieving decoherence is to separate the quantum system under consideration into a preferred subsystem and the remainder, often called the *environment*. Such a separation corresponds to an expression of the system's Hilbert space $\mathcal{H}$ as

$$\mathcal{H} = \mathcal{H}^{\mathcal{S}} \otimes \mathcal{H}^{\mathcal{E}}, \tag{14}$$

where $\mathcal{H}^{\mathcal{S}}$ is the Hilbert space of the subsystem and $\mathcal{H}^{\mathcal{E}}$ describes the environment. This decomposition is actually quite general, and it corresponds to the expression of the configuration space of a classical system as the Cartesian product of subspaces carrying different degrees of freedom; for example, a single particle with spin moving in space can be described as the product of two subsystems, namely the spatial and spin degrees of freedom, each with its own Hilbert space. Once this separation is defined, decoherence is achieved by considering only histories composed of projections $P^{\mathcal{S}}$ onto the subsystem and by allowing the subsystem and environment to interact, which generally greatly accelerates the process of decoherence. This second factor leads to the use of open systems theory for decoherence, as the time evolution of the subsystem is normally far from Hamiltonian even if a Hamiltonian describes the full system.

Some of the most popular model systems for illustrating decoherence, as described by Stamp [12], can be classified as follows. The first and most famous are harmonic oscillators coupled linearly to baths of similar oscillators, typically in a thermal state with some simplified spectrum of frequencies; the preferred oscillator exhibits decoherence in its position basis due to interactions with the bath. The influence functional technique, due to Feynman and Vernon [13], was applied to this system by Caldeira and Leggett [14], and their calculation is still widely cited as an illustration of the efficiency with which environmental interaction produces decoherence. Caldeira and Leggett were at some pains to argue in their work that this model was considerably more generic than it would seem at first sight; the need for such arguments should be clear, given that their intent is to illustrate what should be one of the most widespread phenomena in nature (responsible for the entire classical world, after all). A second set of calculations has focused on finite-dimensional quantum systems coupled to spin baths [15]; the rationale was that such systems should do a good job of representing the low-energy dynamics of condensed matter systems, in which generally only a few low-lying energy levels should be excited. A third mechanism is "third-party decoherence" [12], in which correlations between subsystems which suppress interference arise not because of any interaction between them, but due to their mutual interactions with a third subsystem. (The particular example Stamp gives is of a particle passing through a two-slit interferometer which loses spatial interference because its spin interacts with the slits, even though its spin and spatial degrees of freedom are uncoupled.)

These sorts of calculations have two features in common which are worth mentioning. First, almost all of them are performed not on decoherence functionals, but on *reduced density matrices* describing only the preferred subsystem; for example, Caldeira and Leggett actually showed that the reduced density matrix of the preferred oscillator diagonalizes rapidly in the position representation for a variety of initial states, while its diagonal elements remain relatively unchanged. There have been exceptions, such as [16], but the majority of calculations in the literature have considered not histories but reduced density matrices or related objects. Given that I have argued for the importance of decoherence functionals in establishing classicality, it is worth asking how much light density matrix calculations shed on decohering histories. Second, a large fraction of the calculations have either assumed, or made other simplifying assumptions to guarantee, that the subsystem's evolution is described by a master equation of some sort, often Markovian. When Sudarshan and I discussed this state of affairs, he expressed his concern that such approximations were far more restrictive than these authors perhaps realized, and that they cast some doubt on their claims to have illustrated generic features that give rise to the classical world. These issues are the subject of the next two sections.

5. Decoherence functionals vs. reduced density matrices

Particularly since the histories approach to decoherence puts so much emphasis on diagonalizing a functional, it is easy to misunderstand the importance of diagonalizing reduced density matrices in that approach (at least it was easy for me to misunderstand it). Every density matrix, reduced or not, is of course always diagonal in some basis simply because it is self-adjoint. The goal of the reduced density matrix approach is this: Given a full system with initial state ρ, reduced initial state $\mathrm{Tr}_{\mathcal{E}}\,\rho$, and full time evolution operator $U(t)$, the time evolution of the reduced density matrix

$$\mathrm{Tr}_{\mathcal{E}}\,\rho \rightarrow \mathrm{Tr}_{\mathcal{E}}[U(t)\rho U^{\dagger}(t)] \tag{15}$$

sometimes has the property that the result diagonalizes rapidly in a particular basis, far more rapidly than the diagonal elements in that basis undergo change, and it stays diagonal over time. Further, this behavior is largely independent of the initial state ρ. The states making up such a basis, possibly with additional properties, are called "pointer states" [17], and the idea behind the reduced density matrix approach to decoherence is that such bases serve to define a preferred class of observables in which classical behavior emerges in appropriate systems. This is actually a very sensible idea, and in fact it deals directly with the fundamental issue raised by Dowker and Kent [7], which is that in general decoherence of histories depends far too sensitively on the exact choice of projection operators in the histories and the times at which they are placed. If anything is obvious about classicality, it is that it's robust; one does not have to look at the positions of a planet orbiting the sun at a carefully orchestrated sequence of times in order to see different alternate trajectories failing to interfere. Thus the obvious question to ask is whether histories composed of projections onto pointer states decohere. To see whether this is so, let me consider a representative decoherence functional

$$D(\alpha,\beta) = \mathrm{Tr}\left[P^{\mathcal{S}}_{\alpha_n}(t_n)\cdots P^{\mathcal{S}}_{\alpha_1}(t_1)\rho P^{\mathcal{S}}_{\beta_1}(t_1)\cdots P^{\mathcal{S}}_{\beta_n}(t_n)\right]. \tag{16}$$

Let me also make this expression a bit more explicit by expressing the trace as the composition of two partial traces, over the environment and subsystem, and let me also leave the Heisenberg picture and write the time evolutions explicitly, with the result

$$D(\alpha,\beta) = \mathrm{Tr}_{\mathcal{S}}\,\mathrm{Tr}_{\mathcal{E}}\,[P^{\mathcal{S}}_{\alpha_n}U(t_n,t_{n-1})\cdots P^{\mathcal{S}}_{\alpha_1}U(t_1,t_0)\rho U(t_0,t_1)P^{\mathcal{S}}_{\beta_1}\cdots U(t_{n-1},t_n)P^{\mathcal{S}}_{\beta_n}]. \tag{17}$$

Now let us suppose that the time evolution in Eq. (15) does in fact diagonalize the reduced density matrix in some basis, and further that the projections $P^{\mathcal{S}}_{\alpha_i}$ project onto that very basis. Can we conclude that this functional (or at least its real part) vanishes if any $\alpha_i \neq \beta_i$? Not obviously, for the simple reason that the time evolution is repeatedly interrupted by projections before the partial trace over the environment is taken, and while the partial trace commutes with those projections, it does not commute with the time evolution. This is all obvious, of course, and this problem is discussed in some detail by Kiefer in Chapter 5 of [18]. An important reference in that book is work by Paz and Zurek [19], whose solution, with notation modified for consistency with other work shown below, is as follows. Suppose there exist operators $M(t_2,t_1)$ such that for any operator A acting on the full system,

$$\mathrm{Tr}_{\mathcal{E}}[U(t_2,t_1)AU(t_1,t_2)] = M(t_2,t_1)\{\mathrm{Tr}_{\mathcal{E}}\,A\}; \tag{18}$$

in other words, suppose that the time evolution of Eq. (15) can be expressed as an operator acting on the reduced state only. In that case Eq. (17) can be written

$$D(\alpha,\beta) = \mathrm{Tr}_{\mathcal{S}}\,[P^{\mathcal{S}}_{\alpha_n}M(t_n,t_{n-1})\{\cdots P^{\mathcal{S}}_{\alpha_1}M(t_1,t_0)\{\mathrm{Tr}_{\mathcal{E}}\,\rho\}P^{\mathcal{S}}_{\beta_1}\cdots\}P^{\mathcal{S}}_{\beta_n}]. \tag{19}$$

Now it's obvious that if the times t_i are sufficiently well separated for the decoherence mechanism to do its work, the result of each $M(t_i,t_{i-1})$ will be diagonal in the pointer basis and the $\alpha_i \neq \beta_i$

terms will vanish. But is it reasonable to hypothesize such an operator M? For some time I struggled with this issue, because the justification offered by Paz and Zurek seemed to rely excessively on special assumptions, such as a Markovian master equation, of the type with which Sudarshan expressed such concern. However, recent work by Sudarshan and coworkers has resolved this issue rather neatly, which brings me to my final topic.

6. Decoherence functionals and the theory of open systems

Despite all my concerns with master equations and such, it was shown recently by Jordan, Shaji, and Sudarshan [20] that time evolution of operators in a subsystem of a full quantum system experiencing unitary evolution can always be expressed in a form very similiar to Eq. (18) with no assumptions whatsoever. Suppressing the time indices for the moment for convenience, the expression is

$$M(\mathrm{Tr}_{\mathcal{E}}\, A) = L(\mathrm{Tr}_{\mathcal{E}}\, A) + K \tag{20}$$

where

$$L(\mathrm{Tr}_{\mathcal{E}}\, A) = \mathrm{Tr}_{\mathcal{E}} \left[U \left(\mathrm{Tr}_{\mathcal{E}}\, A \otimes \frac{\mathbb{I}^{\mathcal{E}}}{\dim \mathcal{H}^{\mathcal{E}}} \right) U^{\dagger} \right] \tag{21}$$

and

$$K = \mathrm{Tr}_{\mathcal{E}} \left[U \left(A - \mathrm{Tr}_{\mathcal{E}}\, A \otimes \frac{\mathbb{I}^{\mathcal{E}}}{\dim \mathcal{H}^{\mathcal{E}}} \right) U^{\dagger} \right], \tag{22}$$

where I have reexpressed their results using my own notation. Several comments are in order concerning this form. First, the whole thing may look like simple sleight of hand (after all, this is simply Eq. (15) with Eq. (21) added and subtracted), and it would be if this form had no advantages; but Jordan *et al.* prove several significant results about this form and develop useful expressions for calculating with it in [20], so its apparent simplicity is deceptive. Second, although Sudarshan has always emphasized that the most general time evolution of a density operator can always be cast in linear form, the form shown here is instead affine, and that is the source of some of its charming features. Notice that the linear term depends only on the time evolution operator U and not on any correlations between the subsystem and the environment, while the contribution from the correlations has been isolated in the affine term. Third, this means that in fact M is not a function of only $\mathrm{Tr}_{\mathcal{E}}\, A$, but A in its entirety, as of course it must be; the idea here is not to neglect the correlations but to put them in a more tractable form. Finally, since this expression is available generally, it clearly follows that the $M(t_i, t_{i-1})$, while always defined, will not in general form a semigroup (as this would imply a master equation), and forming a group is almost always out of the question, as that would imply Hamiltonian evolution of the subsystem.

Casting $M(t_i, t_{i-1})$ in this form makes one point particularly clear. While it is typically assumed that the time evolution of the preferred subsystem obeys a master equation, is Markovian, or is otherwise special, all of those assumptions are for computational simplicity only; none of them are needed to express decoherence of histories. Stamp forcefully expresses a concern, hinted at previously in this paper, that the work which has been done up to this point on decoherence has not considered sufficiently general situations to convince the general population of physicists that truly generic mechanisms for producing the classical world have in fact been unearthed [12]; and while I don't wish to endorse Stamp's position fully, I do agree that the model calculations performed up to this point don't quite convey the flavor of a universal phenomenon. This leads me to the fundamental point of this paper: I suggest that a study of decoherence functionals using the formalism of Jordan *et al.* would give a broader and more general picture of decoherence as a phenomenon than the sorts of calculations largely performed up to this point have done. Jordan *et al.* state that the aim of their paper is "to simply describe the Schrödinger picture before making approximations to it" (such as the introduction of master equations), and I suggest that such a perspective would benefit the study of decoherence, which in many ways is so much more mature than it was a decade ago.

This maturity shows itself particularly in the research program of Zurek and coworkers, recently summarized in [17], in which it is recognized that decoherence is only one of many ingredients that characterize classicality. The larger picture is that the classical world is characterized by particular observables that carry several properties:

- The corresponding eigenstates are highly resistant to environmental dephasing, while their superpositions are not.
- Thus they are selected by the details of dynamics, not our arbitrary choices; it is dynamics that tells us what we can potentially observe.
- These observables imprint their values highly redundantly on the environment, so it is possible to deduce their values by sampling only a small fraction of the environment (knowing the position of an object after sampling only very few of the photons reflected from it is an obvious example).
- Finally, histories composed of projections onto eigenstates of these observables exhibit decoherence.

Many of these issues have been studied in explicit models in a master equation context, and I suggest that further study using this more general open systems perspective would be of great value.

However, the results of Jordan *et al.* require extension in certain directions to be useful here. The most obvious is that Eqs. (21) and (22) require the dimension of the environment's Hilbert space to be finite, as they involve the introduction of a "state of complete ignorance" of the environment, which is defined only in the finite-dimensional case. While this is adequate for spin bath calculations, an infinite-dimensional version is clearly required for other applications. Perhaps this is easily done by replacing the state of complete ignorance with a thermal equilibrium state, but that remains to be demonstrated.

7. Conclusions

Establishing that quantum mechanics has the correct classical limit has proved to be a surprisingly challenging problem; some time passed before the full extent of the problem was even clear. Since the recognition that decoherence was an important contributor, however, various research programs have made progress on this issue. Here I have summarized results due to Griffiths, Omnès, and Gell-Mann and Hartle, and I have argued that "decoherence" or "consistency" in their sense is necessary for establishing one aspect of the classical limit, namely the validity of classical statistics. I have also pointed out that decoherence is actually only one ingredient among many for demonstrating this validity in certain regimes. After describing their formalism, I tried to bring it into contact with approaches to decoherence that consider not histories but reduced density matrices of selected subsystems, and I found that the connection seemed to depend on making certain approximations in the quantum theory of open systems, such as the existence of Markovian master equations. I also suggested that this might cast some doubts on claims that truly universal mechanisms explaining the emergence of classicality have been found. Then I described a way to analyze histories based on recent work by Jordan, Shaji, and Sudarshan that seems to justify this connection while avoiding the usual approximations, thus offering the possibility of demonstrating decoherence in a much broader range of situations. I suggest that this work offers a new path forward for performing calculations that can illuminate many different aspects of the classical limit, including but certainly not limited to decoherence of histories, with considerably greater generality.

Acknowledgments
Clearly I am indebted first and foremost to E. C. George Sudarshan, to whom this paper is dedicated. Knowing my interest in foundational issues in quantum theory, he introduced me to Gell-Mann and Hartle's early articles on decoherence, and our analysis of their work and others' has influenced my understanding of these issues long after I moved on to other topics in my day job. I'd also like to acknowledge Thomas Jordan, with whom Sudarshan and I wrote [8]; he has written many clear and useful articles on foundational issues in quantum theory, and he continues to do so to this day. I also appreciate very helpful conversations with Wojciech Zurek, who changed my views on some of the issues described here and also pointed me to some important current references. Finally, I should also mention that this work was supported by the U. S. Department of Energy under contract DE-AC52-06NA25396.

References
[1] Griffiths R B 1984 *Journal of Statistical Physics* **36** 219
Griffiths R B 1986 *Fundamental Questions in Quantum Mechanics* ed L M Roth and A Inomata (New York: Gordon and Breach)
Griffiths R B 1986 *New Techniques and Ideas in Quantum Measurement Theory* ed D M Greenberger (New York: New York Academy of Sciences)
Griffiths R B 1987 *American Journal of Physics* **55** 11
Griffiths R B 1993 *Physical Review Letters* **70** 2201
Griffiths R B 1996 *Physical Review* A **54** 2759
[2] Omnès R 1988 *Journal of Statistical Physics* **53** 893, 933, 957
Omnès R 1989 *Journal of Statistical Physics* **57** 357
Omnès R 1990 *Annals of Physics* **201** 354
Omnès R 1991 *Journal of Statistical Physics* **62** 841
Omnès R 1994 *The Intepretation of Quantum Mechanics* (Princeton: Princeton University Press)
[3] Gell-Mann M and Hartle J B 1990 *Complexity, Entropy, and the Physics of Information* (SFI Studies in the Sciences of Complexity vol 8) ed W Zurek (Redwood City: Addison Wesley)
Gell-Mann M and Hartle J B 1990 *Proceedings of the 3rd International Symposium on the Foundations of Quantum Mechanics in the Light of New Technology* ed S Kobayashi, H Ezawa, M Murayama, and S Nomura (Tokyo: Physical Society of Japan)
Gell-Mann M and Hartle J B 1990 *Proceedings of the 25th International Conference on High Energy Physics* ed K K Phua and Y Yamaguchi (Singapore: World Scientific)
Gell-Mann M and Hartle J B 1993 *Physical Review* D **47** 3345
Gell-Mann M and Hartle J B 1994 *Proceedings of the NATO Workshop on the Physical Origins of Time Asymmetry* ed J J Halliwell, J Perez-Mercader, and W Zurek (Cambridge: Cambridge University Press)
[4] Wigner E P 1963 *American Journal of Physics* **31** 6
[5] Aharanov Y, Bergmann P G, and Lebowitz J L 1964 *Physical Review* **134** B1410
[6] Omnès R 1992 *Reviews of Modern Physics* **64** 339
[7] Dowker F and Kent A 1995 *Physical Review Letters* **75** 3038
Dowker F and Kent A 1996 *Journal of Statistical Physics* **82** 1575
[8] Chisolm E, Sudarshan E C G, and Jordan T F 1996 *International Journal of Theoretical Physics* **35** 485
[9] Griffiths R B 1998 *Physical Review* A **57** 1604
[10] Griffiths R B 2003 *Consistent Quantum Theory* (Cambridge: Cambridge University Press)
[11] Bell J S 1990 *Physics World* **3** 33
[12] Stamp P C E 2006 *Studies in History and Philosophy of Modern Physics* **37** 467
[13] Feynman R P and Vernon F L 1963 *Annals of Physics* **24** 118
[14] Caldeira A O and Leggett A J 1983 *Annals of Physics* **149** 374
Caldeira A O and Leggett A J 1983 *Physica* A **121** 587
Caldeira A O and Leggett A J 1985 *Physical Review* A **31** 1059
[15] Prokof'ev N V and Stamp P C E 2000 *Reports on Progress in Physics* **63** 669
[16] Dowker H F and Halliwell J J 1992 *Physical Review* D **46** 1580
[17] Zurek W H 2003 *Reviews of Modern Physics* **75** 715
[18] Joos E, Zeh H D, Kiefer C, Giulini D, Kupsch J, and Stamatescu I-O 2003 *Decoherence and the Appearance of a Classical World in Quantum Theory* 2nd ed (Berlin: Springer-Verlag)
[19] Paz J P and Zurek W H 1993 *Physical Review* D **48** 2728
[20] Jordan T F, Shaji A, and Sudarshan E C G 2006 *Physical Review* A **73** 012106

Journal of Physics: Conference Series **196** (2009) 012026 doi:10.1088/1742-6596/196/1/012026

Extended gravitational action and novel consequences

K P Sinha

INSA Honorary Scientist, Department of Physics and NMR Research Centre, Indian Institute of Science, Bangalore 560 012

E-mail: kpsinha@gmail.com

Abstract. It is found that the inclusion of higher derivative terms in the gravitational action along with concepts of phase transition and spontaneous symmetry breaking leads to some novel consequence. The Ricci scalar plays the dual role, like a physical field as well as a geometrical field. One gets Klein-Gordon equation for the emerging field and the corresponding quanta of geometry are called Riccions. For the early universe the model removes singularity along with inflation. In higher dimensional gravity the Riccions can break into spin half particle and antiparticle along with breaking of left-right symmetry. Most tantalizing consequences is the emergence of the physical universe from the geometry in the extreme past. Riccions can Bose condense and may account for the dark matter.

1. Introduction

The concept of the superfluid vacuum state connected with the cosmological constant was suggested several decades ago in collaboration with Sudarshan [1,2,3]. These researches involved the study of non-singular cosmological models, symmetry breaking and phase transitions at various stages of the evolution of the universe [4-7]. In the context of cosmology the geometry of space-time (whether flat or curved) of four or higher dimension is intimately connected with the interplay of matter and energy.

2. Formalism and results

In what follows, we present our recent results arising out of our study on extended gravitational action which contains higher derivative terms. The Einstein-Hilbert action.

$$S_g = \int d^4x \sqrt{-g}\, (\tilde{R} / 16\pi G_N + \Lambda) \qquad (1)$$

contains only the first power of Ricci scalar $\tilde{R}$, $1/16\pi\ G_N = M^2_p$, M_p = Planck mass, G_N is the Newtonian constant. In natural units ($\hbar = c = 1$) $16\pi\ G_N$ has the dimension of $(\text{length})^2$. Λ is the cosmological constant and g is the determinant of the metric tensor $g_{\mu\nu}$. Varying the action yields the Einsten field equation

$$R_{\mu\upsilon} - \tfrac{1}{2}\ R\, g_{\mu\nu} + \Lambda g_{\mu\nu} = -8\pi G_N\, T_{\mu\upsilon}\ , \qquad (2)$$

where $R_{\mu\upsilon}$ is the Ricci tensor and $T_{\mu\upsilon}$ the energy-momentum tensor. Here we have taken into account other physical fields also $S_m = \int d^4x\, L_m$, L_m being the Lagrangian density for fields other than gravity.

The above Einstein field equation had great success in explaining low energy phenomena involving gravity. However, in the high energy regime the higher derivative terms such as R^2, $R_{\mu\upsilon}R^{\mu\upsilon}$, and $R_{\mu\upsilon\rho\sigma} R^{\mu\upsilon\rho\sigma}$ become important. They play an important role in the context of renormalizability of the gravitational theory and stringy corrections in the string theory.

A generalized version of the extended gravitational action has the form [7-12]

$$S_g = \int d^4_x \sqrt{-g}\, [R^{\sim} / 16\pi G + \Lambda) + \alpha\, R_{\mu\upsilon}R^{\mu\upsilon} + \beta R^2] \quad , \qquad (3)$$

where α, β are dimensionless coupling constants and G may have value other G_N to account for strong grativational interaction in the early Universe [1]. We confine to lengths larger than L_P (the Planck length) to avoid unitarity problem. For the extended action the field equation is obtained by demanding the condition

$$(1/\sqrt{-g}\,)\delta S_g/\delta^{\mu\upsilon} = 0 \qquad (4)$$

One finally gets

$$R + [32\pi G(\alpha + 3\beta]^{-1}R = - \Lambda/(2\alpha + 6\,\beta), \qquad (5)$$

which is the dynamical equation for R in curved space with a potential

$\Lambda R / (2\alpha + 6\beta)$ as a source term.

$$\imath = \nabla_\mu \nabla^\mu \equiv 1/\sqrt{-g}\;\; \delta/\delta x^\mu\, (\sqrt{-g}\; g^{\mu\upsilon}\, \delta/\delta x^\upsilon)$$

In the absence of Λ, (5) can be written as

$$\imath R + m^2_R R = 0 \qquad (6)$$

$$\text{with } m_R = [32\pi G\, (\alpha + 3\beta)]^{-1/2} \qquad (7)$$

This shows that in the high energy regime R behaves like a scalar material field with its mass given by (7). The structure of (6) is exactly like Klein-Gordon (K – G) equation and from geometry we have made transition to physical field whose quanta are massive. We have named the quanta "Riccion". Note the mass depends on the coupling constants. To avoid ghost problem it is necessary to take $(\alpha + 3\beta) > 0$ [ghost fields are not associated with physical particles].

Compared to the usual K – G equation (where the mass dimension of the field is 1) the field dimension of R is 2. This can be circumvented by multiplying (6) by η (where $|\eta|$ = some number and dimension of η is $(\text{mass})^{-1}$. Thus with $R^{\sim} = \eta R$

$$\imath\, R^{\sim} + m^2_R\, R^{\sim} = 0, \qquad (8)$$

An important point to note is that the above formulation involves the modification of Einstein's theory of gravity in the high energy regime. The Hawking-Penrose theorem is violated and the model should be singularly free and there will be bounce at t = 0. It is found that [7]

$$a^2 = a^2 \sinh\, [(t - t_c)\, \sqrt{T_c}/6\eta + 0.89] \qquad (9)$$

t = time, a the radius of the universe, T_c, t_c critical temperature and time. The model universe expands initially according to (9) and for t sufficiently larger than t_c

$$a \approx a_c \exp [(t - t_c) \sqrt{T_c}/24\eta] \quad (10)$$

giving an exponential expansion in conformity with the inflationary model.

With the expansion the temperature falls and when T is well below $M_P\ [8\pi\ (\alpha + \beta)]^{-1/2}$ Einstein's theory of gravity takes over with R behaving like a geometrical field.

So far we have considered four-dimensional space-time only. We extend the model to (4 + D) dimensional action for R^2 gravity. The action is now [11]

$$Sg = \int d^4x\, d^Dy \sqrt{|g_{4+D}|}\ [(R_4 + D)/(16\pi G_{4+D}) + \alpha R^2_{4+D} + \beta\, \imath_{4+D}\ R_{4+D}]$$

$$\imath_{4+D} = (\sqrt{|g_{4+D}|})^{-1}\ \partial/\partial x^M\ (\sqrt{|g_{4+D}}\ g^{MN}|\ \partial/\partial x^N) \quad (11)$$

R^2 terms dominate over $R_{4+D}/(16\,\pi\, G_{4+D})$
when the energy mass scale $M \geq [\ 16\,\pi\,(\alpha + \beta\ )\ ]^{-½}\ M_P$

The equation obtained is

$$\imath R + \lambda\, R^2 + m^2_R\ R = 0 \quad (12)$$

$$(R_{4+D} = R_4 = R,\ \imath_4 \equiv \imath\ )$$

$$m^2_R = (D + 2) / (16\pi G\ [4(D+3)\ \alpha + D\beta] \quad (13)$$

and

$$\lambda = D\ \alpha / [\ 4(D+3)\alpha + D\beta] \quad (14)$$

subject to the condition $[4(D+3)\alpha+D\beta] > 0$, to avoid ghost problem. The operator $\imath_{D+4}$ reduces to $\imath_4$ as R is independent of y – coordinates.

Thus one obtains the four dimensional K-G equation for $R^{\sim}$ from higher-dimensional space-time.

$$\imath\, R^{\sim} + \lambda R\, R^{\sim} + m^2_R R^{\sim} = 0. \quad (15a)$$

as a result of spontaneous compactification . (For $\lambda=0$ for D=0, as discussed earlier, the above reduces to the usual K-G eqn.). In fact $\lambda RR^{\sim}$ is a manifestation of higher-dimensional geometry in four-dimensional space-time.

For special reasons we choose $\lambda = ¼$; we have

$$(\imath + ¼\ R + m^2_R\ )\ R^{\sim} = 0 \quad (15b)$$

Now a scalar field can be expressed as $R^{\sim} = \Psi^{\sim}\ \Psi$ (where Ψ is a Dirac spinor, $\overline{\Psi} = \Psi^{+}\gamma^{o}$, Ψ^{+} being the Hermitian conjugate of Ψ, γ^{o} is a Dirac matrix. This amounts to taking Riccion as a composite of a spin-1/2 Dirac fermion and anti-fermion. Using the algebra of Dirac matrices and vanishing covariant derivative of tetrad components we can write (15b) as

$$\overline{\Psi}\ (\gamma^{\mu}\ \nabla_{\mu} - im_R)\ (\gamma^{\nu}\ \nabla_{\nu} + im_R)\Psi = 0 \quad (16)$$

provided $\nabla_{\mu}\ \nabla_{\nu}\Psi^{\sim}\ \gamma^{\mu}\gamma^{\nu}\ \Psi = 0$ (17)

This condition can also reexpressed as

$$\nabla_\mu \nabla_\nu \bar{\Psi}_L \gamma^\mu \gamma^\nu \Psi_R = -\nabla_\mu \nabla_\nu \bar{\Psi}_R \gamma^\mu \gamma^\upsilon \Psi_L \quad (18)$$

where

$$\Psi_{L,R} = [\,(1 \pm \gamma^{5)}/2]\,\Psi, \bar{\Psi}_{L,R} = \bar{\Psi}\,[(1+\gamma^{-5})/2] \quad (19)$$

with

$$\gamma^5 = i\ \gamma^o \gamma^1 \gamma^2 \gamma^3 \quad (20)$$

Equation (18) implies breaking of left-right (L-R) symmetry. Equation (16) involves the Dirac equation for Ψ

$$(\gamma^\mu\ \nabla_\mu + im)\ \Psi = 0 \quad (21)$$

and

$$\bar{\Psi}\ (\gamma^\mu \nabla_\mu - im) = 0 \quad (22)$$

The important point to note is that a Riccion R (with $\lambda = ¼$) can break into fermion and antifremion pair only when L-R symmetry is broken. Thus

$$\bar{R} \rightarrow \Psi + \bar{\Psi} \quad (23a)$$

provided that parity is violated. However, the reverse reaction

$$\Psi + \bar{\Psi} \rightarrow R \quad (23b)$$

is not allowed and hence no back reaction. We have named these spin ½ particles as Riccinos.

The Riccinos are also very heavy particles and can decay further according to

$$\Psi \rightarrow \text{baryon + Photons} \quad (24a)$$

$$\bar{\Psi} \rightarrow \text{antibaryons + Photons} \quad (24b)$$

In the reaction (23a) and (23b) CP invariance is broken. As a consequence there will be excess of baryons over antibaryons. This is in agreement with the observed matter distribution in the early universe. This would suggest that the observed baryons and antibaryons were produced through the decay of riccinos and anti-riccinos with the condition that the process (24b) is slower than (24a).

Note that the above situation of massive Riccions and Riccinos and Antiriccinos appearing out of the geometry of space-time is possible only for curved space-time. For flat space time R = 0.

The above scenario prompts us to speculate that our physical universe emerged from higher-dimensional geometry in the extreme past through decay of Riccions and Riccinos.

Besides the spin-1/2 particles that will be created, other scalar particles can also be created [9, 10]. This arises from the non-minimal coupling of the Riccion field with a scalar field ϕ of the form $L_I = ½\ \chi\ R\ \phi^2$. We have noted earlier the inflation. When R undergoes spontaneous symmetry breaking resulting in a phase transition from the state $\langle R\rangle_{VEV} = 0$ to $\langle R\rangle_{VEV} = \pm\ \sigma$ (σ is the spontaneous symmetry breaking mass scale). As long as $T > T_c$, $\langle R\rangle_{VEV} = 0$. But as T falls, $\langle R\rangle$ tunnels through temperature barrier ($T = T_c$) and acquires $\langle R\rangle_{VEV} = \pm\ \sigma$. The scalar field ϕ is mass less for $\langle R\rangle = 0$ and acquires mass in the state $\langle R\rangle = \pm\ \sigma$. Thus a large number of particle-antiparticle pairs are created during the exponentially expanding stage in the state $\langle R\rangle = \pm\sigma$.

The scalar particles, in the cooling stage of the inflating universe, will undergo Bose-Einstein condensation below a temperature given by $T_{BEC} = (3.31)/(g_{deg.}{}^{2/3})\ (\hbar^2)/m_R\ (N/V)^{2/3}$, g_{deg} = degeneracy factor. These massive scalars in the universe play the role of dark matter and those massless as dark energy [13]. With $m_R \approx 100$ GeV, $(N/V) = 10^{48}$, T_{BEC} = a few degree K.

3. Concluding remarks

The model presented involves many novel features:
The TRINITY (Geometry, Matter and Energy) describing the universe are inter-convertible.

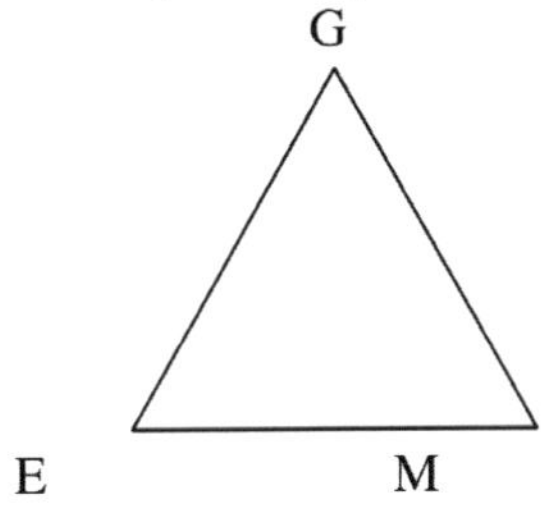

(G E M Model)

Einstein showed that $E = mc^2$, the interconvertibility of energy and matter. The GEM model shows that geometry (of the spacetime), energy and matter are interconvertible to each other. In keeping with the superstring theory, the geometry of the Universe may comprise higher than (3 space + 1 time) dimensions. The extra dimensions are curled up and do not manifest to direct observation. In the context of the above model it is appropriate to note Einstein's remark. "The geometry of space-time is physical, much like matter". The geometry may indeed have lattice structure and other attributes of matter e.g. charge, mass, spin, isospin, hypercharges, charm arise from the topological defects of the lattice.

Acknowledgement:
Grateful thanks to Dr. A. Meulenberg and Dr. S.K. Srivastava for fruitful discussion.

References:

[1] Sinha K P, Sivaram C and Sudarshan E C G 1976 Found. Phys. 6, 65, 717
[2] Sinha K P and Sudarshan E C G 1978 Found. Phys. 8, 823
[3] Sinha K P, Sudarshan E C G and Vigier J P 1986 Phys. Lett. 114A, 298
[4] Satyaprakash B S, Lord E A and Sinha K P 1984 Phys. Lett. 105A, 407
[5] Satyaprakash B S, Goswami P and Sinha K P 1986 Phys. Rev. D33, 2196
[6] Satyaprakash B S and Sinha K P, Pramana 1988 30, 15
[7] Srivastava S K and Sinha K P 1993 Phys. Lett. B397, 40
[8] Sinha K P and Srivastava S K 1994 J. Ind. Math. Soc. 61, 80
[9] Sinha K P and Srivastava S K 1995 Pramana, 44, 333.
[10] Srivastava S K and Sinha K P 1996 Int. J. Theoret. Phys. 36, 135
[11] Srivastava S K and Sinha K P 1997 Mod. Phys. Lett. A. 12, 2933
[12] Srivastava S K and Sinha K P 1998 *Aspects of gravitational interaction*, Nova Science Publishers, Inc, New York
[13] Srivastava S K and Sinha K P 2006 *Riccion as cosmic cold matter candidate*, sent to Phys. Lett B

Journal of Physics: Conference Series **196** (2009) 011013 doi:10.1088/1742-6596/196/1/011013

E C G Sudarshan: a Natural Philosopher

Ranjit Nair

Centre for Philosophy & Foundations of Science, New Delhi

Email: director@cpfs.res.in

It is a great privilege for me to be present here and I thank the organizers for their generosity, which has made me feel less of an interloper in the company of our honoree's students and long-standing research partners. This is a privilege matched only by my good fortune in being able to survey the rich corpus of E.C.G. Sudarshan's writings in the course of preparing his *Selected Scientific Papers* for publication, which contains 57 papers on the occasion of his 75th birthday.

Some personal reflections are in order here. I learnt about George Sudarshan when I was at school in Trivandrum in the early seventies and we had a bunch of highly gifted students for whom nothing except physics made any sense. The example of one our own making waves in international science was more important to us then than the waves in the sea which caressed the Kerala shoreline, which excited everyone else.

It was at a time shortly after the moon landings, which were universally admired as a feat of technological wizardry, although nobody asked whether going all the way to the moon and back was going to fill hungry stomachs of which there was no short supply on any continent. We led a ship to mouth existence, until the green revolution in India, which was a remarkable turnaround that impressed everybody. It was also a time when Indian space research had commenced in the coastal village of Thumba where a launch pad had been set up and every Wednesday we could set our clocks at 8 p.m. when a sounding rocket soared skyward.

That the Indian space programme began in Trivandrum meant a lot to us when we were students, since they had a bunch of highly motivated physicists and mathematicians in addition to engineers. My personal introduction to the space physicists came about as the result of an essay I sent while I was in high school, to the prime mover behind India's space programme, Vikram Sarabhai, who, much to my surprise, sent on my piece to a physicist in the Space Science and Technology Centre, who contacted me. Along with a couple of friends, I was allowed to sit in on their pre-Ph.D. courses in fluid dynamics. Our larger group of science enthusiasts had banded themselves into a science society which churned out a stream of physicists and mathematicians while it lasted for over a decade. Many of them went on to become professionals in the area and almost all of them are in the United States.

Personally I had the opportunity to meet Professor Sudarshan at the Centre for Theoretical Studies he established in Bangalore, in the Indian Institute of Science campus. This was shortly before my departure to Cambridge University. It was a wonderful experience, although I did not heed his suggestion that after a year in Cambridge I should go to America. The Centre was far ahead of its time, since inter-disciplinarity was not received wisdom then as it is today. I recall George and his able student and colleague N. Mukunda conduct an international symposium with great élan. The Centre was lodged in an unprepossessing building, which had earlier been used as

Journal of Physics: Conference Series **196** (2009) 011013 doi:10.1088/1742-6596/196/1/011013

a dispensary, but now what it dispensed was high-quality science. In the seminars held at the Centre, there would be the occasional traditionally attired pandit to represent the Indian tradition.

In India, philosophy as an academic subject suffered the misfortune of neglect, which was not exactly 'studied', since the least studious got in. In the student's list of priorities, philosophy's only consolation was that it was a few notches ahead of Sanskrit. It was and still is viewed with suspicion by die-hard physicists who assume that it is either trivial or irrelevant. This was quite unlike the West where some of the best and brightest pupils take up philosophy as evidenced by a statistic some years ago which showed that the average IQ (if you assume the credibility of such an index) of philosophy students was the highest, followed by physicists.

Traditional Indian philosophy had a poor rating among students and their parents, since it was not only philosophy but was also written in Sanskrit, which was associated with the indigenous religious traditions, making it as politically incorrect as it is possible to be. In this climate, Sudarshan's interweaving of Sanskrit quotes with his physics treatises was treated by some as a deplorable fad, not mitigated in the least by his own Christian origins. The neglect of Indian traditions by the intelligentsia lay the field open to rabble-rousers, who found it rather easy to stand the tradition on its head if need be, to underwrite their own brand of patriotism.

Admittedly, all narrative is 'constructed', but it is not to be disparaged on that account, since we cannot even conceive of a world without narrative. In the great Indian narrative, the Indian tradition was polymorphous. Ashoka the Great in the 3rd century BC carved on pillars in various places across India the message that if anyone disparages another faith to exalt one's own, he did a disservice to his own faith which would be seen in a poor light. The tradition of tolerance of diverse views was celebrated as a central Indian trait by the anti-colonial nationalists.

The first Indian Enlightenment is associated with the ministry of Gautama Siddhartha, who was the Buddha (literally the 'enlightened one'). The most recent was during colonial times when pioneers like Raja Ram Mohan Roy, a scholar of Sanskrit, Arabic, Persian and Bengali actively strove for the introduction of modern science.

European Indologists like Raymond Schwab prophesied a new renaissance in Europe with the diffusion of Indian knowledge, similar to the one sparked off by the return to Europe of Greek thought that followed the fall of Constantinople. It has been suggested that a renaissance of sorts did indeed occur through Arthur Schopenhauer who had read the Latin translation by Anquetil Duperron of the Persian translation of the Mughal Prince Dara Shikoh. Schopenhauer was pretty much at a tangent to the 'mainstream' of European thought which was cluttered by the heavy, lumbering and oh-so-prophetic deliverances of Hegel, a philosopher whom Schopenhauer heartily despised.

The rise of philosophy of science in Vienna, in a professional manner begins with Ernst Mach, an experimental physicist, who was the first person to occupy a chair in the philosophy of science. Like a conductor checking tickets on a bus, he questioned Newtonian ideas of absolute space and of absolute time, by asking for their empirical license. Mach and Boltzmann were involved in passionate debates about the reality of atoms, famously denied by the former, for whom all that was needed is to tweak some knobs in phenomenological thermodynamics. Ludwig Boltzmann's delight at replacing Mach did not cure him of the depression which drove him to suicide, having realized that the alignment of time's arrow with entropy increase that he purportedly established, was fatally flawed.

In the second half of the 20th century, the philosopher of science who won the allegiance of scientists was clearly Karl Popper, who strove to demarcate science from non-science, by decreeing that science dealt with statements that were falsifiable. Even Stephen Hawking who is otherwise disdainful of philosophy ('when I hear of Schroedinger's cat I reach for the gun') approves of Popper.

In his biography of Gell-Mann ('Strange Beauty') George Johnson says that Gell-Mann was fascinated by the name Sudarshan, which was explained to him as 'one who is good to look at'. It means more than that. 'Darshan' translates also as philosophy (cognate with the Greek 'theoria') and Sudarshan could be read as 'good philosophy'. As a natural philosopher, Sudarshan takes to heart the stirring words of Dirghatamas in the oldest of the Vedas, the Rgveda, in which he asserts 'ekam sat vipra bahudha vadanti' which translates as 'the truth is one though the wise speak of it variously'.

In the Upanishads, the last sections of the Vedas, the philosophy of the unity of existence and of the self is articulated through dialogue and argument between rival thinkers. The Vedanta school which Sudarshan is closest to, interprets the Vedas in the light of this foundational insight. The trope of unity is a powerful one in physics, in which the programme of unifying the fundamental forces of nature occupies the attention of some of the world's best minds.

There are different approaches to the issue of unification and among 20th century physicists, Paul Dirac is George Sudarshan's hero. In his autobiographical notes appearing in the *Selected Scientific Papers,* Sudarshan recalls the ease with which he was able to communicate with Dirac, whose sparse utterances were legendary. Harish-Chandra, a student of Homi Bhabha, who had Dirac as his thesis supervisor, gave his work to Dirac to read. The latter would read and return them with nary a word exchanged. Someone asked Dirac why he was so cruel and the latter replied that he did not want to say anything that would break the spell of Harish-Chandra's inspiration. Beauty for Dirac embodied universality, elegance, clarity and economy among other virtues.

Students of Dirac testify that the latter lectured from his book since he believed he had laid everything out with a precision and clarity that could not be surpassed. Dirac took to heart the Keatsian epigram 'Beauty is truth, truth beauty'. Sudarshan's ability to engage Dirac in extended conversations was a result of having carefully pored over the latter's papers, which he greatly relished, beforehand. One may perhaps draw the inference that Dirac's reticence did not apply so far as the exploration of his own work was concerned and that he could open up to a gifted graduate student.

Like Dirac, Sudarshan believes in grasping the nettle when it comes to the interpretation of physical theory by taking mathematics at its word. For instance, Dirac did not hesitate in asserting that in a double slit experiment with photons, the photon passes through both slits at once and interferes with itself. Similarly, having produced a first order wave relativistically covariant wave equation for the electron, Dirac was dismayed to find that it had excess baggage, the negative energy states, which he threw in the sea, only for it to float occasionally as positrons, inaugurating the realm of anti-matter.

The moral inculcated by Dirac was to take the equations seriously and not be afraid to push it to the limit if needed and to boldly go where angels fear to tread. Thus Sudarshan worked out (a) the V-A theory of weak interactions, going against the evidence of four experiments performed by established groups and thus discovering a fundamental law of nature, the third since Newton and Maxwell (b) the diagonal coherent state representation for all optical fields that was the basis of quantum optics although it gave negative probabilities for typically quantum effects, (c) a theory of faster-than-light particles that was consistent with the first signal principle since the light barrier could not be breached on either side, (d) developed the Wigner-Moyal phase space distribution where the negative probabilities could be interpreted as a typically quantum signature, (e) developed the quantum Zeno effect which showed that repeated measurements inhibit decay in unstable quantum systems. Such examples abound in Sudarshan's remarkable *oeuvre*.

Wolfgang Pauli once remarked about Dirac apropos his atheism 'There is no God and Dirac is his prophet'. Sudarshan parts company here, since, *mutatis mutandis*, he combines a belief in the correspondence of his equations with real-world phenomena, with an idealistic monism derived from the Advaita school of philosophy, as did Erwin Schrödinger before him. Fundamental theories, which are sacrosanct for workaday physicists, have to be probed deeply enough to understand what they do say and what they do not. The post WW-II generation of physicists began by assuming that quantum electrodynamics was the fundamental paradigm of particle physics. Sudarshan disagreed and attempted to extend mathematical structures (indefinite metric, rigged Hilbert spaces etc). By the '60s quantum field theory seemed to be getting nowhere and in a fit of abstemiousness, physicists took to the S-matrix approach which was concerned solely with scattering amplitudes and not about the fundamental constituents of matter. The tide turned in quantum field theory's favour in the late sixties, vindicating Sudarshan's faith in particle physics.

Similarly, the 'received view' is that quantum theory reduces to classical theory in the limit of $h \to 0$, which is in fact never the case. One of the approaches to measurement that Sudarshan developed involved a formal representation of a quantum system interacting with a classical apparatus, an attempt to embed within a consistent mathematical description, the desiderata of measurement. Unlike other physicists who were pacified by the nostrums on offer, Sudarshan saw the problem as something that could be formally solved. Similarly with the arrow of time, the connection of spin and statistics and so forth, where the standard practice is to attribute a solution to someone or the other and not seriously bother about questions of consistency and comprehension.

Where we stand at present, we must learn from Sudarshan that it is a moral duty to examine flaws rather than to trumpet the assertions of the virtues of various theories.

Let us draw inspiration from the illustrious career of our friend and hero George Sudarshan and wish him many more years of productive and rewarding research. Thank you.

Journal of Physics: Conference Series **196** (2009) 011013 doi:10.1088/1742-6596/196/1/011013

Reflections on Sudarshan's seven quests

Journal of Physics: Conference Series **196** (2009) 012027 doi:10.1088/1742-6596/196/1/012027

Reflections on Ennackal Chandy George Sudarshan

V. V. Raman

Department of Humanities, Rochester Institute of Technology, Rochester, NY

Email: vvrsps@rit.edu

Abstract: Talk presented to celebrate Sudarshan's 75th birthday.

It is a privilege for me to speak in this august assembly of distinguished physicists and scholars on this joyous celebratory occasion about a man of the stature of Professor George Sudarshan. It is also a personal pleasure for me to be doing this since I have regarded George as a good and esteemed friend for many decades now.

However, this legitimate honor is also a challenge: On what scientific aspect of this creative physicist can I dare to speak to this audience after listening during these past two days to thoughtful presentations on a variety of topics to which George has made significant contributions? Many of his germinal ideas, papers and visions have been explored and extended by fellow specialists in the fields who have assembled here.

I am even less equipped to talk about his interest in Malayalam literature, for, to put it in positive terms, my expertise in Malayalam is of the same depth as my scholarship in the literatures of Mandarin and Mongolian.

So, as one who has followed Sudarshan's career both as caring friend and as interested observer I will reflect on some matters in general terms. I will need no experimental set up for this, no theory to place them in a deductive framework, and no mathematics to give them calculational confirmation.

I recall with fondness the summer course that he gave at Brandeis in 1961. It was here that I met Sudarshan for the first time. There he was, a young and dashing physicist, displaying absolute mastery over his subject. He spoke with a wit and eloquence that was subtly enriched by his mellifluous Kottayam accent. He expounded to the eager post-docs in the room the latest results of Quantum Field Theory. I sat there in utter awe of this young man who began his lecture with a quote from Ernst Mach. I knew right away that he was not just another number-churning theoretical physicist. He went beyond formulas and calculations, and probed into the epistemological roots of the subject. He reminded me of the lines of Alexander Pope:

Let such teach who themselves excel,
And censure freely who have written well.

There is a level at which we speak about space and time, in which our notions of matter and energy are relevant and useful. But as we all know, these reveal considerable complexity when we scratch beneath their definitional meanings, not simply in their properties and attributes, but in the very core of the notions themselves. That is something Sudarshan is always aware of.

I like to compare physics to a swimming arena with a gently sloping floor that gradually merges into the oceanic expanse. At school we barely wet our feet in the waters, then some move a little farther, and most physicists swim in regions of reasonable depth. It is given to but a few to be drawn into the depths of the sea where they explore never-before noticed niches and discover things that are not there in the shallower bodies of water. Sudarshan belongs to the class of physicists who go behind the veils of the knowledge mines, and discern therein gems and nuggets that is for all to admire and make jewelry with.

But there are also other aspects of George Sudarshan that make him unique even in this distinguished assembly. For he is a man who began speaking Malayalam when he was barely a toddler. Now, how many in this room can claim to have done that? How many in this audience even know that Malayalam is the only language with a polysyllabic name, which is also a palindrome. In the jargon of physics, writing it out is a perfectly reversible process.

Among the many great personages who have come from that region of India, known as Kerala, perhaps the greatest is Adi Shankaracharya. For more than a millennium now that most illustrious thinker has enjoyed the highest respect, reverence, and pan-Indian reputation. In the extraordinary reverence he has received and in the sheer volume of commentaries on his work, he may be called the Aquinas of the Hindu world.

And we have in our midst today another prestigious personality from Kerala. He already shines bright in the firmament of his native land, and he too, like Shankara, is claimed by India as a whole. He has surpassed even the metaphysical master, for he is part of America too, and is an eminent citizen of the realm of international science. And it is safe to will here and now that in histories yet to be written of the Kerala world, George Sudarshan's name will stand out among those in whom his people will take personal and patriotic pride.

Sudarshan is also unique in that he was born in a Christian family, and later became a Hindu. I am told that when he was quite young, he read the Bible from cover to cover. I am not sure if this is what prompted him to embrace Hinduism. I must point out that conversion from Christianity to Hinduism is an allowed, and not a forbidden transition, but its probability is relatively small compared to the reverse processes. In any case, this has made him all the richer, culturally speaking.

Everyone here has a story to tell about how he or she was drawn to physics. You may like to know Sudarshan's. Back in India in those days, college students stayed home, and not in dorms. When George's older brother went to college, his physics text was left inadvertently one day on a common table. It piqued George's curiosity, and he began to read it. A statement in that book intrigued him: that the derivation of the formula for the period of a simple pendulum was beyond the scope of the book. Being who he is, George decided that this secret should be hunted down and revealed to his full satisfaction.

Some people in their later years say regretfully, "When I was young, I always wanted to be somebody. I wish I had been more specific." George knew right from the start that he wanted to be a physicist, and there is no other honest way of becoming a physicist without being able to derive the formula for the period of a simple pendulum.

When, in 1659, Christian Huygens derived the formula for the pendulum period, little did he realize that one day that formula would entice a young lad in distant Kerala into the intricacies of physics, much less that this lad would some day carve out seven quests in fundamental physics. No causal chain in the Laplacian deterministic world could have connected the swinging pendulum to Sudarshan, the physicist. It is legitimate to ask: Was George's looking into the book pure chance or pre-ordained? Sophocles wrote that the dice of Zeus make only lucky throws. It is an ancient truth that, in matters human chance encounters and people's whims are what direct the course of events in the world.

George went to Chennai as an undergraduate. He received his bachelor's and master's degrees from there. Sudarshan went way beyond the prescribed physics texts for doing well in the exams. He was fascinated by books like Eddington's *The Nature of the Physical World*, even as he plodded through the volumes of Arthur Haas's *Introduction to Theoretical physics*. While in

Chennai he inevitably acquired a speaking knowledge of Tamil, a sister language of Malayalam. I still enjoy conversing with him in Tamil sometimes.

After winning a first class degree from Madras University, he got a fellowship at the famous *Tata Institute of Fundamental Research* (TIFR) which was then, and still is, the most prestigious center for physics research in India. With great vision, Jawaharlal Nehru, the first prime minister of modern India, had generously supported its establishment at the recommendation of Homi Bhabha of Bhabha-Heitler scattering fame.

Young George shone brilliantly at the TIFR. He joined the team, which was then studying multiple coulomb scattering of cosmic ray particles, and he participated in devising a method for determining the mass of K-mesons. On the side, he read through and made critical comments on Von Neumann's foundations of quantum mechanics.

Robert Marshak from the University of Rochester was among the many physicists who visited TIFR during those years. He came there to give a series of lectures on Meson Physics. George took meticulous notes, and when Marshak saw these, it was love at notes' sight. These so impressed the master of mesons that he invited the young note-taker to Rochester for doctoral work. In those days, as far as graduate schools in the United States were concerned, outsourcing meant, not sending the job to India or Korea, but getting people from there right into the labs and universities here.

So, after some hurdles, Sudarshan moved from Bombay to Rochester where, as is now common knowledge, he did outstanding work. His mind is strong. But it was drawn to the weak interaction.

We have all heard the story to the effect that when Einstein explained his theory of relativity, only four people in the world understood it. Now there is a good physicist who, by this measure, is even greater than Einstein, because when he explained not just relativity, but anything, only two people understood it. One was the physicist himself, and the other was George Sudarshan. This physicist also discovered that there was a direct correlation between a person's intelligence and his ability to understand him. So he formulated a new scale, called OIQ: the IQ that is reflected in one's capacity to decipher Okubo's talk. It is said that on this scale also, George scored higher than anybody.

Another of George's fellow students in Rochester in those days was Tullio Regge. No he is not a Pole, but an Italian physicist. I have heard that Tullio and George wanted to be excused from the physics lab course because they felt it would be useless for them. Professor Marshak said there was a very simple way of determining the validity of that proposition: He asked them to take the course, and at the end of it they could empirically verify if it had been useless or useful. I suppose if it turned out to be useless, Marshak promised not to give them any unnecessary credit for the course.

Sudarshan was energetic and creative, original and insightful. He published on a whole range of subjects, and after he became a professor, he gathered graduate students and research associates in droves. His reputation grew, his work and worth was recognized by more and more people. He became director of research institutes. As the years rolled by, his students became physicists in their own right, affiliated to institutions in many parts of this country and in many countries of the world, and occupying prestigious positions themselves.

There is another aspect of Sudarshan I would like to touch upon. Already in his school days young George was fascinated by colorful Indian epics. Aside from their story content, these grand narratives, which are part of the rich lore of India, carry meaning and message. In some of the George took at college, there were allusions to the Gita and the Upanishads, and he began to explore these classical texts of Indian thought. In his college years he was an active member in a group where Indian philosophy was discussed, and this drew him further into the subject. His learning in these matters was enriched by extensive readings, as also, in later years through exchanges with the likes of Swami Agehananda Bharati.

In the Indic world few professors or practitioners of science have much attraction for the philosophies of the tradition, let alone its theologies. By and large, Indian physicists are content and creative in their various technical fields, and like specialists elsewhere, they would rather not venture into metaphysical domains where sound bites often matter more than substance. Leaving aside the complexities of Sanskrit grammar, to most practitioners of the hard sciences, such works belong to a different age. Some have even argued that too much affiliation to past visions of the world could restrain our march into the future. That is why modern scientists, especially in their problem-solving moods and modes, consciously keep away from philosophical and religious matters.

Yet, though religion is something that involves rites, and rituals, prayers and worship, devotional songs and festivals, all of which have little to do with science, many good scientists in the Hindu world do go to temples and do their prayers and periodic penitence, sing heart-felt bhajans and chant mantras. But generally, they keep all these exercises separate from their scientific commitments.

Unlike in the Western cultural matrix where religion and science are in a state of perennial tension, like a couple on the verge of divorce, arguing whether to live or not to live together, in the Hindu world, by and large, science and religion cohabit, to borrow Gould's phrase, as two as *Non-Overlapping Magisteria*.

From the Hindu perspective, beneath all the cacophony in the human world, and beyond the turmoil even in the stellar realms, there is an underlying cosmic harmony: a peace that transcends the conflicts and contradictions that mark the world of experience. This serene and silent world order or *rita*, as it is called, is revealed to the religious aspirant as the spiritual undercurrent that weaves the world of reality, and it is revealed to the physicist as the laws of nature. Who can touch or feel gravitation or the electro-weak principle, or the symmetry groups that orchestrate the song and dance of the universe?

Aside from such spiritual, poetic, and experiential reflections, there is much in classical Indian thought that is analytical, logical, interesting and insightful also. In epistemological subtleties, some classical Indian thinkers show extraordinary keenness. That is why Sudarshan, the physicist who is also a reflective thinker, delves into Indian spirituality and appreciates its relevance to culture in the modern world. He is one of the very few Indian scientists I know of who has attended, participated in, and contributed to Indian philosophical conferences as well. Sudarshan's public lectures on science and society are replete with quotes from and allusions to Sanskrit and Malayalam literature. For his knowledge and wisdom in these matters he was honored a couple of years ago with a prestigious recognition from the International Vedanta Conference.

Because of all this he was invited into the inner circles of some nationally and internationally know god-men. He gracefully accepted their invitations, and his name added prestige to their organizations.

With all that, it is important to remember that Sudarshan is faithfully wedded to the physicist's framework. Just as he can recognize cracks in the foundations of the scientific worldview, unlike some enthusiasts, he can also distinguish metaphysics and meaning from mathematics, and metaphor from measurable scientific truths. He once flew to Switzerland to observe attempts at mantra-induced levitation, which is based on the hypothesis that anti-gravity fields come into play at the bottom of squatters in meditation. But he declared that there was little empirical evidence of this even in the heights of the Swiss Alps where the acceleration due to gravity has smaller values. He is also cautious and critical when people link S-matrix to sutras in Buddhism or loftily write on quantum leaps and the field theory of consciousness.

Sudarshan's knowledge, skills, and scientific leadership have been in demand in India, in the United Sates, and elsewhere too. He has received several local, national, and international honors, such as *honoris causa* from many universities, First Prize for Third World Scientists, and the

Padma Bhushan, the highest honor – the equivalent of the knighthood, one might say – bestowed by the Government of India.

Back in India we call the celebration of the 75th year Platinum Jubilee. For this special joyful occasion when so many of his friends, well-wishers, and peers have gathered here to celebrate his work and honor him, as a tribute to George Sudarshan I have prepared a Power-point presentation reviewing his life and achievements.

In the Tamil tradition, there is a grand celebration when a person reaches the age of 83. Why 83? Because when you are 83 years old, there have been a thousand full moons since your birth. Dear George, we hope to see you celebrate that landmark too.

George Sudarshan

We've come here on this happy day
For lots of talks and fun,
To honor in a special way,
Our friend George Sudarshan.

Who's this man, some may ask,
Whom we honor thus?
Well, my friends, it 's now my task
To tell you why this fuss.

A simple man is how he looks
With a twinkle in his eye.
This author of some excellent books
Is modest and sometimes shy.

He's written scores of articles
And treatises on physics:
On elementary particles
And on quantum optics.

George was born in Kerala
The land of dance and beaches
From here came also Shankara
Who Vedanta teaches

This man who hails from Kottayam,
Before he heard of Bohr or Bohm
First spoke the language MALAYALAM
Which is a palindrome.
Like that great Vedantic sage
George moved from his place of birth.
And became of the modern age
A physicist of great worth.

When George was still a growing lad
He read the Bible whole.
But science and math he also had
Where he was to play a role.

He spent some years in Chennai where
He was an undergraduate.
To Mumbai next he went from there
To the TIFR great.

Here his genius really took
Off in different ways:
From commenting on Von Neuman's book
To theories on cosmic rays.

In Mumbai he also chanced to see
Men like Paul Dirac,
Tomonaga, Wolfgang Pauli
And of course Bob Marshak.

When Marshak George's notes did see.
He wanted both notes and him.
But George in Bombay had to be,
'cause that was his boss's whim.

After a year or two when the Indian chief
Could him no longer pester,
Sudarshan got his great relief
And went off to Rochester.

There he was a student bright.
This was clear for all to see.
His answers in tests were mostly right,
So he got his Ph.D.

There were of his colleagues plenty
The (not Pole) Regge Tullio,
And the mass-formula celebrity
Well-known as Susumu Okubo.

From relations which were by Sudarshan shown
A new result would flow:
Linked to the mass formula known,
As that of Okubo.

In fifty-seven was the conference
With all the greats of the day.
It was like a remembrance
Of the Congress at Solvay.

Bethe, Lee, Oppenheimer,
Yang and Murray Gell-Mann,
Wu, Dyson, Julian Schwinger,
Weisskopf, Salam and Feynman.

To the famous Rochester conference
With a paper Sudarshan came.
Though the paper made a lot of sense
George hadn't yet made a name.

The insight in his thesis
On the interaction called weak,
Was profound but on this
He couldn't at that meeting speak.

Others physicists of renown
Published a like result, they say.
This is the famous theory known
As the interaction V minus A.

If Sudarshan's paper had first come
At that meeting of the physicists wise,
He could have received, so say some
The coveted Nobel Prize.

"Elementary Particles" came out shortly
By Marshak and Sudarshan.
A book that was rendered promptly,
Into French, Russian, and German.

Of his lectures on Mechanics
Good notes Mukunda took.
In the literature of physics,
These became a solid book.

Sudarshan's work was systematic
On SU-3 and their representation.
He worked on functions analytic
And on their continuation.

Emil Wolf looked very sad.
George asked if he could something fix.
With his partial coherence Wolf was mad.
Whence George's work in quantum optics.

In the spring of Nineteen-Eighty
He met Glashow and Edward Witten;
And other physicists mighty
Who had major papers written.

Snow White had seven dwarfs, it says.
Some birds have seven eggs in nests.
Every week has seven days,
Sudarshan has his seven quests.

The seven quests he has been in
Are: weak interaction (this makes sense);
Statistics and of course spin
Quantum optical coherence.

Another relates to Einstein's note:
Naught faster than light can move on.
To this Sudarshan and Bilaniuk wrote
A paper on the tachyon.

Baidhyanath Misra and Sudarshan showed
That an unstable quantum entity
Can stay for good if kept observed:
They called this the Quantum ZP.
There is symmetry and invariance:
In all of physics, that's true.
But the physics of open systems
Is in quantum mechanics too.

He has written on quantum systems,
And on the hot electron too;
On Dirac brackets, abstract theorems,
And on the dynamics of Nambu.

He's written on density matrices
On beta, not alpha decay.
And on quantum optics
All his work can't be told today.

He has interesting things to tell
On the mysticism of Capra.
He wrote a whole book to dispel
The field-theories of Chopra.

Sudarshan has taught in many places
From Syracuse to Austin, Texas.
His students have a variety of faces
From all races and both sexes.

In the Math Institute in Madras
On the director's chair he sat.

'Cause of misunderstandings crass,
That was the end of that.

George led research institutes
He inspired students galore.
He was drawn into some disputes
But he did countless things more.

He has wanted as teacher and as student
To push the bounds of physics forward.
This he did, wherever he went,
Not just in the halls of Harvard.

Some seminal ideas of Sudarshan
Have sure influenced others,
But his papers to explicitly mention
Not everyone always bothers.

When the full history is written some day
All this will be brought to light.
And for sure in this way
The records will be set right.

No matter what theme Sudarshan takes,
And at philosophers' meetings too.
Witty comments he often makes,
As few other thinkers do.

Lecturing gives him lots of kicks
He has many a subject taught.
Also on matters beyond physics
He has often wisely thought.

He had friends like the Indic fighter
Agehananda Bharati;
Raja Rao, the writer,
And Jiddu Krishnamurti.

His friends include Feinberg
Mukunda, Marlan, Rothman,
And of course Steven Weinberg,
Hagen and V. V. Raman

With everyone he moves with ease,
Mahesh Yogi was in his range.
With Satya Sai he's been at peace,
Though some find this rather strange.

He has received many honors,
In India and beyond.
He has given many lectures
Of which all are fond.

He likes science fiction,
And reads Castaneda even,
And some Malayalee writers
Like M. T. Vasudevan.

Besides books from Kerala
Sudarshan has read a lot
On Upanishads and Vedanta
And other aspects of Indic thought

He likes detective stories,
Especially on TV,
But few things he likes more
Than mushrooms in hot curry.

In his garden he is happy
With all the plants he's got.
He is very glad to see them,
And not himself, go to pot.

Many are certain his works deserve
Not one but two Nobel Prizes.
But there's doubt they this observe
In the Committee that advises.

But George forgets not the T-O-E
Taught by Shankaracharya:
That everything we hear and see
Is a dream of the Cosmic Maya.

We celebrate George today,
Who has had a life-long alliance
With India, the USA,
And the whole world of Science.

Such is the man, simple and great.
Who is here with his wife.
Both of them we congratulate:
And wish more joys in life!

Journal of Physics: Conference Series **196** (2009) 012028 doi:10.1088/1742-6596/196/1/012028

Ennackal Chandy George Sudarshan

Swadesh M Mahajan

Institute for Fusion Studies, The University of Texas at Austin

Email: mahajan@mail.utexas.edu

Both compelled and powered by his ever-enquiring yet itinerant intellect, George Sudarshan's wanderlust made him a fugitive in the realm of ideas and laws of nature. The fateful wanderings led to successes so thrilling and far reaching that mortal men dare not dream of.

Seven is a jolly good number for successful quests; seven denotes both mystery and aesthetics- for what could be more sonorous than the alliteration

Sudarshansya sapta safala sadhana. Ladies and Gentlemen, you will agree with me that when *recited* in Sanskrit even simple and innocent statements become deep, magical and authoritative.

George, this illustrious gathering of savants has regaled us with their own achievements that were stimulated by the fruits of your manifold quests.

These achievements are impressive, indeed. Hearing them and conflating them with my own understanding of your work, I can hear the following distinguishing notes in the symphony that constitutes your science:

When you entered a field, you fundamentally changed it

And when you did not, you left indelible marks of your visit- these marks stayed long after you had strayed off to distant lands

When you had one of your frequent epiphanies, those who had the good fortune to be around were, forever, inspired. Your generous sharing of thoughts and ideas catalyzed the emergence of thoughts and ideas in several generations of physicists.

Many would say that the ability of seeding fertile minds is what makes a great teacher-Add to it kindness and care and one gets the perfect prescription for honing young minds- you should rejoice and we all do that you have honed so many.

The history of Physics, of course, is a grand illustration of the profound catalytic stirring that great ideas cause in their wake. It is said that after quantum mechanics was formulated, anyone familiar with Greek alphabet could get a Nobel Prize. There is one exception, however, that proves the rule- I did learn my Greek alphabet well and all I got for my troubles was the ability to read Greek alphabet.

I have already talked a bit much. In the Indian tradition, it is considered somewhat impudent for the young- relatively, that is- to prattle endlessly about the learned elders. They should show their respect by performing appropriate rituals and not by chatter- Ritual is, indeed, what I am leading up to. I begin by reminding you that 75 the year is a year of major phase-transition – one graduates from the stage of being a *vanaprasthin,* a forest dweller, to become a *sanyasin*, an itinerant wanderer and mendicant. Incidentally for those who may not be aware, Sudarshan has spent all twenty five years of *vanaprastha* in the exalted and salubrious forest called Austin- much of the time with his favorite consort whose resplendence today transcends both classical and quantum descriptions.

I have nothing against a homeless itinerant wanderer- some of my best friends are hopelessly homeless. And being an itinerant will suit George fine- but there is a catch! When you become a

Journal of Physics: Conference Series **196** (2009) 012028 doi:10.1088/1742-6596/196/1/012028

sanyasi you must renounce everything, and I mean everything- thinking and pursuit of knowledge included.

Now a curious thing happened on the way to the phase-transition. A group of trouble makers who call themselves Friends of George Sudarshan –FOGS, in short, do not approve of this exalted state of dynamical cessation. They remonstrated and demonstrated before the Creator- I am not privy to their methodology but they sure got their way and extracted a mighty concession- Folks, for Sudarshan, the phase transition is indefinitely delayed. So George- by virtue of the authority vested in me by FOGS and the blessed Sarasvati, I enjoin you to continue creating knowledge – It is hoped that before the next hearing, you would have created a totally new branch of science.

By the way you FOGS - Roger, Prashant and Alaka- you have competition now – a brand new Californian band called DOGS- the disciples of George Sudarshan is in town- beware there is much melody in their tunes too.

Yes, indeed, I promised a ritual and have gone on prattling. But you will soon see why this digression wasn't purely wanton. After the most amazing victory in the Creator's august court, the FOGS decided to revolt against the establishment in Stockholm- If Stockholm was not going to do its job well, they were going to create a Stockholm in Austin. Honorable scholars, investigators, and friends, FOGS will, now, present a robe of honor to

George Sudarshan - just when he is so happy - surrounded by his larger family, biological and intellectual.

This robe is a symbol both of defiance and of joy- we honor the man and his Himalayan intellect -- we do not have the Nobel money to go with the honor --- and Bhamathi you don't get to dance with the king- but in this context I feel sorrier for the king; he does not get to dance with you!

Barring such minor details, this little ceremony of ours has all the nobility

that should grace institutions that honor higher thinking, invention and discovery. We honor and salute you for your science and for your humanity.

Most of all we thank you for having been so kind and wonderful, bearing so many wanton disappointments and still creating -your enthusiasm for science, for creativity, for the dissemination of all the wonderful things you do has never faltered- In a profound sense you are truly transcendental- you are beyond the constraints of your local surroundings.

Ladies and Gentlemen, I present to you Professor Ennackal Chandy George Sudarshan. Once he begins talking, my incarnation will be over and I will not come back. The last words will be his. He will talk for as long as he wishes and will answer only those questions that he wishes to indulge.

Journal of Physics: Conference Series **196** (2009) 012029 doi:10.1088/1742-6596/196/1/012029

Happy recollections of five decades in physics

E.C.G. Sudarshan

Physics Department, The University of Texas at Austin, Austin, TX

Email: sudarshan@physics.utexas.edu

These two days of discussions on physics with special attention to my scientific quests gives me great gratification. So many of my friends who have distinguished themselves as physicists have put their minds to present such excellent expositions. The contributions that I have made over the last half century, when handled by these experts, show their interconnection through the paths chosen by them, and also seem to show the interconnection between these ideas.

1. The Vector-Axial Vector Theory of Weak Interaction

When I joined the University of Rochester as a graduate student, I had to undergo a written qualifying examination, and I had to review various subjects like quantum mechanics, nuclear physics, electromagnetic theory, and classical mechanics. The study of nuclear physics gave me a chance to review the evidence for the relativistic form of the beta decay interaction. It was clear that the data was inconsistent, but the popular vote seemed to be for an S-T interaction. But this gave no chance for combining the muon decay with the beta decay. After the Ph.D. qualifying examination, Robert Marshak asked me to look into the problem of beta decay and muon decay interaction. The decay of the charged pion could be only axial vector or pseudoscalar interaction. So to make possible a Universal Fermi interaction, the beta decay had to contain A (axial vector); but if that was so, the companion interaction should be V (Vector). It then has a great formal similarity with the electromagnetic interaction, which was pure vector. So my conclusion was that in view of the inconsistency of the data, on angular correlation and decay rates, there should be some experiments that were wrong. I wanted to present this result at the Rochester Conference on High Energy Physics in the spring of 1957; it would have resolved the problem of weak interaction. But I was not given the chance. By the time I got to present it, I had talked to Murray Gell-Mann about what I wanted to present at the Rochester Conference in Spring; finally my manuscript was presented by Robert Marshak at a conference in Italy[1]. While this was going on, Richard Feynman and Murray Gell-Mann wrote a paper[2] with the assertion that the beta interaction was V-A, but with no analysis of experiments. This was fifty years ago, but there are still people who call it the Feyman-Gell Mann theory. There are even textbook writing "experts" who are innocent of this knowledge. (On one occasion when I went to McGill, I wanted to tell the sory to the author of a standard book on Nuclear Physics. Even though he and I were present at a dinner, he bolted his food and made himself scarce.)

The vectorial nature of the leptonic interaction current suggests similarity between electromagnetism and weak interactions. It was suggestive to later investigators that gauge couplings may be involved. Once this beta interaction was understood, the other weak interactions became understood one by one. The most remarkable development was the electro-weak unified theory of Weinberg[3] and Salam[4].

Journal of Physics: Conference Series **196** (2009) 012029 doi:10.1088/1742-6596/196/1/012029

In his opening on the V-A-theory[1], Steven Weinberg pointed out that the discovery of V-A gave the key to future developments of gauge theories of electron weak unification. When we studied the symmetries of the V-A interaction and the isotopic spin selection rules and considered the SU(2) characterization generated by the weak electron-neutrino currents[2], little could one imagine that the theory would suggest the gauge theory of electro -weak interactions. With the participation of Susumo Okubo, it gave us an opportunity to find the solution of the singular integral equation[3] for Λ decay. Ashok Das and Sandip Pakvasa surveyed the present status of the theory of weak interactions and the "coming of age" of neutrinos.

2. Symmetries

Symmetries were the tool for understanding much of atomic and nuclear physics. In atomic physics, the interactions were well known, but for complex spectra one had to emphasize symmetry aspects. In nuclear physics symmetries provided the vehicle for the computations. But when it comes to particle physics and field theory, symmetries gave us mass and cross section sum rules. For example Heitler showed that charge independence gives the sum rule for pion-nucleon scattering:

$$\sigma(\pi^+ p \rightarrow \pi^+ p) + \sigma(\pi^- p \rightarrow \pi^- p) = \sigma(\pi^0 p \rightarrow \pi^0 n) + \sigma(\pi^0 p \rightarrow \pi^0 p)$$

On the other hand, Lie algebras were used to study the mass and electromagnetic properties of observed particles . Marshak, Okubo, and Sudarshan[5] found a charge independence relation between the magnetic moments of the three Σ hyperons. Okubo used the seminal method of the Wigner-Eckart theorems to discover the mass formula for SU(3) multiplets. Macfarlane and Sudarshan[6] found results for transition magnetic moments of $\Sigma^0\Lambda$ and their transition masses in relation to the magnetic moments and mass differences within the multiplet. In the meantime, Levinson, Lipkin and Meshkov, and our group in Rochester were leapfrogging through the question of combining the internal symmetries with relativistic invariance raised by McGlinn followed up by Han, Sudarshan, and finally by O'Raifeartaigh[8].

Symmetry is not only in quantum particle physics, but also in Classical Mechanics. Currie, Jordan, Mukunda, and Sudarshan[9] studied the consequences for classical dynamics, and one result was the No Interaction theorem for relativistic classical Hamiltonian dynamics. Later, in an extended collaboration led by Marmo and Wang on nonholonomic constraints we found ways of going beyond the no-interaction limitation. The reviews of Sydney Meshkov, Moo-Young Han, N. Mukunda, and Thomas Jordan commented on these topics authoritatively, and brought out the vitality of symmetry ideas. It was a major milestone when Susumu Okubo[5] discovered the broken SU (3) mass formula. When SU(3) was introduced into particle physics, the formalisms got new applications

3. Spin and Statistics

From the works of Stoner[11] in magnetism, followed up by Pauli[12] in atomic spectra, one arrived at the basic principle of atomic structure, namely the Pauli principle. At about the same time, Bose had shown that photons obeyed a different statistics. Many people consider that Pauli proved the spin-statistics theorem for relativistic quantum field theory[13], but the main consequences are in the non-relativistic domain of atomic structure and the specific heat of dielectrics; so a non-relativistic and a more complete proof was necessary.

For more than forty years[8] I have held that we must have a simple and direct derivation of the spin statistics connection, not restricted to relativistic quantum theory. Some time ago, Neuenschwander[9] also raised the same question.

My attempts to do this, beginning in 1968, have been elegantly summarized by Shaji[14,15]. Luis Boya showed how to extend it to even theories with higher space dimensions, David Finkelstein related it to super-conductivity, and Ian Duck reviewed the whole domain in his spontaneous poetry. Tom Imbo and I explored the spin-statistics connection on manifolds with nontrivial topology of the configuration space. Unfortunately, a health emergency prevented him from being present here .

4. Quantum Coherence Optics

I am again reminded of my youth spent in Rochester. I trace my love of optics to Professor M . A. Thangaraj, who taught me optics in my undergraduate class in India ; but the development in coherence optics was annotated to me by Emil Wolf and Leonard Mandel. While there were statements made about "Quantum Coherence" that cannot be spanned by "Classical Optics, only classical coherence optics embellished with quantum labels was treated in " Quantum Optics" by other authors in which only classical distribution functions were postulated . In 1963 I gave the first and only formulation of optical coherence in quantum field theory, first in a short paper in Physical Review Letters[12?] and later in an extended paper (of about a hundred pages) in the Journal of Mathematical and Physical Sciences[12]. Chandra Lal Mehta, who was at Rochester, in the early sixties, contributed significantly to the development of "quantum" optics[14]. He has reviewed for us the theory and its history. The quantum aspect of coherence leads to anti bunching[15] and negative intensity correlations[16], and this was demonstrated experimentally by the work of Jeff Kimble and Leonard Mandel . He also did remarkable experiments in "Squeezed Light,"[17] which is an extreme case of quantum optics. Kimble's "Assessment" gives us a firsthand account of nonclassical light and its experimental confirmation. In his presentation, Rajiah Simon traces the relevance of nonclassical light in Information Science. He also puts on record the history of the discovery of the Diagonal (Coherent) Representation. Simon's analysis of false claims should convince any intelligent scientist. It is indeed a treat to hear the field being reviewed by four people who were pioneers in quantum optics and who continue to enrich the subject. Little did I realize when I wrote down the Diagonal Representation, that it would have such ramifications, or that false claims would be made to confuse the physicists who have not done their homework.

5. Quantum Zeno Effect

In classical physics we have the physics of discrete particles as well as those of waves. For these waves the "intensity" or "density" or any physical property is proportional to the square of the amplitude. But it is the amplitude that obeys equations of motion and in particular, the change from one wave amplitude to another is linear in time. Hence transition probabilities should be quadratic in time. On the other hand, in the decay of radioactive nuclei the rate is constant, and the probability grows linearly with time. How come we can get a constant rate (and linear growth of probability)?

There are two ways of obtaining this. One is best illustrated by Dirac's calculation of the rate of decay of an excited atom. Here we take into account a large continuum of energy levels of the final states; the integrated probability of transition is well approximated (for times not too short) by an expression linear in time. This is generally the accepted method to deal with an unstable system. Another method is to deduce a Markovian (probabilistic) evolution by resetting the phases at a very short time[29], and this leads to the Kossakowski[30] semigroup . In either version, we get a constant rate (and the radioactive law).

But what about the exact time dependence for very short times when the transition rate is proportional to the square of the time? If we were to consider this as a measurement and reset the system at this time interval, we could proceed to determine at a later time whether the system has decayed or not. So the survival probability of decay over a time t is given by and is the limit of $P(t) = (P(t/N))^N$. For large N, the survival amplitude of the original state approaches unity, and hence the decay is inhibited. This is the Quantum Zeno effect[20] of Misra and Sudarshan. Prashant Valanju was the first to demonstrate this effect by an analysis of the experimental data on hadron-nucleus collisions for his doctoral thesis.

Journal of Physics: Conference Series **196** (2009) 012029 doi:10.1088/1742-6596/196/1/012029

The presentations by Marmo, Pascazio, and Chiu discussed various aspects of Quantum Zeno Effect. Clearly this effect provides a mechanism for stabilizing qubits and controlling decoherence.

Giuseppe Marmo and Saverio Pascazio have put the theory in a proper and rigorous mathematical framework including the Trotter formula, and showed its manifold manifestations[19]. Charles Chiu reviewed this effect with particular attention to excited atom decay and nucleon-nucleus collisions[20] and exhibited the details of the true evolution of a metastable state[21]. He also connected it with Aharanov-Vardman's[22] work on the technique of realizing a simple Feynman path experimentally. Wayne Itano critically reviewed the work of his group on the experimental verification of QZE. Their experimental[23] results agree with the work of Raizen and collaborators[24] in Austin and yield sufficient experimental verification of the Quantum Zeno Effect.

6. Tachyons

The story of superluminal motion is an interesting one. In 1959 summer I wrote a brief paper on particles traveling faster than light and sent it to Physical Review. It promptly came back with a referee report which said it was "all wrong," and the Editor regretted that in view of this report, they cannot publish it. I requested a second referee whose report said it was quite correct, but all the results were "well known," and so publication was not recommended. I requested a third referee, since the first one did not say what was wrong and the second referee said "it all well known," but did not indicate where. The third referee was the most gifted. He said, "I have read the manuscript, and the two referee reports. I agree with both of them." I requested that the Editor read the paper and the confusing and contradictory reports. As was to be expected, I did not even receive the courtesy of a reply. But when someone else wrote a paper several years later (which was clearly inconsistent), the same Editor published it[25](and did not send it to me refereeing)! Eventually Olexa-Myron Bilaniuk and I rewrote the article and got it published in the American Journal of Physics. It generated a lot of correspondence and discussions. The revolutionary idea came to many people's attention due to the excellent article in "Discovery" by David Freedman[26] . Erasmo Recami has been the most eloquent exponent of superluminal motion. He showed that many interesting consequences result from superluminal motion. He has reviewed the experimental searches so far, and gave an exposition of the visual effects of such motion, which are quite characteristic. Olexa-Myron Bilaniuk gave his simple but elegant exposition of this unusual physical system. Samir Bose addressed the question whether special relativity theory forbids faster than light particles and its possible relevance in general relativistic Cosmology.

7. Open Systems and Dynamical Maps, Irreversibility

In the course of being a research guide, I have learned new things from my younger colleagues. I feel at ease not only with Fiber Bundles, but also with manifolds and homotopy . But I feel uneasy when people talk about information and entropy, not because others do not make sense, but my feeling that information and order are encoded and we do not have the decoder. The next session involved one of the most challenging research areas, beginning with the question of how can we get irreversibility from reversible equations of motion[28]. After much study and critical examination of the various trendsetters, I am still of the belief that the question is open. For a many particle system the trajectories are quite complicated, and resemble the strands of cooked spaghetti[29]. (See Fig. 1.) There seems to be no pattern and no predictability about the various trajectories. But if we insert a fork and pull out a few strands, they are almost parallel near the fork (Fig. 2). One also recognizes that in both directions the strands end up in complicated patterns. Seeing regularity at the present time does not say that it was regular at either end. If the trajectories belong to a confined space, it is likely that approximately the same regularity

Journal of Physics: Conference Series **196** (2009) 012029 doi:10.1088/1742-6596/196/1/012029

reappears (Fig. 3). The reason that regular strands get tangled up is not cussedness of the universe, it is our arbitrary imposition of a particular order that is described by the dynamics.
We are also aware of arbitrary choice of 'order' when we consider the motion of a quantum particle in a potential. We identify the "in" solution (Fig. 5) with 'order' in the far past, but disarray in the future. But there is an equivalent set of 'out' states with order in the far future but disarray in the far past, and even at the present (Fig. 6).

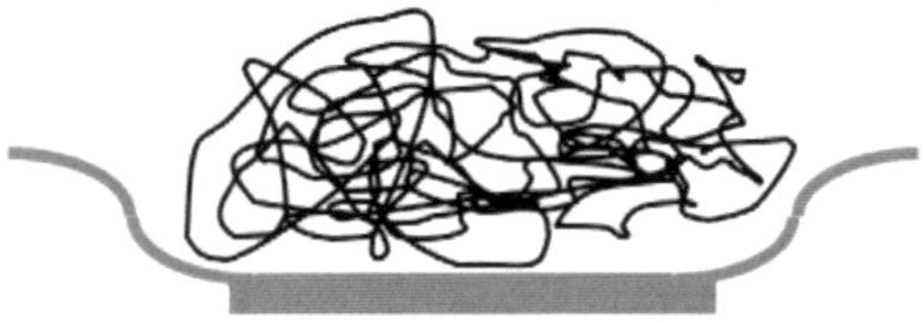
Fig. 1

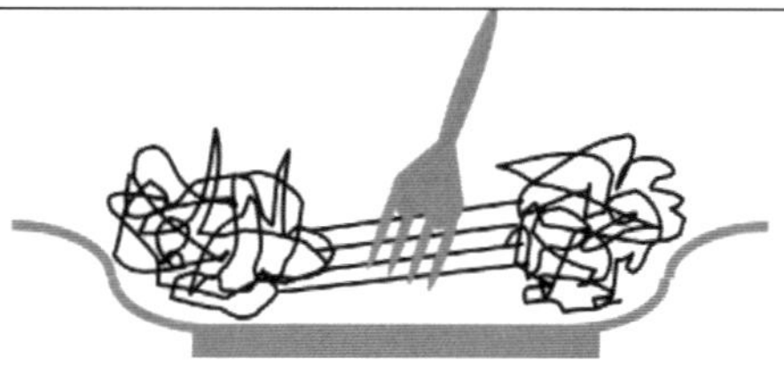
Fig. 2

Fig. 3

Fig. 4

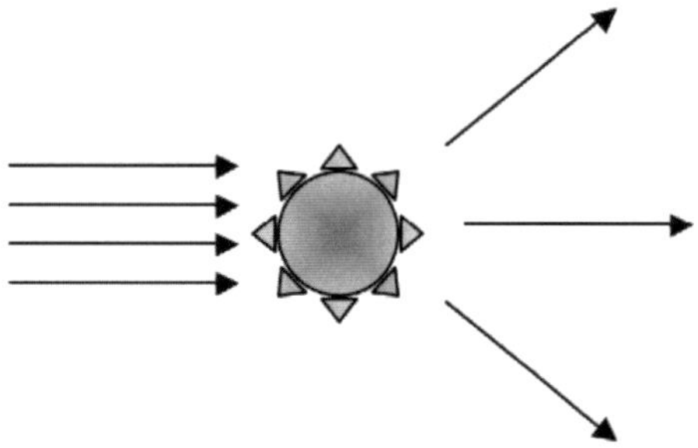
Fig. 5

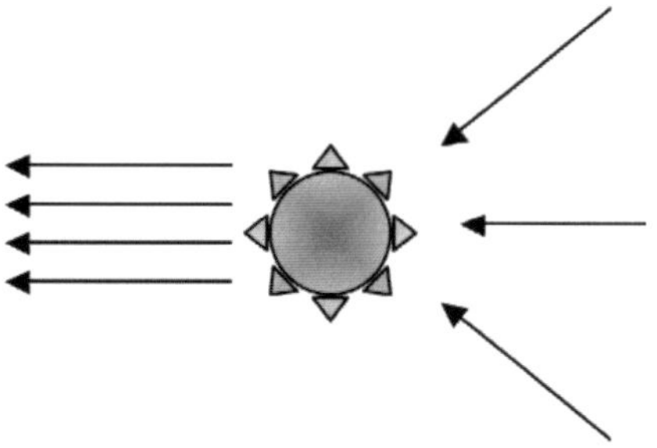
Fig. 6

Classical stochastic processes involve a non-negative probability distribution obeying a linear equation of motion. This is a Markov process. The stochastic propagation in time is then given by the solution to a partial differential equation akin to a non-negative stochastic matrix (or structure kernel). It is interesting that for a discrete probability vector stochastic evolution is more natural than any continuous evolution in time.

What about Quantum Mechanics? In this case, probabilities are bilinear in the amplitude. So we must work only with these quantities. But these are precisely the elements of the density matrix ρ. Only non-negative, unit trace density matrices are acceptable; and the quantum stochastic process must map these matrices into other density matrices. The quantum density matrices constitute a convex set

$$\rho = \mathrm{Cos}^2(\theta)\,\rho_1 + \mathrm{Sin}^2(\theta)\,\rho_2$$

is an allowed density matrix for all values of θ . Since the stochastic super matrices takes them into the convex set, stochastic matrices constitute a convex set.

The implication of this requirement on the stochastic matrices is seen best by constructing a super matrix B , , by defining

$$B_{rr',ss'} = A_{rs,r's'}$$

For the finite dimensional case the eigenvectors $v(\alpha)$ of B give

$$B_{rr',ss'} = \Sigma_\alpha\, \mu(\alpha) v_{rr'}(\alpha) v^*_{s\,s'}(\alpha)$$

for the case all $\mu(\alpha)>0$, we could consider the stochastic map in the form

$$B_{rr',ss'} = \Sigma_\alpha C_{rr'}(\alpha)\, C^*_{s\ s'}(\alpha)$$

These cases with all $\mu(\alpha) > 0$ are called "completely positive" maps.

Every such map can be obtained by embedding ρ in a larger super density matrix R, by writing

$$R = \Sigma\ \rho_{rs}\ X\ \tau_{r's'}$$

and then evolving R by a unitary super matrix U where

$$U_{r\ r'}^{1\ \alpha} = C_{rr'}(\alpha)$$

and taking the trace over α.

This theory of dynamical maps has been extended, refined (and plagiarized). It has also been applied to quantum computing and quantum tomography. The maps provide a method to derive the Kossakowski stochastic semigroups[28], Asorey's masterly presentation discussed these matters and some newer developments. Hegerfeldt discussed quantum jumps in quantum optics. Chisolm gave an exposition of the treatment of decoherence. This work on stochastic mapping is being pursued with Chelikowsky , Shaji, Jordan, Rodriguez, Asorey, Kossakowski, and Marmo.

It was a gratifying experience to have so many of my former collaborators and former students review these developments spanned by five decades.

8. Causality versus Reversibility

Perhaps our ideas of casuality, the requirement that cause precedes effect is dependent on this concept of order. A collimated beam of particles gets scattered by a potential. But what then of the 'out' states? A potential seems to be able to take a diffuse beam and make them collimated. Does it jibe with our idea of cause and effect?

As mentioned earlier, we take a bunch of (cooked) spaghetti strands, pick them up and put them on a separate plate, you may find that somewhere along their length more or less the same regularity may appear. This kind of recurrence is explored on general grounds (Fig. 4).

It is interesting to recognize that when tachyons are involved emission and absorption could change depending on the focus of the observer. This is an aspect of the Reinterpretation Principle that Recami and Bilaniuk discussed.

9. Reincarnation of Scientific Discoveries

In Indian tradition, such recurrences are part of the cosmology, except that it would then be called "reincarnation."

But to my dismay, this happens to scientific discoveries also. When Jagadis Chandra Bose[30] designed and demonstrated long distance propagation of radio waves in England , it was rediscovered by Giulio Marconi, and Marconi was awarded the Nobel Prize and the credit for the discovery, even though Marconi did it after Bose and Aryabhatta'a result for the power series for arctangent is assigned the name Gregory's series[31]. In my own work, my work on quantum stochastic processes was rediscovered (with Greek and Latin characters interchanged) some seven years later[32]. Andréj Kossakowski discovered the generic semigroup structure, but many people ascribe it to only Lindblad, perhaps because he did it later[33].

Some of these developments must have been clear to you from live presentations by Manuel Asorey. The presentations of Gerhard Hegerfeldt and Eric Chisolm were so clear that I need not elaborate on their clear expositions.

10. In Gratitude and Appreciation: Beyond Seven Quests

I am grateful to Rodger Walser, Mary Ann Rankin, Juan Sanchez, and John Markert, as well as to Austin Gleeson, Swadesh Mahajan, and V. V. Raman for their kind words of praise and encouragement. I need also to thank Alaka Valanju, Gopalakrishnan Bhamathi, Prashant Valanju and Rodger Walser for their tireless efforts in bringing about this Seven Quests Symposium.

These seven quests by no means exhaust the scientific topics that I am interested in. Stanley Deser reminds me of our work with Gilbert on 3- and 4-point function in quantum field theory[34]. Similarly the work on axiomatic field theory, quantum mechanics in dual spaces and work on inconsistency of charged spin 3/2 fields and topics like Fiber Bundles in Classical mechanics[35], the structure of the Dirac brackets, relativistic interactions using covariant constraints, theory of rays in quantum optics and radiative transfer[36]; with the scattering of charged particles in nuclear emulsion[37], multiple meson production, the theory of weak measurements and measurement theory, and several other topics are not covered by the seven quests. But the essential point to be brought out is that all these are inter-related. For example, the study of the convex set of statistical states[38] led to the discovery of the incompleteness of the Wightman axioms[39].

Even more important is the gratifying recognition that most of the young scientists who studied with me for their doctoral work have gone on to be excellent and productive scientists.

Substances and process are but two aspects of reality; and in quantum theory we see this identity even more directly. This wisdom should translate into an understanding of the process of measurements, and ultimately, to unite scientific theory and experiment with aspects of human consciousness. This is the ultimate frontier[40].

Recently I came across my scientific genealogy (published in these proceedings). It is indeed gratifying to see Hans Bethe, Arnold Sommerfield, Felix Klein, Carl Friedrich Gauss, as well as Joseph Fourier, Simeon Poisson, Joseph Lagrange, and Leonard Euler along the chain. These intellectual giants continue to influence scientists, some of them through me!

References

[1] Sudarshan E C G and Marshak R E 1957 *The Nature of the Four-Fermion Interaction*, Proc.Padua - Venice Conference on Mesons and Newly Discovered Particles, Reprinted in Kabir P K, *The Development of the Theory of Weak Interactions*, Gordon and Breach, New York, George Johnson *Strange Beauty*, Sudarshan E C G and Marshak R E 1958 *Chirality Invariance and the Universal Fermi Interaction*, Phys. Rev. **109,** 1860-1862

[2] Feynman R P and Gell-Mann M 1958 *Theory of Nuclear β Decay*, Phys. Rev. 109,193

[3] Weinberg S 1967 *A Model of Leptons*, Phys. Rev. Lett. 19,1264

[4] Salam A 1968 Proc. Nobel Conference, Stockholm

[5] Marshak R E, Okubo S, and Sudarshan E C G 1957 *Consequences of Charge Independence for the Magnetic Moments and Masses of the Λ Hyperon*, Phys. Rev. **106,** 599-601

[6] Macfarlane A J and Sudarshan E C G 1964 *Electromagnetic Properties of StableParticles and Resonances*, Nuevo Cimento **31,** 1176

[7] Levinson C A, Lipkin H J, and Meshkov S papers are cited in Meshkov's presentation in these Proceedings

[8] References to the work of McGlinn, Han, Sudarshan, and O'Raifeartaigh are given in Han's presentation in these Proceedings

[9] Jordan T, Currie D, and Sudarshan E C G 1963 *Relativistic Invariance and Hamiltonian Theories of Interacting Particles*, Rev. Mod. Phys. 35, 350-375

[10] Marmo G, Mukunda N, and Sudarshan E C G 1984 *Relativistic Particle Dynamics – Lagrangian Proof of the No-Interaction Theorem*, Phys. Rev. D 30, 2110-2116

[11] Stoner E C 1924 *Distribution of Electrons Among Atomic Levels*, Phil. Mag. **48,**719-736

[12] Pauli W 1920 *Uber den Zusammenhaug des Abschloress der Spektren*, Z. F. Physik **31,** 765-782

[13] Duck I M and Sudarshan E C G 1997 *Wolfgang Pauli and the Spin-Statistics Theorem*, World Scientific, Singapore

[14] Shaji A, These Proceedings

[15] Boya L J and Sudarshan E C G *Spin Statistics Relation in D>3 Space Dimensions*, paper

under preparation
[16] A systematic account is given in Born M and Wolf E, *Principles of Optics*, Pergamon Press, New York
[17] Batchelor G K 1982 *The Theory of Homogenous Turbulence*, Cambridge University Press
[18] Sudarshan E C G 1969 *Theory of Partial Coherence*, J. Mathematics and Physics (IIT/Madras) 3,123. Reprinted in *Selected papers of E C G Sudarshan 2006* Ed. Ranjit Nair, Scientia, New Delhi, 2006. See also Klauder J and Sudarshan E C G, *Foundations of Quantum Optics*, Benjamin, New York. Dover edition published in 2006
[19] Hanbury-Brown R and Twiss R Q 1956 *A Test of a New Type of Stellar Interferometer on Sinus*, Nature 178, 1046
[20] Mandel L 1963 Prog. In Optics, II (Ed. Wolf E) Netherlands, Amsterdam
[21] Sudarshan E C G 1968 *Equivalence of Semiclassical and Quantum Mechanical Descriptions of Statistical Light Beams* Phys. Rev. Lett. **10,** 217
[22] For misrepresentation/misunderstanding see Glauber R J 1963 Phys, Rev,**131,** 2766, especially the footnote to Section X
[23] References in Kimble's Contribution in these Proceedings
[24] Mehta's contribution to these Proceedings
[25] Simon R, these Proceedings
[26] Scully M and Zubairi M 2001 *Quantum Optics in Phase Space*, Wiley-VCH
[27] See, for example, Dirac P A M 1958 *Principles of Quantum Mechanics*, Oxford University Press, 4 th ed
[28] Rau J 1963 *Relaxation Phenomena in Spin and Harmonic Oscillator Systems*, Phys. Rev. 129,1880-1888
[29] Kossakowski A 1972 *On Quantum Statistical Mechanics of Non-Hamiltonian Systems*, Reports on Mathem Phys, **3,** 247-274
[30] Bilaniuk O M P, Deshande V K, and Sudarshan E C G 1962 *Meta Relativity*, Am. J. Phys. 30, 718
[31] Sudarshan E C G, Matthews P M, and Rau J 1961 *Stochastic Dynamics of Quantum Mechanical Systems*, Phys. Rev. **121,** 920-924

Journal of Physics: Conference Series **196** (2009) 011014 doi:10.1088/1742-6596/196/1/011014

Congratulatory mesages to Sudarshan on the occasion of the 2006 Sudarshan Seven Science Quests Symposium at The University of Texas at Austin

Abstract. Some of the many messages sent to George Sudarshan by those who could not attend this symposium for various reasons.

Girish S Agarwal
Noble Foundation Chair and Regents' Professor, Oklahoma State University

In appreciation of George Sudarshan

"It is my great honor and privilege to send this message on George's 75th birthday. I have known George from my student days at Rochester however a little later I had a chance to know the man whom I admired. Over the years I came to know him very closely. As widely recognized his scientific contributions are of the highest category. The ones which had immense impact on the quantum optics community, though many times the community did not give credit where it richly deserved, are— diagonal coherent state representation; positivity of master equations and a general framework for open systems , which has gained new importance in connection with quantum information science; quantum Zeno effect; weak measurements. The work on diagonal coherent state representation broke totally new ground and it became the basis for new directions of research in 80's and continues of great importance in connection with studies trying to understand quantum entanglement and non classicality. My own work was influenced by discussions with George and I would always look forward to an opportunity to have a stimulating discussion with him.I salute him on this occasion and wish him good health in years ahead so that the community could benefit from his ideas and from interactions with him."

Michael Berry
Melville Wills Professor of Physics (Emeritus), University of Bristol

"You have much to celebrate, and I greet you on your birthday symposium, unfortunately not in person but in spirit, from a remote wetland in Argentina. Although I'm not a high-energy physicist, we have common interests in other areas (perhaps I should rather say that we share uncommon interests). In phase space representations of quantum mechanics, your applications to optics have been seminal. And your persistence, against the grain, in arguing that that the spin-statistics theorem can be understood without relativity, is something I have taken up and developed; we don't agree about the details, but the aim is the same. I recall how much I have enjoyed our encounters - all too few - and your kind personal hospitality in Madras (as that city was called then).

This is not the occasion to recall my efforts to get your housekeeper to provide toilet paper (while you were away from the house) - well, perhaps it is the occasion, but I still won't tell the story.

With my affectionate wishes for many more years of productive engagement with our beloved physics."

Roland Omnes
Professor of Theoretical Physics (Emeritus), Faculté des sciences at Orsay, Université Paris-Sud XI.

"Professor Sudarshan's contribution to theoretical physics has earned him the respect of all his colleagues in the world. Transcending his Seven Quests, I wish to salute through him the unique quest of all devoted scientists, which is to understand better and investigate further the wonderful depth of the Laws of Nature and rejoice in their beauty. This is worth a life long dedication for the happy people who met them, Sudarshan's dedication particularly, which is celebrated in this meeting. I join my voice with joy to the homage he highly deserves, with my best greetings for his 75th birthday."

Yasushi Takahashi
Professor of Physics Emeritus, University of Alberta

"It is my great pleasure and honour to send this message on occasion of the symposium "Sudarshan: Science 7 Quests". Needless to say, George is an internationally respected theoretical physicist, philosopher and great organizer.
I have known George since 1963 when he organized a very stimulating symposium in Rochester attended by small numbers of active theoretical physicists in the field of high energy physics. The symposium was extremely informal , stimulating and of atmosphere like a family gathering. Through this symposium, I got to know George and other active physicists personally. The success of that symposium was totally due to George's friendly personality.
Since then, I was excited whenever I saw George's works: They always consist of clear analysis of the problem and very interesting idea and argument to save the problem. This style had been kept from the famous V-A theory of weak interaction to the quantum mechanics of open systems which will be discussed in this symposium. I am sure this symposium" 7 Quests" will be successful as much as that of 1963 symposium. Last but not the least, I firmly believe that George continues to be playing the leading role in theoretical physics for years to come, I wish him many more years of productive research and good health."

Wojciech Zurek
Laboratory Fellow, Theory Division, Los Alamos National Laboratory

Best wishes on this important occasion! Over the years I have benefited from various advances you have pioneered starting with the dynamical maps and theory of open quantum systems, via Zeno effect, and, above all, through your contributions to quantum optics including the Sudarshan-Glauber representation. Along with Johnny Wheeler, you were an inspiration ever since I arrived in Austin as freshly minted University of Texas student three decades ago, and I am sure I will continue to learn from you in the future!

With admiration,
Wojciech

Journal of Physics: Conference Series **196** (2009) 011014 doi:10.1088/1742-6596/196/1/011014

Guests at the Sudarshan Seven Quests Symposium

1. Anderson, Phillip W.
2. Acharya, Raghu
3. Agarwal, Girish
4. Aharonov, Yakir
5. Arima, Akito
6. Arnowitt, Richard
7. Arons, Michael
8. Ashtekar, Abhay
9. Asorey, Manolo
10. Aspect, Alain
11. Bacry, H.
12. Balasubramanya, Murley
13. Barger, Vernon
14. Berry, Michael Victor (Sir)
15. Bialynicki-Birula, Iwo
16. Bilaniuk, Olexa-Myron
17. Bohm, Arno
18. Bose, Samir
19. Boya, Luis
20. Braunstein, Sam
21. Brink, Lars Elof Gustav
22. Bryan, Ronald
23. Capri, A Z (Tony)
24. Chaturvedi, S.
25. Chew, Geoffrey
26. Chisolm, Eric
27. Chiu, Charles
28. Chou, Kuang-chao
29. Cohen – Tannouji, Claude
30. Cooper, Leon N
31. Das, Ashok
32. Datta, Kamal
33. Davies, EB
34. Dell'Antonio, Gianfausto
35. Deser, Stanley
36. d'Espagnat, Bernard
37. Duck, Ian
38. Durand, Loyal
39. Durr, Hans-Peter
40. Dyson, Freeman
41. Eberly, Joseph
42. Edwards, S F
43. Emch, Gerard Gustov
44. Finkelstein, David
45. Freedman, David
46. Gardiner, Crispin William (Cris)
47. Ghosh, Partha
48. Ghosh, Rupa
49. Glashow, Sheldon
50. Gleeson, Austin
51. Gorini, Vittorio
52. Greenberg, Oscar (O.W.)
53. Gruber. Bruno
54. Gustafson. K.
55. Hagen. C. Richard (Dick)
56. Hammermesh
57. Han, Moo-Young
58. Harte, John
59. Hegerfeldt, Gerhard
60. Heinzen, Daniel
61. Hiida, K
62. Home, Dipankar
63. Hulet, Randall
64. Imbo, Tom
65. Itano, Wayne
66. Jordan, Thomas
67. Joseph, Babu
68. Josephson, Brian
69. Kailath, Thomas
70. Kembhavi, Ajit
71. Kenkre, V.M.
72. Khanna, F C
73. Kim, Y. S.
74. Kimble, H. Jeff
75. Kolb, Ed
76. Koshiba, Masatoshi
77. Kossakowski, Andrej
78. Lamb, Willis
79. Lee, Tsung-Dao
80. Leutwyler, H
81. Lewis, Marc
82. Lindblad, Göran

83. London, George
84. Luippold, Dave
85. Ma'nko, Margarita
86. Macfarlane, Alan
87. Mahajan, Swadesh
88. Mahanthappa, Kalyana T
89. Man'ko, Vladimir I
90. Marmo, Giuseppe
91. McGlinn, William D
92. Mehta, C.L.
93. Meshkov, Sydney
94. Misra, Baidyanath
95. Mukunda, N.
96. Nair, Ranjit
97. Narlikar, Jayant
98. Nicolis, Grégoire
99. Nilsson,
100. Okubo, Susumu
101. Omnes, Roland
102. Padmanabhan, T.
103. Pakvasa, Sandip
104. Pappu, Rama Rao
105. Pascazio, Saverio
106. Pasupathy, J.
107. Pati, Jogesh , Paul
108. Babu, Piron
109. Constantin, Polkinghorne
110. Preskil, John
111. Puligandla, Rama
112. Puthoff, Harold
113. Raizen, Mark
114. Rajaraman, R.
115. Raman, V.V.
116. Ramond, Pierre
117. Rauch, Helmut
118. Recami, Erasmo
119. Rechenberg, Helmut
120. Rivera, Juan
121. Roman, P
122. Rothman, Tony
123. Sanchez, Juan
124. Schechter, Joseph
125. Schleich, Wolfgang
126. Schrieffer, John Robert
127. Schweber, Silvan
128. Scully, Marlan
129. Semenov, G
130. Shaji, Anil
131. Simon, R.
132. Singh, V.
133. Sinha, K.P (Krityunjai Prasad)
134. Srinivasan, S.K.
135. Srinivasan, V.
136. Stapp, Henry P
137. Stevenson, Paul
138. Sukhatme, Uday
139. Takahashi, Yasushi
140. Tasaki, Shuichi
141. Tata, Xerxes
142. Taylor, J.C.
143. Thouless, D.J
144. Tilma, Todd
145. Ting, Samuel C C
146. Tiomno, Jayme
147. Tollaksen, Jeff
148. Townes, Charles
149. Tripathi, Vikram
150. van der Merwe, Alwyn
151. Van Kampen, N G
152. Vigier, J P (Died 2004)
153. Vittiello, G.
154. Voisin, Jacques
155. Wali, Kameshwar
156. Walther, Herbert
157. Weinberg, Steven
158. Weiner, Richard M.
159. Werner, Reinhard
160. Williams, Randy
161. Willis, Wayne
162. Wolf, Emil
163. Yang, C N
164. Zaccaria, Francesco
165. Zurek, Wojciech

Journal of Physics: Conference Series **196** (2009) 011014 doi:10.1088/1742-6596/196/1/011014

Sudarshan's Students and past research associates

1. Acharya R (Professor Arizona State Univ)
2. Al Hendi (Riyadh, Saudi Arabia)
3. Arons M E (Professor Long Island Univ.)
4. Bardakci K (Professor U C Berkeley)
5. Burcu A (Financial Services)
6. Byrd M (Southern Illinois Univ, Carbondale)
7. Chiang C C (Prof, Academia Sinica, Taipei)
8. Chisolm E (LANL, Los Alamos)
9. Currie D G (Professor Univ of Maryland)
10. Gousheh (Teheran, Iran)
11. Han M Y (Professor Duke Univ.)
12. Hans Bebie (Director, Institute of Exact Sciences)
13. Hans Hofer (Experimental High Energy, Bern)
14. Hans Schatz (Institute of Exact Sciences, Bern)
15. Jordan T (Professor Univ of Minnesota)
16. Khalil M A (Environment Research Institute , Portland)
17. Kuriyan J G (UCLA/ now in private enterprise)
18. Mac E (not known)
19. Mims M (Private Enterprise)
20. Mukunda N (Professor IISc , Bangalore)
21. Pinski G (Psychiatrist)
22. Schechter J (Professor Syracuse Univ)
23. Shaji A (Univ of New Mexico, Albuquerque, NM)
24. Tilma T (Japan)
25. Underwood J (Professor, Austin Community College, Austin)
26. Urjit Yagnik (Professor, IUCAA, Pune, India)
27. Valanju P M (Research Scientist, Institute for Fusion Studies, Univ of Texas at Austin)
28. Varma S (Financial services)

Current Ph.D. students at University of Texas at Austin:

1. Aik-Meng Kuah
2. Antonia Chimonidou
3. Dixit K
4. Jantzke P
5. Modi K
6. Rodriguez C

Past research associates and their current status:

1. Bohm A (Professor Univ of Texas at Austin, Austin, Tx)
2. Chiang C C (Professor Academia Sinica, Taipei)
3. Cziffra P (LRL, Livermore, Ca.)
4. Fr. Jacques Voisin (Louvain Univ, Belgium)
5. Gardiner C (Professor, New Zealand)
6. Gleeson A (Professor Univ of Texas at Austin, Austin, Tx)
7. Gundzik M (Private Enterprise)
8. Gyuk I (Professor Univ of Wisconsin/Milwaukee)
9. Han M Y (Professor, Duke Univ)
10. Hsu L P (unknown)
11. Hurley W A (not known)
12. Imbo T (Professor Univ of Chicago , Chicago , Il)
13. Macfarlane A J (Professor Cambridge Univ, U.K.)
14. Maiella G (Professor University of Naples, Naples, Italy)
15. Mehra J (Professor ICTP, Trieste, Italy)
16. Misra B (Univ. of Thessaloniki, Greece)
17. Moore R M (not known)
18. Mukunda N (Professor IISc, Bangalore, India)
19. Nandi S P (Professor Oklahoma State Univ.)
20. Nelson C A (Professor, SUNY, Binghamton)
21. Pakvasa S (Professor, Univ of Hawaii)
22. Papstamatiou N (Professor Univ of Wisconsin/Milwaukee)
23. Raman K (Professor Weslyan Univ, Middletown, CT)
24. Rechenberg H (Professor Max Planck Institute, Munich, Germany)
25. Ryan C (Deceased)
26. Schnitzer H J (Professor Brandeis Univ.)
27. Sherry T N (Professor, Mathematical Physics, NUI, Galway)
28. Simoni A (Professor University of Naples, Naples, Italy)
29. Takasugi E (Professor, Osaka Univ, Japan)
30. Tata X (Professor Univ of Hawaii, Hawaii)
31. Valanju P (Research Scientist, University of Texas at Austin)
32. Wang K H (not known)
33. Yun S K (unknown)
34. Zaccaria F (Professor University of Naples, Naples, Italy)

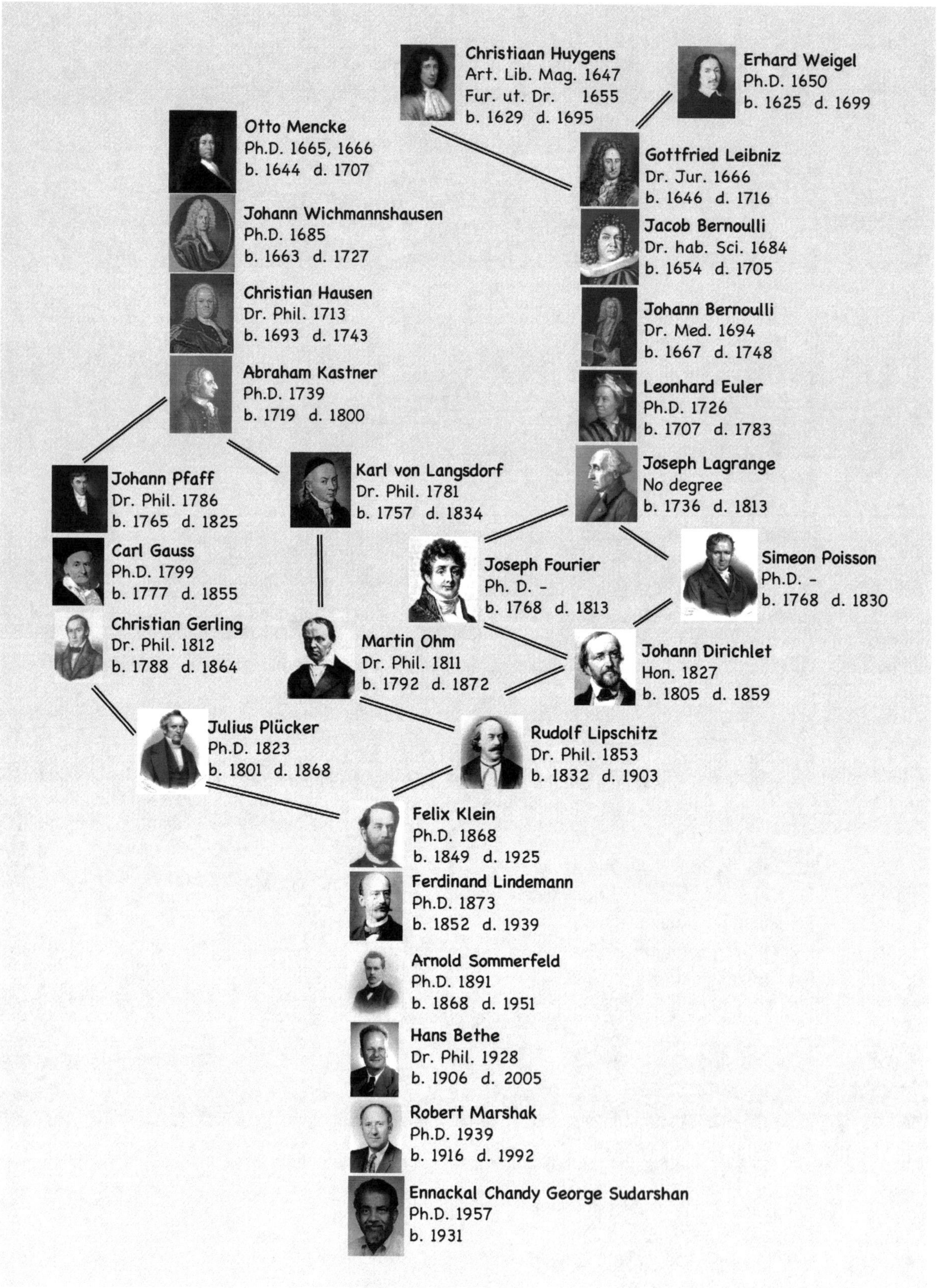
Christiaan Huygens
Art. Lib. Mag. 1647
Fur. ut. Dr. 1655
b. 1629 d. 1695
Erhard Weigel
Ph.D. 1650
b. 1625 d. 1699
Otto Mencke
Ph.D. 1665, 1666
b. 1644 d. 1707
Gottfried Leibniz
Dr. Jur. 1666
b. 1646 d. 1716
Johann Wichmannshausen
Ph.D. 1685
b. 1663 d. 1727
Jacob Bernoulli
Dr. hab. Sci. 1684
b. 1654 d. 1705
Christian Hausen
Dr. Phil. 1713
b. 1693 d. 1743
Johann Bernoulli
Dr. Med. 1694
b. 1667 d. 1748
Abraham Kastner
Ph.D. 1739
b. 1719 d. 1800
Leonhard Euler
Ph.D. 1726
b. 1707 d. 1783
Johann Pfaff
Dr. Phil. 1786
b. 1765 d. 1825
Karl von Langsdorf
Dr. Phil. 1781
b. 1757 d. 1834
Joseph Lagrange
No degree
b. 1736 d. 1813
Carl Gauss
Ph.D. 1799
b. 1777 d. 1855
Joseph Fourier
Ph. D. -
b. 1768 d. 1813
Simeon Poisson
Ph.D. -
b. 1768 d. 1830
Christian Gerling
Dr. Phil. 1812
b. 1788 d. 1864
Martin Ohm
Dr. Phil. 1811
b. 1792 d. 1872
Johann Dirichlet
Hon. 1827
b. 1805 d. 1859
Julius Plücker
Ph.D. 1823
b. 1801 d. 1868
Rudolf Lipschitz
Dr. Phil. 1853
b. 1832 d. 1903
Felix Klein
Ph.D. 1868
b. 1849 d. 1925
Ferdinand Lindemann
Ph.D. 1873
b. 1852 d. 1939
Arnold Sommerfeld
Ph.D. 1891
b. 1868 d. 1951
Hans Bethe
Dr. Phil. 1928
b. 1906 d. 2005
Robert Marshak
Ph.D. 1939
b. 1916 d. 1992
Ennackal Chandy George Sudarshan
Ph.D. 1957
b. 1931

www.ph.utexas.edu/fogs/symposium
For more information, email Professor R.M. Walser at quest@physics.utexas.edu.

Nilsson | Weinberg | Okubo | Das | Pakvasa | Aharonov | Meshkov | Han | Mukunda | Jordan | Glashow | Boya | Duck | Finkelstein | Imbo | Shaji | Gleeson | Nair | Mehta | Kimble | Simon | Birula | Marmo | Itano | Chiu | Freedman | Recami | Bilaniuk | Bose | Omnes | Kossakowski | Hegerfeldt | Chisolm | Mahajan | Raman | Sudarshan

INTERNATIONAL ADVISORY COMMITTEE Chair: Ranjit Nair *New Delhi*; Yakir Aharonov *Tel Aviv*, Luis Boya *Zaragoza*, Jan Nilsson *Stockholm*, Sheldon Glashow *Boston*, Austin Gleeson *Austin*, Jeff Kimble *Pasadena*, Gerhard Hegerfeldt *Goettingen*, Swadesh Mahajan *Austin*, John Markert *Austin*, Giuseppe Marmo *Naples*, Sidney Meshkov *Pasadena*, Roland Omnes *Paris*, V.V. Raman *Rochester*, Erasmo Recami *Milan*, Steven Weinberg *Austin*

ORGANIZING COMMITTEE Chair: Rodger Walser; **Advisor**: George Sudarshan; G. Bhamathi, Murley Balasubramanya, Vijay Mahajan, Mel Oakes, Harold Puthoff, Alaka Valanju, Prashant Valanju, Ardyce Walser

SPONSORS University of Texas at Austin: Office of Vice President for Research, Department of Physics, and Center for Electromagnetic Materials and Devices, and MetaMaterials LLC, Austin, Texas

Journal of Physics: Conference Series **196** (2009) 011014 doi:10.1088/1742-6596/196/1/011014

Sudarshan: 7 SCIENCE QUESTS

SYMPOSIUM

सुदर्शनस्य सप्त शास्त्र साधना

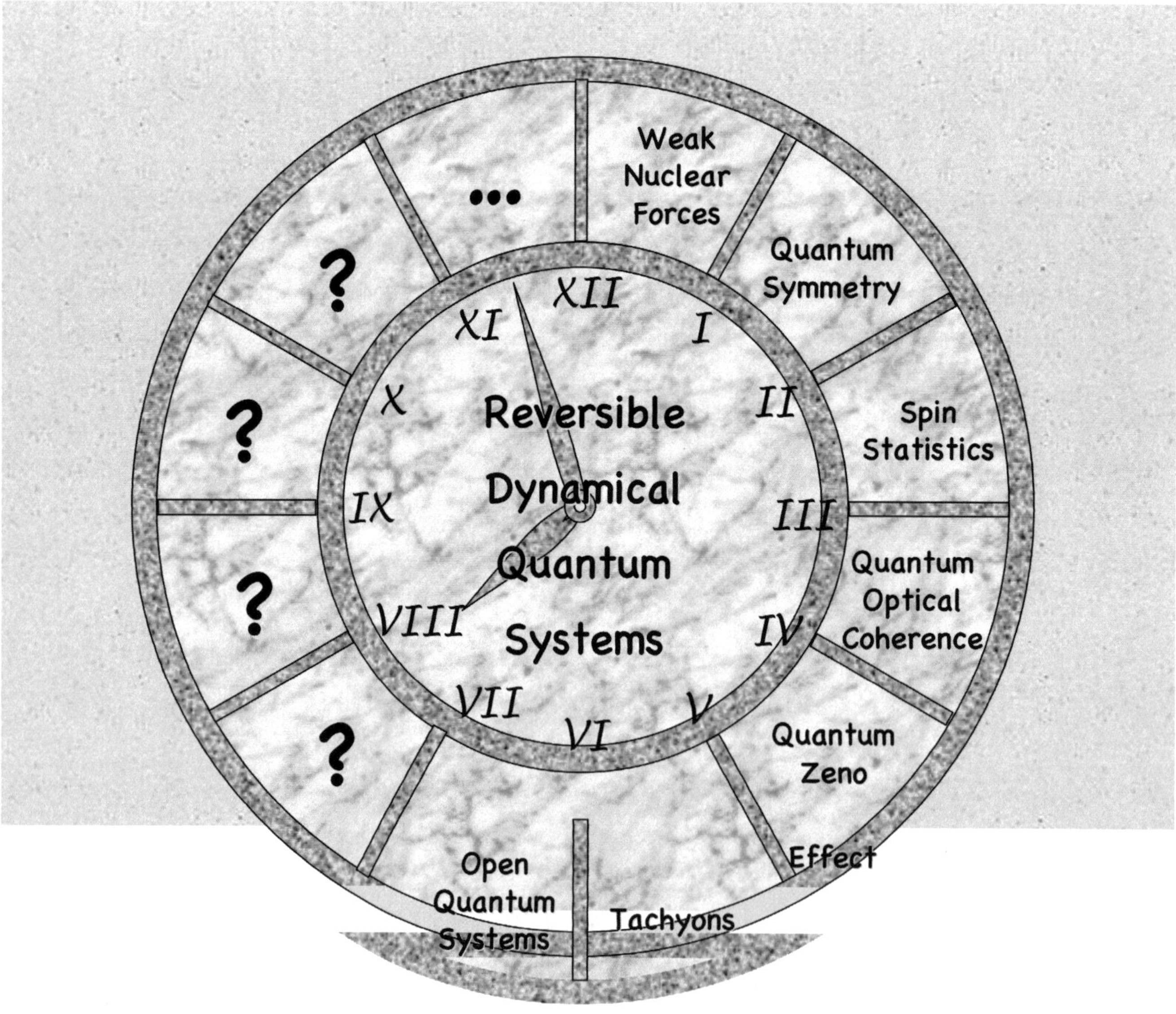

AUTHOR INDEX